LIFEBOAT

LIFEBOAT

John R. Stilgoe

University of Virginia Press Charlottesville & London

University of Virginia Press
© 2003 by the Rector and Visitors of the University of Virginia
All rights reserved
Printed in the United States of America on acid-free paper

First published 2003

9 8 7 6 5 4 3 2 1

LIBRARY OF CONGRESS CATALOGING-IN-PUBLICATION DATA
Stilgoe, John R., 1949–
   Lifeboat / John R. Stilgoe.
      p.   cm.
Includes bibliographical references and index.
   ISBN 0-8139-2221-6 (alk. paper)
   1. Lifeboats. 2. Shipwrecks. I. Title.
VK1473 .S75 2003
910.4ʹ52—dc21

                        2003001882

FOR HAROLD TUTTLE

CONTENTS

PREFACE

LIFEBOATS AWAIT DISASTER. UNNOTICED AND UNREMARKED, OFTEN deliberately ignored by cruise-ship passengers determined to frolic, somehow invisible to bystanders waving bon voyage or watching tugs nudge a rusted freighter against a wharf, lifeboats whisper of the uncertainty of ocean travel. All but the officers and crewmen charged with their care dismiss them as mere canvas-protected appendages, insignificant structural details dwarfed by hull and funnel and derricks, vestiges of archaic government regulation. Lifeboats remain perhaps the least known of all small boats, intriguing neither yachtsmen nor naval architects, and certainly not marine historians.

Lifeboats acquire character only when ships sink. Shipwreck confers on every lifeboat individual identity and significance beyond measure.

Only in disaster do passengers—and sometimes even officers and crewmen—discover that lifeboats are oceangoing vessels themselves. Something small and insignificant, almost invariably beneath notice, becomes a miniature city afloat in a vast wilderness.

Lifeboats make passages. When help fails to arrive, lifeboats proceed somewhere, and even now must often proceed under the oars and sails they have carried for more than a century. Life rafts merely drift.

I sail a ship's lifeboat built in 1935. Why I bought it and restored it and sail it, I do not know exactly. Maybe a premonition of disaster prompted my 1995 purchase and my beginning this book, and after the terrorist attacks in New York and Virginia and over Pennsylvania—which occurred after this book arrived at its publisher—I shiver at the premonition. Somehow a long-standing scholarly interest in the fragility of contemporary American living angled me toward a ship's lifeboat as a way into a complex subject.

What I know from boyhood salts this book too. Fishermen use the word *vessel* a lot, for they go to sea in craft smaller than ships, and fisherman vocabulary figures strongly in the conversation of those who live around the small harbors near my home. Fishermen and oceangoing mariners distinguish between a voyage—an out-and-back trip to nowhere in particular, usually in search of fish—and a passage, a journey from one port to another. Mariners take passage in or aboard ships and boats, not on them, and while they tend to define ships as vessels making ocean passages, they respect old distinctions between ships and schooners, brigs and barks; they instantly identify and name the now rare lugsail rig that drives my lifeboat; almost never do they attach "the" to the name of a vessel; for them it is *Lusitania*, not *the Lusitania*. Ships and boats lack gender, too, however much the vessels may have pronounced character traits that suggest some sentient identity, benevolent or otherwise. Only when something important happens, when mariners must pay a vessel the closest attention and respect they can pay, do they nowadays speak of a vessel as feminine. Male and female alike, mariners speak of sailing toward a particular port, not to it: they know that the sea sometimes thwarts intention. None of this is affectation. The language of this book is the speech of the south shore of Massachusetts Bay and the margins of Cape Cod I have known since birth.

It is the language of tales, too. Anyone who sails a ship's lifeboat attracts all sorts of attention, some of which produces stories of shipwreck and lifeboat passage making. I rely on published sources here because I seek to understand the lifeboat in broad cultural terms, while recognizing that any inquiry into United States, Canadian, and British manuscript lifeboat-passage accounts pushes a responsible scholar into oral history as well. I have no desire to create an oral-history project focused on shipwreck and lifeboat passage making, especially in World War II. But as I mention in the conclusion to this book, oral history proves worthwhile. Once national authorities determined that World War II tankers should carry steel lifeboats, none paid attention to seamen anxiously pointing out that such lifeboats became skillets in the blazing oil and gasoline that covered the sea around torpedoed tankers—and that the traditional wooden lifeboats smoldered but pulled free of the flames. Their view remains as unpublished today as ever, as hard to find as information about how well modern inflatable rafts survive fire. So while I depend on published sources I also keep in mind what I learn on summer days when my lifeboat nestles against some wharf on which old men sun themselves and remember.

Libraries and archives contain no narratives by castaways who died aboard lifeboats. Exposure, especially in wartime in the winter North Atlantic and Arctic oceans, killed many people aboard lifeboats, although the boats themselves often fetched up on beaches. Heat and thirst killed many castaways in the tropics. So the sources I use are those produced by survivors, and usually—although not always—by officers. Men wrote most of the narratives, but among the best is a little-known account by a very tough-minded woman indeed. And survivors seem to have taken few photographs. Cameras went into lifeboats only in an Alfred Hitchcock film. Once lowered and away, lifeboats vanish from the visual record only to reappear—rarely—after meeting a ship or making land.

Three-quarter-inch planking sheathes my lifeboat. It is a thin lapstrake skin indeed that separates me from a sea so cold that even in summer immersion means near immediate hypothermia. Lately I wonder about the thin veneer of contemporary American civilization, the ease with which fire extinguishers and automobile jacks and lifeboats disappear into a limbo masked by a shimmer of high-tech gadgetry and official assurance of safety. I have no wish for intimacy with disaster, but my acquaintance with it predates the terrorist attacks of September 11, 2001. My acquaintance confirms my hunch that lifeboats reward scrutiny.

ACKNOWLEDGMENTS

BOYD ZENNER, ACQUIRING EDITOR AT THE UNIVERSITY OF VIRGINIA Press, supported this book from the moment she heard of it. Her efforts on behalf of it deserve the highest commendation.

My friends and colleagues have known about my lifeboat thinking for years. Thomas Armstrong, Robert Belyea, William Bunting, James Fox, John Goodman, Stephen Jones, James Mitchell, Edward Murphy, Laura Nowosielski, Michael Rubin, and Harold Tuttle all contributed to what follows here.

Martin Robinschaugh, Canadian Consul in Boston, made determined efforts to decipher the worn incisions on the stem of my lifeboat.

At Harvard, librarian Kenneth Carpenter preserved an obscure seafaring collection from the disorders of book relocation and library renovation. Another librarian, David Cobb, routinely produced charts of unfrequented seas. To them I owe a special debt.

Claudia Jew, Photographic Archivist at the Mariners' Museum, made an extraordinary effort to find images of lifeboats.

Ruth Melville did wonders as a copy editor. I cannot praise her efforts too highly.

My wife, Debra, scuba dives from the lifeboat she first saw as a hulk. It is not the best diving platform, but she knows its value on the North Atlantic coast. I cannot thank her enough for the tremendous efforts she has made on behalf of this book.

LIFEBOAT

Embowered in shrubbery, the lifeboat I had purchased sight unseen loomed from vegetation that disguised the bulk and weight that immediately frustrated the boat haulers. (Photo by author)

INTRODUCTION

ALL DAY THE GALE HAS WORSENED, THE WIND GROWING STRONGER hour by hour after a bleak and still dawn, curling in upon itself in a jarring, ceaseless, implacable shuddering. Rain came a little after eight, fitful at first, then driving sideways, changing to sleet around noon, masking the solid gray overcast blasting past a few hundred yards above. Overhead raced only solid, pockmarked cloud, a seamless gray darker than the gray of warships, the gray of snow to come. By early afternoon, sleet had become snow, the tiny-flaked snow that heralds a prolonged gale, the three-day storm coastal New Englanders call a "no'theaster," what tourists and television announcers mislabel "nor'easter." Snow, fine and driving, darkening everything, destroying distance, squeezing shut the sun, welds to wind, becomes a single force raging from somewhere, maybe the Gulf of Maine, maybe Cape Farewell in Greenland, maybe everywhere in the North Atlantic.

And in the bottom of the lifeboat, trouble. Crouched sideways against the ribs and planks exposed by the floorboards removed and jammed against the starboard side, I lie against the massive wooden keelson, socket wrench dropped against the garboard strake, the lowest starboard plank. Chilled now despite the one-piece quilted cold-weather suit my wife bought for me only a few weeks ago, I feel the numbness in my hands. Gloves get in the way of the wrench and drift bolts, and icelike hex nuts slide from the slippery fabric. I strip off the gloves, shove myself sideways and further down, and look again. In the gloom I peer at the massive bolts, then close my eyes against the gale and the repair I cannot finish now, and marvel at Newfoundlanders who overbuilt this boat so long ago, before World War II, long before I first breathed. No three-quarter-inch-diameter bolts for those cold-sea men. No seven-eighths drift bolts either. In the gathering dark I confront the undeniable truth. From Newfoundland so long ago this wooden lifeboat went forth with keel and keelson attached with one-inch-diameter bolts. The socket wrench might lie, but the micrometer cannot. My selection of galvanized and stainless-steel and silicon bronze hex nuts includes none sized to Newfoundland conscience. I squint at the bolt exposed in the dark, shift my weight from hip to shoulder, shiver with the cold slicing up my forearms and stabbing into my knees, the cold riding the gale. Why am I doing this, I ask myself, grunting as I pull myself upright, one hand on the gunwale, the other scrabbling for the socket wrench, upsetting the box of seven-eighths-inch galvanized nuts. My gloves topple from the keelson into the bilge. Why am I in a lifeboat, cold, cramped, half-deafened by the gale?

In the kitchen, hot tea does nothing to help answer the question. Not at first. I wrap my fingers around the mug, inhale the steam, shiver slightly, and look out at the gale, wincing at the snow, the early afternoon dusk. Through the kitchen window I see the barn dwarfed by a Norway spruce, the barn sheltering the lifeboat from the gale, the barn fading in twilight and ripping snow. I shake my head in wonder. I am restoring an artifact that lacks both general and specific history, but one with a peculiar power to fix attention on disaster.

No one loves lifeboats. No one studies them either. Only rarely does anyone write about them at all, and then not at length. Somehow they remain essentially invisible, ranked in rows to which cruise-ship passengers pay no attention, strung necklacelike, each under a stout canvas cover, each hidden from possibilities implicit in lifeboat drill. Obsolescent, almost everyone shrugs, in this age of airborne rescue, supplanted by self-launching rafts that inflate automatically and self-activating electronic homing beacons that summon long-range aircraft and rescue ships. But lifeboats linger in fraying memories, in a handful of stark narratives, now and then in some boatyard corner, worn out from war, then worn out again from conversion to pleasure boats. Like the boatyard lifeboats resisting decay, bluff bows awash in bayberry and beach rose, the need for lifeboats endures.

History scarcely stains my boat. What I know is merely alongshore hearsay moored in vagueness. In 1968 a Coast Guard patrol found the boat adrift east of Stellwagen Bank and towed it into Woods Hole on Cape Cod, not to preserve it from loss, but because the vessel showed signs of onetime occupancy. Somewhere some ship had sunk, or at least had reason to launch a lifeboat, and in its experienced, sophisticated, and somewhat plodding way, the Coast Guard understood the boat as evidence. Whether the boat figured in some unreported accidental sinking or some marine crime like barratry—the deliberate misuse of a ship by its crew—it could not at once determine. The lifeboat might have been launched to pick up a man fallen overboard or by some drunken or mutinous crewmen. Its occupants might have been rescued by a freighter bound for Europe or Africa, and the joyful radio message never received. The numbers chiseled in its stem demanded the attention of the Canadian Board of Trade and perhaps whispered of some Cold War near ignition. Investigation determined nothing, but time determined that the nameless lifeboat cluttered the Coast Guard wharf. A boatyard owner bought it at auction, found it heavy and a bit tricky to maneuver, and sold it to a marine painter known as a collector of nautical artifacts, the set pieces of his paintings. The artist found it a bigger-than-usual artifact that dwarfed his back lawn. In time he found it something to sell to a man from Massachusetts chasing a dream or nightmare, pinning down a thought, fixing an image as he fixed the boat, replacing planks and ribs and gunwales, worrying out an idea, remembering not so much a long-ago fear as a long-ago awareness, something about growing up to be a man alongshore.

Lifeboats have intrigued me since boyhood. They stabbed harsh reality deep into the tramp-steamer dream, but not until adolescence did my interest focus.

At Christmas time, 1963, the great liner *Lakonia* caught fire in the Atlantic, about 180 miles north of the island of Madeira. We learned of it at breakfast, eating hot oatmeal, listening to WBZ broadcasting from Boston through the white radio nestled against the flour canister on the kitchen counter. Passengers taking to the boats. *Women and children first.* I had no time to find its position on a map, and on the school bus no one cared much about the news, nor did anyone in my freshman homeroom. I remember my keen disappointment that no one cared, that the assassination of John F. Kennedy overwhelmed anything else, even *Lakonia* afire. That evening, newspaper and television news remained sketchy. From somewhere the American air force had arrived, dropping supplies, and merchant ships had found their way in the dark to an ocean liner blazing from stem to stern. A freighter had begun collecting survivors in lifeboats.

After dinner, by the wood-burning stove, my boatbuilder father mentioned the 1934 burning of the liner *Morro Castle,* for him and my mother an enduring mystery bound up in sabotage, in newsreel footage, in photographs of lifeboats coming ashore on the New Jersey coast. I had never heard of *Morro Castle,* and a year after the Cuban missile crisis I found it hard to imagine a fleet of liners running regularly between Havana and New York. As *Lakonia* burned, I learned of one more 1930s mystery that I added to my list of before-the-war oddities. The 1929 stock-market collapse, the Lindbergh baby kidnapping, the *Hindenburg* zeppelin explosion at Lakehurst,

New Jersey, the disappearance of Amelia Earhart, all seemed romantic yet sinister, bound up in the coming of war. Decades before I first read about the Manhattan telephone exchange failure during the stock-market crash or the innuendo swirling around the refusal of Charles Lindbergh to ask the FBI for help, I heard in my parents' voices, and saw in their faces, their continuing doubts that newspapers had told them much at all. Even as I obeyed high school teachers and guidance counselors and studied physics as my Cold War patriotic duty, I struggled to read about the odd events framed by the 1929 stock-market crash and the 1941 bombing of Pearl Harbor which my history teachers dismissed as mere sparks in the Great Depression gloom. Neither my small-town public library nor private James Library collections offered much about the 1930s, and *Morro Castle* afire became one more tale about which I learned only bare facts and hints.

Innuendo lingered like the smudge of an Atlantic liner long over the horizon or like the acrid, oily smoke from the burning, abandoned *Lakonia*. It lingered like the accusations against the *Lakonia* captain and crew that exploded a day or so after the rescue. My father and other men, all taciturn, salt-marsh Yankees, alluded to the July 1956 sinking of the liner *Andrea Doria* off Cape Cod. Rammed in the fog by the Swedish liner *Stockholm*, *Andrea Doria* capsized and sank in a glare of publicity. Aboard the liner *Ile de France*, which had departed New York some ninety minutes before *Stockholm* and put about in a wild dash to the scene, a *Life* magazine photographer caught not only the rescue and the sinking but the glimmerings of scandal. When the first lifeboats reached the French liner, the photographer and other passengers found them filled with *Andrea Doria* hotel staff, rather than passengers. Women and children had not gotten away first; a Boston photojournalist, flying above the gashed liner, recorded the portside lifeboats all in their davits. The *Andrea Doria* sinking effectively marked the end of ocean-liner travel. Immense Constellation airliners thereafter replaced Atlantic steamships, and jet aircraft quickly supplanted the propeller-driven Connies. By the early 1960s few Americans cared about shipwreck, chiefly because few Americans traveled by sea. In those years I realized that many adults, including my high school teachers, did not want to talk about the failure of seamen to rescue passengers.

Lakonia proved that failure beyond measure. In the midst of fighting the fire, seamen had no time to follow the captain's mangled order to launch lifeboats, and stewards and passengers struggled to lower boats they quickly found unseaworthy and unprovisioned. The bodies rescuers found floating in the calm sea around *Lakonia* were those of passengers abandoned by the crew, dumped into the water by toppling lifeboats, killed by lifeboats falling from paint-jammed davits, drowned when they jumped over the side to be dashed against the steel hull of the ship. What *Lakonia* and *Andrea Doria* taught came down to something simple and dark. Every steamship *passenger* ought to know how to lower and crew a ship's lifeboat.

Lakonia reinforced in 1963 what several authors had explained in the middle to late 1950s. In 1955 Walter Lord published his best-selling history of the 1912 foundering of *Titanic*, probing as much at the failure of the social compact as at the failure of watertight compartmentalization. His grim *A Night to Remember* heralded other books: *The Last Voyage of the Lusitania*, by A. A. Hoehling and Mary Hoehling; Thomas Gallagher's *Fire at Sea: The Story of the Morro Castle*; Alvin Moscow's *Collision Course*; A. A. Hoehling's *They Sailed into Oblivion*. Except for Gallagher's, all the books appeared as paperbacks, and after about 1960 I bought them all and read each more than once, being especially transfixed by the lurid cover of *The Last Voyage of the Lusitania*. Taken together, these authors make it clear that shipwreck sorely tests the rules of humanity most people unthinkingly accept.

Airline travel had become so commonplace in the late 1960s that concern about passenger safety at sea struck many people, especially my college friends, as archaic. In the limbo between the end of trans-Atlantic liner service and the beginning of Caribbean cruising aboard small liners, questions about survival during shipwreck elicited only mild interest. Throughout the 1970s, as Hollywood produced a surfeit of disaster films involving airline crashes and blazing skyscrapers, my graduate student friends seemed less and less inclined to pay attention to fire drills. Disaster and the possibility of disaster bored them.

In subsequent years as a university professor I slowly realized that many undergraduate women know almost nothing of how past generations of women have managed in chaotic, often violent situations. National-emergency situations demanding military conscription cause many young women to reassess their understanding of military conscription, especially after they learn that the Selective Service form provides space for women to register. My women's studies colleagues tell me that the specter of the female conscript defeated backers of the Equal Rights Amendment. Maybe the specter of physical harm and the collapse of social order continues to deter women students from imagining themselves tossed into chaos.

Whenever I mention the word *lifeboat*, even in conversations utterly and completely unrelated to anything social or political, the word somehow connotes *women and children first*. Only the most ruthless thinkers among my feminist friends embrace the questions ideology begged after 1965. Why not put the children in a lifeboat crewed by strong men able to row and bail and steer? Should the strongest not carry the children to safety, away from mothers? Why not train everyone, including women, how to swing out davits and lower lifeboats? Why not let everyone heft a twelve-foot-long oar, then embark on physical fitness training like that John F. Kennedy envisioned? Bit by bit, however, I learned that such questions cause trouble in universities committed to certain sorts of equality but not others.

By the 1980s travelers avoided these issues to such a degree that they no longer read the airliner safety card, glancing about to locate emergency exits, count the seats between themselves and egress, think about *getting the women and kids out.* No one seemed concerned that the Boston-to-Denver plane often circled far into the Atlantic before heading west, that the Dallas-to-Boston flight routinely skimmed Boston Harbor before touching down, that the airliners carried no life jackets or life rafts, but only seat cushions usable for flotation. How do passengers get the seat cushions out of a sinking airliner, say an airliner that slides from the end of an icy runway into Boston Harbor? If women must remove high-heeled shoes before using the inflatable escape slides, what happens to them if the wrecked plane squats in a pool of flaming jet fuel? What do high-heeled shoes do to inflatable rubber life rafts, come to think of it? I started musing on such questions around the year 1971, when I began flying regularly out of Logan Airport in Boston. Logan runways extend into the sea in ways that give first-time passengers spectacular shocks as they look from windows and see whitecaps just beneath the wings. Those runways helped launch this book long before terrorists destroyed the World Trade Center towers.

A smug security grounded in credit cards and expense accounts and well-regulated white-collar office work, an unwillingness to imagine disaster away from the cinema or television or paperback thriller, an unswerving faith in electronic technology, maybe a secular fatalism too, all subtly infected most airliner passengers. And by then no one cared about liners, let alone about lifeboats. Few people took passage anywhere on a steamship anyway. If anything happened, "they" would come and rescue everyone.

But suppose *they* do not come? Suppose the radio operator fails to send an SOS or shout "Mayday" into the microphone? Suppose the automatic radio beacon fails to turn on when the life raft or lifeboat hits the water? Suppose *they* are too busy, or just do not care? Suppose the ship carries some dirty state secret, and *they* would rather everyone die than have it revealed?

Nasty or perverse or hardheaded in the old salt-marsh Yankee way, such questions lie at the heart of everything that follows here. Unlike the life jacket and life raft, the ship's lifeboat is designed and built to go somewhere safe, not only to endure the chaos of shipwreck and the violence of lowering from davits but to survive in any seas, carrying mostly or exclusively people inexperienced in small-boat handling. Designed, built, and supplied against disasters many ocean passengers deem unthinkable, the lifeboat itself raises many issues demanding scrutiny, some lingering from the early 1940s, questions I blundered into wholly by accident.

I first read Alistair MacLean's extraordinary *South by Java Head,* a 1958 adventure novel set in the days immediately after the 1942 surrender of Singapore, shortly before *Lakonia* burned. Its first-chapter description of the chaotic, dying city choking in the "sulphurous and evil" smoke of burning oil tanks, collapsed buildings, and blazing ships seared deeply into my psyche. For its handful of refugee characters, escape from Singapore involves extreme and immediate peril, woven into larger frameworks involving the collapse of well-ordered ways of life: the appalling, almost instantaneous destruction of a city, then of two ships; the enslaving, illusory freedom offered by the sea; the slow but steady conversion of a lifeboat into a microcosm peopled by accident. A grand sea story by a master storyteller, *South by Java Head* endures as one of the greatest in a narrow, half-forgotten literary genre once designated "tales of pursuit" or "tales of escape." It endures too as a memoir of Allied defeat, demoralization, and disaster, one of the few portals through which an American boy at mid-century might glimpse something of World War II history unmentioned in schoolbooks.

In decades afterward, *South by Java Head* and the *Lakonia* sinking endured at the back of my mind. Cities burn. Civilizations collapse. Then old things count: a wool shirt, heavy pants, thick socks, good boots, well broken in. Fresh water, some food stuffed in a backpack, matches. A sharp knife of high-quality carbon steel that takes and keeps an edge. A reliable compass. And a destination, a goal. But at sea—and sometimes in burning cities—much equipment morphs into a lifeboat.

Lifeboats sail through many accounts of escape from Singapore. Katherine Sim's *Malayan Landscape,* written by a painter determined to stay by her irregular-soldier husband, focuses on her escape south to Singapore: the continual bombing, the fraying of both civil and military authority, the devastating awareness that the great fortress city would fall. "Heavy black smoke from the burning oil tanks in the Naval Base drifted across the Island," she noted in her January 24 journal entry. "The warm, noxious-sweet stench of decay was appalling."[1] In *Singapore to Freedom* (1943), civil engineer Oswald W. Gilmour outlined not only his efforts to maintain the physical fabric of Singapore's roads, bridges, and public utilities—and bury the corpses Sim smelled—until the day before the surrender but also his astonishing escape from a city turned blast furnace. "Oil tanks and other supplies on the north of the island were set alight by our own forces to deny them to the enemy," he recounted. "Great volumes of smoke filled the skies and spread like a cloud over the town and surrounding country. The tropical sun was obscured behind a dense black mantle. It was the most depressing sky I had ever seen or ever want to see, and it remained thus for days." As Gilmour slipped from the waterfront toward the open sea, what astounded him most was the fire and smoke. "The oil islands of Pulau Bukum, Pulau Sambo, and Pulau Saborah in front of us were sheets of flame, rising as if in wrath from the dark blue waters."[2]

For the inhabitants of Malaya in 1940, war seemed half a world away, and the good life of colonialism seemed likely to last forever despite the Japanese invasion of China. Ordinary men and women trusted politicians and the military to protect them from harm, and for decades they had been right to trust. If war came, they reasoned, it would come gradually. They would have time to think, to plan, to act deliberately, and the authorities would direct their activities and provide for them. As H. M. Tomlinson remarked in his 1950 history of the Straits Steamship Company, *Malay Waters: The Story of Little Ships Coasting Out of Singapore and Penang in Peace and War,* "Is there anything better than an elaborate military establishment for encouraging a false sense of security and well-being, except perhaps the bottle?"[3] If the fall of Malaya illustrates anything at the start of the twenty-first century, it is that peace, civilization, the Order of Things are as fragile as *Titanic* slicing itself against an iceberg or *Andrea Doria* the graceful gored by an equally graceful liner.

Until very recently, that fragility scarcely interested most Americans, and it most certainly did not interest my students.

After September 11, 2001, alumni who once endured my remarks on great forest fires and blizzards, the chaos of the Civil War, and the way women rescued themselves and their children from the 1906 San Francisco earthquake began writing and telephoning that what made them uncomfortable as students suddenly made sense. Many in New York pointed out that Manhattan is indeed an island, and that Brooklyn and New Jersey suddenly seemed very far away, and a person who works in New Jersey but has a toddler in day care in Manhattan now saw New York Harbor differently. Several told me they thought of the lifeboat in the pictures hung on my office wall.

So I stand, hands around my mug of Canadian tea, and watch the gale bend the great Norway spruce that blocks my view of the barn. Another squall

Over and over, shipwreck images feature a lifeboat alongside foundering vessels.
(Courtesy of the Mariners' Museum)

of snow, and the electricity dims again. The bulbs burn red for an instant, then brighten. No day for the computer and the World Wide Web. I walk into the study, get the brass boat lantern and shake it to hear the swish of high-grade lamp oil. I leave the brass case filled with NATO-issue lifeboat matches certified to light and stay lit when wet but so expensive I never use them indoors. Military issue once, now military surplus. Maybe lifeboat thinking, at least *my* lifeboat thinking, originates in some vestigial Cold War preparedness frenzy, maybe in hardheaded thinking about the aftermath of unlikely disaster. If a meteor storm or a solar flare instantaneously destroys all computerized equipment and creates long-lasting famine, or if plague rages across North America, the lifeboat in the barn will be of little use, although it might make Greenland in a pinch.

I light the boat lantern, marvel at its nearly watertight design, trim its wick into white light. Maybe I have a lifeboat in the barn because I have never once owned an outboard engine I trusted to start when I wanted it to start. An artifact, the lifeboat in the barn represents both the enduring possibility of things failing and the opportunity to snatch salvation from disaster.

Victors write history. The successful have the time, money, and motive to write their side of events. Failures rarely write, and in any case Americans have little patience with them and their treatises. A lifeboat snug on its chocks whispers of that great potential failure, shipwreck; but a lifeboat sailing across an empty sea or pulling slowly and hesitantly toward land whispers of the possibility of survival after failure. Yet whatever the victory of the lifeboat's passage, the immense failure of shipwreck always overshadows it.

Even the most upbeat of lifeboat accounts, Guy Pearce Jones's 1941 *Two Survived: The Story of Tapscott and Widdicombe, Who Were Torpedoed in Mid-Atlantic and Survived Seventy Days in an Open Boat,* begins with the devastating attack upon the British steamer *Anglo-Saxon* by the German raider *Weser.* Neither its ex–Royal Marine gunner nor the Royal Navy saves the steamer itself, and indeed the Royal Navy fails, as it did in the Great War, to keep German surface ships from the sea-lanes a thousand miles west of the Cape Verde Islands. Moreover, William McFee points out in his introduction to the book that the log of the castaways "might direct public attention to the criminal neglect of ordinary, common-sense provisioning of life-boats carried on cargo vessels."[4] He might have added that the narrative suggests other criminal neglect, say a wheelhouse protected only by a breastwork of concrete building blocks or the master of *Anglo-Saxon*'s failure to specially prepare his lifeboats for a long passage, despite the well-known risks of wartime.

A lifeboat represents the chance of knowledge overwhelming secrecy, something that as a citizen and scholar I value more than knowledge conquering ignorance. The lifeboat makes land some dawn, sails close inshore, then backs through surf to grind onto shingle, sand, or coral, and its occupants stagger ashore and meet the locals, who summon physicians and civil or military authority and who spread the tale of shipwreck. At least usually the tale spreads, as it did across the Bahamas when a farmer and his wife found Tapscott and Widdicombe crawling ashore on Eleuthera Island.[5]

But not always. Sometimes, especially in wartime, the shipwreck remains secret, although the survival of castaways, even the arrival of an empty lifeboat, makes it difficult to keep quiet.

In the decade I have researched this book, tales have come to light of

Princess Sophia, USS *Juneau*, HMT *Rohna*, and other ships sunk in both peace and wartime and nearly lost to popular memory and scholarship, often by deliberate effort that has no place in democratic societies. Only in 1990 did Ken Coates and Bill Morrison publish *The Sinking of the Princess Sophia: Taking the North Down with Her*, an incisive analysis of a 1918 catastrophe in Alaskan waters in which a steamship running blindly at full speed through a howling snowstorm struck Vanderbilt Reef, and all 353 passengers and crew perished. "It is hard to understand why the story of the *Princess Sophia* is not better known in North America, particularly in the Pacific coastal regions and the far northwest," they write, but then again, no one lived to tell the tale.[6] Until Dan Kurzman published his *Left to Die: The Tragedy of the USS Juneau* in 1994, no one had pieced together the interservice stupidities and intra-navy personal rivalries that in 1942 shaped the much-belated rescue of ten men out of the torpedoed cruiser's crew of seven hundred. And only in 1997 did Carlton Jackson reveal one of the best-kept secrets of World War II, the 1943 sinking of a British troopship, HMT *Rohna*, by a German airplane-launched guided missile, one of the first "smart bombs" ever used in combat.

Jackson's book is sobering reading. More than a thousand American soldiers died when the missile sank the ship, and so anxious was the United States government to keep the American public—and perhaps the men of the army if not the navy—from knowing of the newly developed weapon that it attempted to keep the incident secret from the men involved in it. The army dumped the survivors ashore on the African coast and deliberately delayed them from rejoining their units simply because they knew too much about Axis might.

Salvaging any history of ships' lifeboats involves navigating darknesses long unlit, except maybe by burning oil saturating midnight skies. World War I British and American government censorship exasperated and perplexed American newspaper reporters and other citizens who saw lifeboats pulling slowly into New Jersey and Delaware beaches with the news of U-boats just offshore, submarines whose existence the United States Navy at first denied. Perhaps no World War I incident more jarred the American public than the June 4 arrival at Atlantic City of a lifeboat from the torpedoed liner *Carolina*. A festive Masonic Knights of the Mystic Shrine parade along the boardwalk stopped suddenly as its band realized what sort of boat had begun to make its way through the surf. Within moments, Masons, onlookers, bathers, and lifeguards plunged wildly into the surf, and the twenty-nine men, women, and children of the lifeboat arrived on the sand to strains of the national anthem and roaring cheers. Almost instantly, rumors and consternation swept the coast, then the nation. A retired admiral warned that the U-boats might have seaplanes aboard, and the New York City police, trying to forestall aerial attack, immediately forbade all outdoor display lighting. Rumors of secret Mexican U-boat bases frazzled Washington, D.C.[7] This panic was driven by censorship: after the lifeboats began arriving, thoughtful Americans, and especially journalists, no longer trusted War Department news releases.

Throughout World War II, the British government "secret notice" slapped on incident after incident, and the American penchant for secrecy born in the defeats following Pearl Harbor, ensured not only that news media remained ignorant but that decades later scholars might learn only accidentally of events hidden from immediate postwar historians writing officially sanctioned narratives. When two young women found a bullet-riddled ship's life preserver on a Texas beach in 1942 and the *Galveston Daily News* published a photograph of them holding it, the paper made an oblique comment on the usefulness of newly inaugurated Coast Guard foot patrols.[8] On Gulf Coast beaches Texans suddenly realized that the collapse of world order meant Singapore in flames and U-boats striking with impunity half a world away.

Despite American popular interest—especially cinema- and television-initiated interest—in marine disasters, few people outside Germany know of *Wilhelm Gustloff* and *General Steuben,* two liners torpedoed by a Soviet submarine in the last days of World War II. *Wilhelm Gustloff* alone took down seven thousand people, more than the number who drowned in the founderings of *Titanic, Lusitania, Athenia, Andrea Doria,* and *Empress of Island* together. Another three thousand at least were lost aboard *General Steuben,* a graceful old liner once famous on the Hamburg–New York run.[9] Embedded in the little-known but stupendously successful effort of German naval and merchant service forces to transport two million refugees along the Baltic away from the Red Army—an effort that laid the foundation for the economic prosperity of West Germany—the sinking of the two overcrowded liners remains obscure for reasons other than sheer horror and political nightmare.[10] Throughout World War II *Queen Mary* and *Queen Elizabeth* steamed unescorted back and forth across the Atlantic. Having decided that the great speed of these ships secured them from U-boat assault, both American and British governments accepted the risk of a U-boat sinking liners racing eastward, each crammed with seventeen thousand American fighting men.[11] What happened to *Lusitania,* HMT *Rohna,* and *Wilhelm Gustloff* might well have happened to *Queen Mary* or *Queen Elizabeth.*

Lifeboats not only frustrate secrecy but also advertise the enduring vagueness of marine charting. Often they reached uninhabited islands and islets even now infrequently identified on navigational charts and often entirely absent from the atlases and maps used by the educated public. Places like French Frigate Shoals in the Pacific remain largely uncharted today, and ordinarily their absence frustrates almost no one. Yet on June 2, 1942, the shoals figured in an important and eerie coincidence when both the United States and Imperial Japanese navies elected to use them for seaplane refueling. An American seaplane tender arrived first at the anchorage, unknowingly causing Japanese refueling submarines to abort a long-distance aerial reconnaissance mission over Pearl Harbor that would have warned Admiral Yamamoto of the trap history knows as Midway. But who can find French Frigate Shoals on a contemporary globe or map?

Into oblivion, cartographic and otherwise, lifeboats sailed and sail and will always sail, thrusting away from sinking liners and freighters and tankers and cities and nations, nosing out downwind, toward the nearest sea-lanes or the nearest leeward land, the nearest safe place. All too often the castaways succumb to the evils peculiar to lifeboats, perhaps especially to the novelty of the lifeboat passage itself, as so many British seamen did in World War II. "The men were accustomed to work in big ships, where the motion was comparatively slight, the quarters reasonably comfortable, and the work often not actively concerned with real seafaring," Allan Villiers wrote in *And Not to Yield: A Story of the Outward Bound.* "They were good crewmen, but they were not always seamen, and they hadn't an idea how to look after themselves or anybody else when they were cast adrift in a small boat on the North Atlantic or anywhere else."[12] Outward Bound evolved, from lifeboat training and real seafaring efforts, into the educational institution that familiarizes people with storms and electric failure and personal resilience.

In his *Death of a Supertanker,* Antony Trew probes at the modern separation of merchant seamen from the sea at which Villiers and others wondered. Of an ultramodern ship headed for catastrophe he writes, "The windows were shut to keep out the fog, and the atmosphere was pungent with the mustiness of charts and books, the stale smell of deck polish, of coffee and of human bodies, and the indefinable yet unmistakable odor of electronic instruments."[13]

"Newly opened gear is likely to stink," warn the corporate authors of *A Practical Guide to Lifeboat Survival,* a 1990 publication of the Center for the Study and Practice of Survival, in Pornichet, France.[14] The wrench from familiar smells to the stomach-twisting ones of plastic, rubber, nylon, and other materials of contemporary life rafts demoralizes because it announces unseasoned innovation. But better the stink of plastic spray hoods and ponchos than those of vomit, gangrene, and death, all smells the authors of *A Practical Guide to Lifeboat Survival* painstakingly identify after focusing on the stench of fear.

Fear takes passage in every lifeboat, but lifeboats carry some fears better than plastic rafts do. When an oil-scummed harbor burns, when a burning ship glows brick red and spills its flaming cargo onto the sea hundreds of yards in every direction, when the volcano erupts and showers the harbor with sparks, when incendiary bombs explode high overhead, then the lifeboat smolders but proceeds while plastic rafts shrivel and melt. Fire separates the life raft from the lifeboat, but about fire at sea even the most dedicated of contemporary survival experts say little. Fire frightens plastic-raft advocates into silence: the possibility of fire keeps lifeboats aboard ships today, artifacts from the days when torpedoed ships sank in pools of flaming gasoline and whole cities burned and no one came to help.

Definition vexes the casual inquirer. Only in books published in the last decades of the nineteenth century does the word *lifeboat* begin to designate a particular sort of ship's boat, although the term seems to have been used thus in spoken English a few decades earlier. Certainly by the 1890s, mariners writing autobiographically used lifeboat to identify ship's boats used in rescue work thirty or forty years earlier, and they meant boats other than shore-based rescue craft. While Noah Webster defined the lifeboat in 1826 as "a boat so constructed as to have great strength and buoyancy, for preserving of lives in cases of shipwreck or other destruction of a vessel at sea," he clearly meant a shore-based surfboat. The definition endured essentially unchanged through the 1890 edition of *Webster's International Dictionary;* not until the 1934 edition did lexicographers add a secondary definition, one focused on ship's boats kept ready for emergency use.

Tracing definitions of lifeboat illuminates not only the lack of attention dictionaries paid to marine terminology in an era of great change but also the significance of shore-based lifeboats in the minds of landsmen. But war permanently skewed usage. Peacetime tourists watched shore-based lifeboats in drills and rescues, but wartime readers learned of ship-launched lifeboats in one newspaper story after another. Postwar dictionary publishers focused on what the 1966 Funk and Wagnalls *Standard Collegiate Dictionary* defined as "a boat constructed and equipped for saving lives at sea in event of shipwreck, storm, etc.: especially such a boat carried on board a larger vessel."

On Christmas Day 1940 the German pocket battleship *Admiral Scheer* and a captured refrigerator ship rendezvoused with the raider *Thor* and three

Kriegsmarine supply ships some six hundred miles north of Tristan da Cunha. Two days later, the German cruiser *Komet*, having left its Baltic base for the Pacific via the Arctic Ocean and Bering Straits, shelled the phosphate plant at Nauru Island upon which depended the agricultural productivity of Australia. And in February 1941 two German raiders and their prizes rendezvoused near Saya de Malha Bank in the Indian Ocean. The German surface warships wreaked havoc among merchant ships, sinking and capturing vessels not only on regular sea-lanes but in regions as unfrequented as the Antarctic whaling grounds.

U-boats worsened the situation. When U-159 fired a torpedo on November 5, 1941, in South African waters, the vast explosion atomized the targeted vessel, slightly injured three U-boat crewmen, and startled the Cape Point lighthouse keeper three hundred miles away. Postwar South African historians suspect U-159 hit the United States ammunition ship *La Salle,* but perhaps it also destroyed the Greek ammunition ship *Aegeus,* since both ships vanished without trace at the same time.[15] German and Japanese warships turned up just often enough in remote areas that even late in the war any unusual inshore activity precipitated near panic. When United States Army tugboat ST250, its compass temporarily defective, signaled for a pilot one night off Limón in 1944, Panamanian officials scrambled to drag a century-old cannon to the harbor parapet, evacuated women and children, and prepared to defend their city against coordinated German assault.[16] Nowadays only military historians and thoughtful mariners know anything of the reach of Axis naval forces and the terror that reach inspired, let alone the loneliness of so many wartime castaways.

No mariner thinks ocean travel perfectly safe. Mariners know what journalists seem to ignore, the regularity of cruise-ship fires for example. In 1980 *Prinsendam* burned and sank off Alaska, all of its 524 passengers and crew remaining in lifeboats for thirteen hours before rescue ships arrived—just ahead of a fierce storm. Fourteen years later, *Achille Lauro* caught fire for the second time at sea (it had suffered two other fires twice when docked), and 975 people abandoned the liner in its lifeboats.[17] Foreign-flag cruise ships experience small fires frequently, and any of these holds the potential for disaster. The world cruise-ship industry is a flag-of-convenience business governed by safety regulations far less stringent than those Congress imposes on United States vessels. Cruise ships are often built of flammable material that produces poison gas once afire, and crewmen may speak a language alien to both ship's officers and passengers. In the end, the passengers themselves may have to launch lifeboats in the midst of black smoke and panic.

Despite all sorts of modern rescue systems, passengers and crew in the end must depend on lifeboats and await rescue, not by helicopter or even by warship, but by the nearest ship, usually a cargo vessel some distance away. Off the coast of Somalia, the crew and passengers of *Achille Lauro* discovered that their lifeboats floated in an ocean nearly bereft of ships. Most small nations, especially those so remote and "romantic" that ecotourism cruise lines provide the only comfortable means of visiting them, lack anything in the way of ocean-rescue services. Not until the arrival of the bulk carrier *Hawaiian King*, a cargo vessel fitted to carry no more than fifty crewmen, did the *Achille Lauro* castaways see any hope of rescue, and the thousand castaways spent a full twenty-four hours aboard the cargo ship until more help arrived. *Achille Lauro* sank in desolate seas like those crossed by so many cruise ships steaming toward deserts and Antarctica, where ships' lifeboats remain the only genuine hope of salvation, maybe more so now than a few years ago.

"We have seen and heard reports that when a ship gets into a bad storm and gets into trouble, the first thing to go is satellite communications," the Canadian Coast Guard warned five months after the International Maritime Organization phased out Morse code on February 1, 1999. The high-tech Global Maritime Distress and Safety System that replaced the public and private networks of commercial radiotelegraphy does not reassure the Canadians, whose territorial seas are some of the most dangerous in the world. "Morse code was right there 'til the ship flipped under the water."[18] But Morse code is gone, companies like Globe Wireless no longer transmit and receive, the radio listeners are unemployed, and if the Canadians are right, then somewhere soon a ship's lifeboat will make land after a long and unnoticed passage.

What follows here, then, is a lifeboat narrative, but not a narrative of a particular passage. Instead it is a careful if incomplete look at an artifact that makes most people uncomfortable and at the environment peculiar to that artifact. Here is a sort of natural history of lifeboats.

A chapter on maps follows this introduction because lifeboat passages all too frequently thrust castaways, even castaway mariners, into regions steamships rarely traverse. A lifeboat sails toward safety, but only rarely is the safety found the original destination of the foundered ship: more often, safety lies in some unfrequented place. What is missing from charts leads on in chapter 2 to what went missing from seamen decades ago, about which Villiers and others anguished in the earliest years of Outward Bound, and

1

MISSING

In the summer of 1909, the Blue Anchor Line passenger and cargo liner *Waratah* went missing off the Cape of Good Hope. An almost new, medium-size ship, *Waratah* exemplified the sort of vessel that routinely called at ports punctuating very long passages between continents. On July 26, after discharging cargo and taking on coal, *Waratah* left Durban, on the east coast of South Africa, on a three-day coastal passage for Cape Town, the second stop on its Port Adelaide-to-London run. At four in the morning of the second day out, a day of splendid weather, *Waratah* overhauled a freighter proceeding in the same direction. Then the weather deteriorated rapidly, as it often does in winter off the headland first named Cape of Storms by the fifteenth-century Portuguese. After overtaking *Clan MacIntyre*, *Waratah* passed another vessel, the tramp steamer *Harlow*, close enough to be recognized, but by then visibility had worsened dramatically. Soon hurricane-force winds drove against the prevailing Agulhas Current, creating a very dangerous cross sea chopped by the shallowness of the ocean and by proximity to land. "We experienced a great storm—I had never met with anything as bad on this coast during my thirteen years in the trade," the master of *Clan MacIntyre* later reported of the force that drove his ship astern while its engines ran full speed ahead. But neither *Clan MacIntyre* nor *Harlow* suffered much in the way of damage or delay, and both arrived at their destinations in good order.[1] *Waratah* vanished.

Insurance-company lexicons omit *vanished*, and instead include a number of other terms still employed precisely by Lloyd's of London, the great marine-casualty underwriting firm that insures almost all the world's shipping. A ship that fails to arrive in port on schedule is first classified as *overdue*, the sailing-ship-era term now accounting for the slight inexactitudes of steam- and diesel-powered navigation and having no specified degree of lateness attached to it. The word appears in notices warning insurers that

which is so fundamentally vital in every castaway lifeboat. Chapter 3 concerns passages of ship's boats in the era before steamship travel. Awareness of those passages heartened castaways embarking on similar efforts in ships' lifeboats in the steamship era, especially during the Great War and World War II, and writers of survival manuals still use them as exemplars.

The naval small boat and coastal lifeboat form the core of chapter 4, because the ship's lifeboat evolved not only from ship's boats like the launches of HMS *Bounty* and the California-bound American clipper ship *Hornet* but also from fishing craft such as dories and peapods, along with naval small craft and the land-launched rescue craft of the British Royal National Lifeboat Institution. A private charity that provides ocean-rescue service around the British Isles now as it did 170 years ago, and helped mold the old United States Life-Saving Service, which Congress made into the Coast Guard in 1915, the RNLI dispatched some of its lifeboats offshore to help evacuate the British army from Dunkirk in France. So traumatic did the Dunkirk evacuation prove that after 1940 there developed, at least in English-speaking nations, a lasting confusion about *lifeboat* designating coast-based vessels or ships' lifeboats. Chapter 5 focuses on the ship's lifeboat itself, what generations of mariners knew and know as the "Board of Trade" boat, the double-ended, wooden, oared, lugsail-rigged lifeboat specified by the British government and accepted and copied the world around. For contemporary Americans—although not for Britons—that boat remains an integral part of cultural memories like the sinking of *Titanic* and Hollywood films like Alfred Hitchcock's *Lifeboat*. Launching lifeboats organizes the whole of the subsequent chapter, for unlike most boats, lifeboats are not ordinarily in water.

Lowering a lifeboat from its davits, a less than easy operation on a sunny day when a great steamship lies at anchor or alongside a wharf and a government examiner springs a surprise examination, turns into an awesome task in storms, after explosion, during fire, especially when a ship lists to one side. So brutal is it that for decades governments have insisted that some seamen be specially trained as *lifeboatmen*. But often lifeboats must be launched not by seamen, not by lifeboatmen, but by passengers, sometimes by landsmen unfamiliar with anything nautical. Unlike lowering a lifeboat, navigating one is often a very long-term task indeed, one that shapes chapter 8. But in this book *navigation* designates more than directing a lifeboat somewhere, using whatever clues the stars and clouds and winds provide. It includes probing the moral wildernesses into which lifeboats inevitably move. Chapter 9 focuses on women, the women of the old cry "Be British:

Women and Children First." Women—or at least special sorts of women—have done stunningly well in lifeboats, often surviving far longer than men, and all too frequently pulling a twelve-foot-long oar, holding a tiller, and providing everything from moral guidance to medical care.

Chapter 10 focuses on the moral order and disorder that suffuses so many lifeboat narratives, the problems of morale, discipline, and cannibalism that lurk in the background of so many alongshore tales. Chapter 11 traces the transformation of tales and memory into cinema imagery, and the post–World War II memory of lifeboats kept alive by the thousands of lifeboats converted into pleasure boats, most with cabins, almost all of which disappeared in the 1970s.

Many subjects are not mine. Here is no history of the United States Coast Guard, no analysis of shipwreck, no detailing of naval architecture, no emphasis on military personnel. Perhaps the last deserves special attention. Men and women under military discipline behave differently, and ordinarily have no access to lifeboats anyway. Sailors are not the seamen who crew merchant vessels, but naval personnel who rarely deal with soldiers, let alone civilians. The navies of the world employ specialized vessels, particularly motor whaleboats, and few warships have space for lifeboats. So while now and then this book analyzes narratives like Ray Parkin's *Out of the Smoke* and John Morrill and Pete Martin's *South from Corregidor,* two accounts of escape from Japanese forces in the early years of World War II, its focus is on civilian castaways, seamen and passengers alike.

Merchant seamen and merchant-ship passengers have written with surprising frequency and at surprising length over the years of their lifeboat experiences. Narratives like *1700 Miles in Open Boats* by Cecil Foster, master of the foundered steamship *Trevessa,* and *Lifeboat Number Seven* by Frank West, *Blue Hell* by Derek C. Gilchrist, and *Standing Room Only* by Elizabeth Fowler, the last three books authored by lifeboat-passage survivors of varying experience, form one of the great channels opening into the subject of this book.

Only rarely do sailors board lifeboats, and all too frequently they do badly, expecting these vessels to perform like naval small craft. Over the decades, naval personnel have acquired the deepest respect for those who bring lifeboats to safety. "We hailed it from the bridge, and were answered by a torrent of Chinese," remarked Nicholas Monsarrat in *Three Corvettes* of a "bobbing black spot" in a gale-swept winter night in the wartime North Atlantic. "I thought, 'God, this is going to be difficult. . . .'" But at a single shout from the Welsh second officer at the tiller the castaway men fell silent, and the lifeboat swept crisply alongside the stopped warship, its occupants

disembarking into military discipline before Royal Navy officers impressed with the multilingual boat handling of the merchant service.[19]

In the dusk running into dark I await the arrival of my wife and our sons, wondering about the gale, the slippery roads, the likelihood of downed tree limbs. I think about another cup of tea, then glower at the useless electric stove.

Out in the barn, snug in its cradle, squats a ship's lifeboat intended to survive far worse gales than this. Out in the barn, snug in its cradle, squats a reminder that things go wrong, and that sometimes no one comes to help.

some partial or total loss may have occurred, and that they should prepare themselves to pay. In a day or two, Lloyd's advertises the overdue ship as *a vessel for inquiry,* meaning simply that the firm actively seeks any information about the overdue ship. If a week of advertising produces no information, Lloyd's labels the ship *missing,* as it has since 1873. *Missing* really means sunk. Only one ship, the square-rigged *Red Rock,* which spent nearly five months of 1899 sailing a thousand miles against prevailing winds in the Coral Sea, has ever been posted missing and subsequently made port.[2] Telegraphy began transforming insurance-company procedures late in the nineteenth century, of course, connecting many ports by landlines and undersea cables and quickening information gathering.

Wireless telegraphy altered matters again early in the twentieth century, enabling delayed steamships to report their positions and forestall what used to be an implacable piece-together-the-puzzle process demanding the best efforts of shipping agents, harbor pilots, and above all masters of sailing ships and steamers. The master might recognize some vessel on the inquiry list as the ship he had heard about anchored for repairs in an unvisited bay, or glimpsed far off course, laboring in heavy weather. But the old terminology endures despite the telescoped notification process.

"On some days there appears the head 'Overdue'—an ominous threat of loss and sorrow trembling yet in the balance of fate," wrote Conrad in *The Mirror of the Sea* in 1905, before wireless telegraphy changed everything. Within three weeks or a month, the names of many overdue ships appeared again in the missing column, in the "strictly official eloquence" that constituted what he knew as the funeral orations of sunken ships. Conrad wondered at length about the connotations of *missing,* determining that in the word "there is a horrible depth of doubt and speculation" about how the ship foundered. "Nothing of her ever comes to light—no grating, no lifebuoy, no piece of boat or branded oar—to give a hint of the place and date of her sudden end."[3] Only ambiguity remained for the readers of the shipping columns, an aching, endless unknown.

A bit of wreckage, especially lifeboat wreckage, however, usually moved a ship in time from the missing column to another, *lost with all hands.* Insurers knew that while a ship might have sunk, its crew and passengers might yet turn up, having sailed in lifeboats across parts of oceans routinely avoided by sailing ships and steamers alike. Moreover, the lifeboats may well have reached islands so remote that the castaways might live for months, even years without anyone knowing. For underwriters of ships and cargoes, *missing* meant an immediate need to pay claims, but underwriters of human lives expected to wait an interval of years before paying, since determined men and women in a well-found lifeboat might turn up alive and well months later.

But *Waratah* vanished in the July storm, and within days three Royal Navy ships, another Blue Anchor ship, and a ship chartered by the Blue Anchor Line began searching, steaming further and further south, finally into the latitudes where sailing vessels tracked the gales east to Australia. Authorities queried every sailing ship arriving in Australia—more than a hundred arrived between August and December 1909—and masters of sailing ships making land in Chile and Peru found consuls and shipping agents asking for information about anything, ship or wreckage or bodies, sighted in the stormy, southernmost latitudes. The search intensified even as it widened, for in the minds of everyone lingered the example of another steamship, *Waikato*.

In June 1899, bound from London to Australia, *Waikato* snapped its propeller shaft about 180 miles south of Cape Agulhas and began drifting helplessly toward Antarctica. For fourteen weeks the liner drifted southward, and while several sailing vessels spotted it, and one even tried to tow it, not until the steamship *Asloun* found it did efficient salvage begin. *Waikato* arrived, under tow, in Freemantle on October 9, sixteen weeks after snapping its shaft, its crew reduced to eating macaroni, sardines, and baby food from the cargo. It had drifted far from steamship lanes into seas still crossed by the last of the great sailing ships.[4]

What happened to *Waikato* happened to other steamers in far southern waters, and a nearly identical incident profoundly affected Conrad. But for the astounding coincidence of a whaling ship spotting it in a snowstorm and matter-of-factly reporting in Australia having sighted a disabled steamer in latitude 50° south, longitude uncertain, the New Zealand–bound steamship almost certainly would have foundered against ice. Conrad knew the second officer of the unnamed steamship, and mused on the incident at length in explicating the changing meanings of *overdue* and *missing*. He decided that the derelict steamship represented a modern-day horror of seafaring. Unlike the sailing ships Conrad commanded, which even if dismasted could be repaired with jury rigs and the setting of a bit of canvas, a steamer minus its propeller becomes something like a drifting log. A ship whose career had been "a record of faithful keeping time from land to land, in disregard of wind and sea," suddenly becomes nothing more than a steel float.[5] Aboard a steamship lacking its propeller and shaft, even the most competent of mariners suddenly had nothing much to do except hope that another ship would sight the derelict.

On December 8 Lloyd's posted *Waratah* as a "vessel for inquiry," and a week later posted the ship as missing. But hope endured, not so much for the ship but for its crew and passengers. On February 25, 1910, the steamship *Wakefield* left Durban for the far southern latitudes and spent five months, much of it in storms, visiting places ordinarily avoided even by the sailing ships running the "easting down" from the Atlantic to Australia. By increments, *Wakefield* touched the Crozet, Heard, and MacDonald islands, then Kerguelen Land, before proceeding to Australia. Its crew found no sign of *Waratah* or its people. When the British Board of Trade opened its public inquiry in London on December 15, 1910, it found itself embroiled in what became a ten-week furor.[6] How, demanded the British public, could an ocean liner disappear without trace? In Britain and the United States, the public mind slipped deeper and deeper into the depths of mystery, dread, and speculation Conrad so acutely identified. Rumors swept through the British Empire: one reporting that a *Waratah* deck chair washed ashore at Coffee Cove on the coast of South Africa, another suggesting that the ship blew up within sight of another liner that ignored its flash, a third intimating that the great seas sucked the liner into a "blowhole" cliff on the African coast. For decades after it vanished, *Waratah* continued to plague the public imagination.

A passenger liner had vanished without trace. In legal terms, it had not sunk with all hands, since no one had seen it sink and no trace of it had ever appeared. Perhaps it had been hijacked, if not to St. Paul's Rocks, perhaps to St. Paul Island.

No incident opens more fully on turn-of-the-century marine mystery than the *Waratah* disappearance. Within a period of six years the world confronted the loss of three great steamships, each with horrendous loss of life, *Titanic, Empress of Ireland,* and *Lusitania,* but each of these catastrophes had witnesses. Among experts, the *Waratah* disappearance diminished to a historical benchmark, the last prewireless big-ship sinking, something about which mariners mused when they doubled Cape Horn in seas already haunted by the Flying Dutchman. But the *Waratah* disappearance engaged the public not only as a mystery but as a sort of gothic romance.[7] After all, its crew and passengers might have taken to lifeboats, might still be living on some remote, unfrequented, indeed unknown island.

Well into the twentieth century, the locating and exploring of islands and islets like St. Paul's Rocks remained the province of interested amateurs. Now and again naval survey ships examined little-visited islands, and in the long decades of peace following the Crimean War warships occasionally probed into seas rarely traversed by merchantmen, but parsimonious governments kept official exploring spotty. While parts of inland Africa and

Asia, and certainly of the Arctic and Antarctic, beckoned well-known and often well-financed explorers, hundreds of thousands of square miles of cold, frequently stormy seas attracted only the interest of sailing-ship masters like Nova Scotian C. C. Dixon, who routinely made very long passages, like one from Philadelphia to Pusan, Korea, a distance of 20,600 miles.

Dixon studied winds and currents to speed his passages and now and then disregarded official government advice and ventured into little-known regions searching for strong, steady wind. He spent years measuring and photographing waves (incidentally debunking tales of hundred-foot-high giants), analyzing icebergs, dropping current-study bottles for the United States Hydrographic Office, bounding the western edge of the Doldrums near Cape St. Roque off Brazil, and examining the extent and biological richness of the Sargasso Sea. "I am not taking any particular credit" for taking ships far from regular routes, he recalled in 1933, "because, speaking by and large, the accepted routes were the safest ones to follow, and a skipper's first and last duty is to protect his ship."[8] But Dixon was all his life a student of "practical meteorology," which revealed the hitherto unknown potential of prevailing winds. He trusted his seamanship and scientific knowledge enough that he sailed routinely in unfrequented regions, and he recorded observations, took soundings, and made sketchy surveys of little-known islands.

He crossed and recrossed the Sargasso Sea and collected writing about it from Plato to Columbus to contemporaneous popular magazine writers. He debunked its supposed dangers, determining that journalists exaggerated its horrors. He studied the coral snakes, the masses of weeds (which he collected with a drogue and calculated to weigh about "10.9 tons to the square nautical mile"), and the channels in the miles of floating weed. He realized that the opening and closing in these channels might have struck sixteenth-century becalmed mariners as something supernaturally insidious.[9] His dozens of passages across the Sargasso provided him with fodder for at least three scholarly articles.

In the Royal Geographical Society's *Geographical Journal* of November 1925, Dixon published a lengthy analysis of the Sargasso Sea that built on two briefer articles outlining his attempts to determine the weight of weed by area. Accompanied by splendid charts, "The Sargasso Sea" is part description and part hypothesis and concludes that the sea is either shifting location or decreasing in area. All of it derives from Dixon's own observations in twenty-five sailing-ship and seven steamship passages across the sea. When becalmed in the Sargasso, Dixon not only put his instruments overboard to test the observations of Darwin and other scientists but recorded what he did *not* see, especially sharks. Frequently he wondered at the differences in point of view resulting from being aboard a modern sailing ship, a sixteenth-century sailing vessel, and a ship's boat.[10] "The Sargasso Sea" remains a model of geographical writing, one buttressed by quantitative data and a healthy dose of skepticism, and it demonstrates that Dixon and his fellow sailing-ship masters recognized that the steamship era offered scant opportunity for sustained scrutiny of the sea.

Running through much of Dixon's thinking is an awareness that point of view changes what mariners see. He knew that the Sargasso seemed far more solid to sixteenth-century seafarers because from their small vessels they could survey such a limited area that the weed seemed endless. From a small boat the channels in the weed look like meandering creeks, and Dixon concluded that anyone steering a small boat along what "are in truth phantom rivers" would be disoriented unless he followed compass bearings. At least once in the Sargasso Dixon found tangible if rotting evidence that something of the old tales endured. He discovered a derelict ship and boarded it in calm weather, but could learn nothing of its identity because so much of the dismasted hulk had rotted. It struck him that the derelict was as grotesque as a fish he had caught in the Sargasso and had preserved for presentation to the British Museum; it put him in mind of the "air of mystery about abandoned ships," especially about those found abandoned but with boats intact.[11] Once at sea in lifeboats, point of view quickly changed and perhaps became as antiquated, and indeed as grotesque, as the odd fish and the derelict he encountered in the Sargasso.

Because abandoning ship thrusts castaways into alien perspectives and often into seas bereft of ships, Dixon felt that his observations of out-of-the-way regions might be useful to mariners in distress. His explorations offer a way into understanding the furor surrounding the *Waratah* disappearance as something involving both turn-of-the-century point of view and turn-of-the-century half-charted seas.

On the slimmest of evidence—a domestic cat prowling around the broken-into emergency food supply on Hog Island in the Crozets in 1910—Dixon argued that the cat proved that some ship had wrecked on the islands and wondered if *Waratah* castaways, or the ship's cat in an otherwise empty lifeboat, might have reached the place.[12] In 1910 educated people agreed with experienced mariners like Dixon that *Waratah* might well have drifted south to the Crozets, but perhaps more important, in 1910 the educated, sea-traveling public and mariners alike knew that lifeboats from *Waratah*, in particular weather conditions, might themselves have fetched up on Hog

Island. In 1910 those knowledgeable about the sea knew of extraordinary open-boat passages made from sinking ships and knew too that most mariners, especially officers, had trained in sail even if they served in steamships.

Training in sail provoked heated controversy for some fifty years after the 1880s, and fragments of the debate shaped public discourse until well into the 1940s. In the early decades of steam propulsion, all officers and men had "come up" in sail, and between 1850 and perhaps 1885, steam-powered vessels typically carried masts, yards, and sails routinely used in favorable winds and in emergencies. Increased engine reliability soon made auxiliary sail obsolescent, then obsolete, and larger engines demanded below-deck specialists, the engineers, stokers, trimmers, and oilers who understood reciprocating steam engines, coal fires, dark bunkers jammed with coal, and, subsequently, oil-fired furnaces. By the end of the nineteenth century a deep chasm divided deck officers and seamen from "the watch below": topside men and even passengers derided "the black gang" as fake seamen unschooled in the ways of sail.

Argument continued for decades about the proper training of merchant seamen, but it flared hottest when experienced master mariners disputed the best training for would-be officers, especially for cadet apprentices. Most men trained in sail took the position Captain D. J. Munro outlined in 1929, recalling his cadet life forty years earlier aboard the clipper ship *Wild Deer*. While Munro spent only four years in sail, he learned a lifelong suspicion of "steamboat sailors," partly from watching the master of *Wild Deer* race a steamship from Australia to London, and win. "No form of training ever yet devised will equal that received by a boy in sail, if courage, confidence, resourcefulness and ability to command are the desired qualities," Munro asserted in a chapter devoted to the issue. The young man training in sail "takes an interest in the sea and sky because the progress of his ship depends so much on their moods."[13]

Munro was no crank. He based his argument not on nostalgia, and not even so much on his own youthful experiences, but on his observations as a Royal Navy officer in World War I, when he served as Commodore of Convoys. In only two instances did officers trained only in steam perform as well as those trained in sail and steam both, he concluded somberly of his decision to promote no more steam-only men. He insisted that his evaluation applied also to officers in the Royal Navy: "Those who trained in the old squadron sailing vessels knew a different—and better—seamanship."[14] Perhaps Munro worked badly with officers whose experience differed from his own, but his analysis of why British merchant officers and seamen should train in sail echoes that of many of his contemporaries who glimpsed something unsettling in the chaotic years of the Great War. Postwar reflection coalesced into new respect for traditional master mariners.

"Captain West was a typical seaman of the old school—rough, masterful, and equal to all emergencies," Rex Clements remembered of the man he served under as apprentice and officer for eight years. "Steamers he held in a large-hearted contempt, but the sea he knew and loved." West trained Clements and his three other cadets thoroughly, earning their respect and lifelong friendship, all the while giving them an attitude they had to mask aboard steamships. "Some of his methods would give a modern liner-officer cold shudders," Clements reported in his 1925 *A Gipsy of the Horn*. "Once when I was taking an azimuth he swept me and the instrument aside with a lordly gesture. 'A handful of degrees don't matter, my boy,' he said, 'take the sun like this'—laying a great hand edgeways on the compass and squinting along it at the sun—'and keep your eyes about you.'"[15] West loved his bark, the four-masted *Arethusa,* and he sailed it well, making swift passages in which he devoted whole days to fishing from the stern. Unlike Dixon, West spent his days examining hammerhead sharks and Peruvian octopi and bonito and every bird from penguins to albatrosses, not just ordering his apprentices to open their eyes but setting them an example of a man who used what he saw and knew of the ocean environment—and at times especially of fish—to sail his bark efficiently.

Clements remembered West as a master of small-boat sailing, and a far better teacher than the second officer, who had no patience with apprentices "who knew as little about stepping a mast in a seaway as a cow does of handling a musket." After one zany mishap in an Australian harbor, when the apprentices infuriated the second officer first by their inability to step a mast and control a lugsail, then by losing an oar overboard, and then by almost crashing into an anchored ship thronged with seamen shouting insults and advice, Captain West took the young men in charge and thereafter taught them not just the rudiments of small-boat handling but the details. He ordered his carpenter to convert the ship's gig from lugsail to yawl rig, then showed his apprentices how to manage the two-masted rig while they ferried him about anchorages and along the Chilean coast in search of stingrays and other out-of-the-ordinary animals. "It was a joy to watch him; in light winds he could keep the gig within four points of the wind, and in heavy weather would have her riding easily when anyone else at the helm would be taking water aboard wholesale," Clements recalled.[16] West made the apprentices learn to row every bit as thoroughly they learned to sail, often keeping them at the oars from six in the morning to eight at night while he explored inlets rumored to shelter rare fish and birds.

In time the training paid, and West rewarded the apprentices by letting them demonstrate their skill to onlookers. At Cape Town, when a gale blew down from Table Mountain and churned the bay into a cauldron, he had them row him back to *Arethusa* while other masters vowed to stay ashore in hotels and collect West's corpse next morning. The teenagers rowed desperately for three hours to cover the mile-and-a-half distance, Clements recounted with awe. "The Italian fishermen at the mole told us the next day they never believed so small a boat could live in such a sea and they never expected to see us again. What possessed the skipper to attempt it I don't know, but it made our boat's crew cock-o' the walk as long as we remained in Cape Town."[17] West instilled in his apprentices a skill that master mariners had begun to lose. Their small-boat training, begun in Australian waters and continued in a dozen harbors, paid its first dividends in Table Bay and made the young men immensely proud of themselves.

As a reward for their three hours' courage and effort, West subsequently took them on longer passages while the bark lay at anchor in Callao Bay off Peru. They took the gig to San Lorenzo, an island that shelters Callao Bay, where West taught them a bit about rare shells and a great deal about the habits of the deep-sea crab. On another day they made a thirty-mile round-trip passage to the Pachacamac Islands, two rocky islands whose shallows contained a variety of octopi, some as big as twenty-five feet. As West demonstrated, none of these posed much danger to men in a small boat, unless they had some tentacles anchored to a rock or other solid object. "I know that instances are recorded of an octopus dragging a man out of a boat, but I think in such cases the octopus must have had a grip on something," Clements decided after years of observation begun under West's direction in the Pachacamacs.[18] By then, of course, Clements had made West's lessons his own, not only those about the habits of deep-sea crabs and octopi. Clements and his three fellow apprentices were at home in ship's boats, able to launch a boat into a wild sea and save life, to sail and row efficiently for long distances, and to land through surf.

West's training methods were rigorous, but he never indulged in physical brutality. As late as the 1920s, young men determined to train in sail encountered coarse teaching methods, even in the schooners trading along the New England coast. In *Wake of the Coasters*, John Leavitt recalls flinging down some ropework during a lesson taught by the mate. After he twice refused to pick up the strands and try again, the mate knocked him cold with one blow, then revived him by dumping seawater on his face. Despite the tiny crew—Leavitt, the mate, and the skipper made up the whole company of *Alice Wentworth*—and what seemed like an easygoing democracy of sleeping accommodation and food, the mate had been a deepwater seaman and he knew mutiny when he saw it. A man of principle and courage—he saved Leavitt's life one winter day when the bowsprit footrope broke and Leavitt plunged into the sea—the mate knew that obedience mattered greatly in the life-and-death instances so common aboard aging sailing vessels, and he punished Leavitt for disobedience.[19] But perhaps Leavitt's flinging down the old-fashioned ropework provoked the mate too, because sinnet, especially the extremely fancy variety made with fourteen strands, had scarcely any place aboard a coasting schooner, especially one in the recession year 1920. Sinnet is decorative, flat ropework, the stuff of clipper-ship cabin-table ornament, the proof that expert, "marlinspike" seamen serve aboard a vessel.[20] When Leavitt threw down his tangled strands, he unknowingly discarded the way of life, especially the way of training but also the learned skills, to which the mate had committed himself years earlier and to which he had remained faithful long after steam supplanted sail.[21]

For centuries, sailing ships carried small boats used to ferry crew and cargo between ship and shore and from ship to ship. Known as *tenders, shallops, yawls, pinnaces, launches, gigs,* and *jollyboats,* and by dozens of other names over the generations, these boats—ordinary, taken for granted, often very roughly used—have scant place in voyage narratives and marine dictionaries. Their maintenance and frequently their handling fell to crewmen, often to bosuns, whose enduring title is merely a contraction of *boat's swain,* the diminutive of the antique synonym for master, *ship's husband.* Yet simply because sailing ships often loaded and unloaded in ports lacking wharves and shore boats, boatmanship remained important if commonplace.

On the coast of California in the 1830s, for example, Richard Henry Dana learned the importance of boatmanship in the hide trade, since the stiff hides had to be ferried from beaches to anchored ships. Much of his narrative *Two Years before the Mast* concerns alongshore boatmanship in small craft he identified by name, "launch, pinnace, jolly-boat, larboard quarter boat, and gig—each of which had a coxswain, who had charge of it, and was answerable for the order and cleanliness of it." From the beginning of his time aboard *Pilgrim,* Dana learned the harsh realities of landing and launching small boats through surf that taxed boatmen of limited experience. "Whalemen make the best boat's crew in the world for a long pull, but this landing was new to them, and notwithstanding the examples they had had, they slued round and were hove up—boat, oars, and men—all together, high dry, upon the sand," he remarked of one high-surf incident.[22] Whaling crews knew a great deal about open-sea boatmanship, he determined, but they knew perilously little about working their boats close inshore.

Months of brutal boat work made Dana and his fellows so expert that eventually they could deliberately wet then terrorize the landsmen they ferried to ships anchored beyond the breakers, but the crew of the brig *Pilgrim* had a lot to learn about small-boat passage making.

Once, when sailing a hired, schooner-rigged open launch to an island a few miles away to pick up firewood, Dana and his fellows encountered unfavorable winds and tide together, then a chilling winter rain.[23] The incident reveals the limitations of Dana's experience. He had learned the hard way how to land and launch boats in surf and how to row heavily laden boats the short distance between shore and anchored ships, but he knew next to nothing about managing a launch all night in bad weather in confined waters. *Two Years before the Mast* endures as a book in which novelty and adventure and coming of age mask the simple fact that Dana sailed in a poorly officered brig crewed by poor-quality seamen, and so learned remarkably little in the way of seamanship and boatmanship. Only his native intelligence and sustained scrutiny made him aware that seamen experienced in some situations proved ignorant in others, and that some ships carried far better boats and more experienced seamen than did *Pilgrim*.

The unevenness in sailing-ship-era boatmanship Dana records worsened by the beginning of the twentieth century, something Conrad explores in his 1900 novel *Lord Jim*. Much of the novel focuses on the panicked abandonment of a passenger steamship by its officers and subsequent lifeboat irrationality, but the hell to which fate consigns one of the officers is to work for one ship chandler after another as a "runner," a white man in charge of a native-crewed fast rowing boat that raced other similar boats to greet the masters of ships anchoring in small Far Eastern ports.[24] Ship chandlers from New York to Penang maintained "parlors" just off their storerooms of marine hardware, often accepted mail addressed to particular masters, and provided other services to encourage shipmasters to purchase paint and additional supplies. Among their services, free boat livery loomed large in the minds of many masters, since it reduced opportunities for crew members deserting and enabled mates to keep aboard full complements for in-harbor loading, unloading, and repair work. Until ports like Penang built wharves, ship chandlers' runners spent whole days racing to anchoring ships, then ferrying masters back and forth, engaging in harbor boat work many deep-sea masters scorned. But however beautiful and fast the small boats created especially for such service—the so-called Whitehall type, which evolved in the 1850s to serve chandlers and ship agents along Whitehall Street in New York, remains one of the fastest open-water pulling boats today—their proliferation cheated seamen out of the training on which Captain West insisted.

Only in retrospect did Clements and other former apprentices realize why West and other tradition-bound sailing-ship masters insisted on being rowed to shore by their apprentices. Not only did the brief passages put the master alone with his apprentices and available for private conversation, but the unsupervised time apprentices spent on wharves waiting for masters to return furthered their education. They were immersed in alien cultures, made to learn languages other than English, forced to examine local small craft, thrust into contact with mariners like the Italian fishermen at Cape Town whose intimate knowledge of particular ports might one day prove invaluable. A ship's boat careering about with apprentices trying to lower its lugsail, even one moving spiderlike across a harbor on a calm day, taught more than any boat supplied and manned by a fawning chandler, and the hours spent idling on wharves provided an intellectual feast for any apprentice willing to look, listen, and learn. But by 1900, as Conrad so well knew, boatmanship had gone the way of royal sails and sinnet, and fewer and fewer officers and men knew anything of the training techniques and skills imparted by masters like West.

"I agree that the ability to do fancy knots and splices has been of no assistance to me when bound up Channel in thick traffic and low visibility," wrote Victor Leslie Making in his 1937 book *In Sail and Steam*.[25] Sinnet, fourteen-strand and otherwise, counts for little in Making's laconic book, but boatmanship most certainly does count.

Unlike so many other old and middle-aged master mariners reminiscing about changes obvious only after the divide of World War I, Making quotes extensively from letters he wrote as a boy and a young man, and the letters reveal his youthful awareness that service in steamships differed greatly from his service in sail. "While on watch I merely walk up and down the bridge for four hours and take sights for finding the ship's position," he had written home decades earlier of his first steamship position. He paced the bridge for four hours, then went below for eight; he enjoyed the food (dinner alone consisted of seven courses every night), delighted in having a cabin boy make up his bed and shine his shoes, and looked forward to operating on a schedule. "A fortnight passage is a bit different from 165 days," he concluded after his first weeks in a tramp freighter. "We shall be a few hours in Port Said, I expect, and the same in Suez. The next ones are Port Sudan, Suakim, Aden, Mombasa, Kilindini, and Zanzibar."[26] For young Making, and for all the apprentices and young merchant-service officers shifted from sail to steam, and of course for all merchant seamen, the world seemed capsized.

Making at first thought that most of what he had learned in sail had lost all value forever: "What good did it do me to know how to cross a royal yard? A steamer had no yards, and only two stumpy masts!" More than most steamship officers reminiscing about the brutal transformation they underwent the day they joined a steamship, Making describes the intellectual paralysis that accompanied his shift from sail to steam. "Five years at sea, four times round the Horn, Third Mate of a wind-bag, Second Mate's square-rigged ticket—and I didn't seem to know one damn thing about anything," he recalled of his first days in a steamship.[27] But Making eventually learned the value of his training in sail.

On December 12, 1910, eighteen months after *Waratah* steamed into oblivion, Second Officer Making sat reading in his comfortable cabin aboard *Parisiana,* a nearly new freighter bound from New York to Australia with a general cargo including twenty thousand gallons of turpentine, fifteen hundred cases of refined oil, and ammunition. Suddenly he heard men shouting "Fire!" Spontaneous combustion had ignited the coal filling number three hold, carried instead of cargo because *Parisiana* steamed along a course with no convenient refueling ports, one that led across the emptiness of the South Indian Ocean. When Making leaped from his easy chair, *Parisiana* stood roughly 2,800 miles from Cape Town and 1,900 miles from Perth, about the worst possible place for disaster, especially in prewireless years, for it lay north of the prevailing-wind route followed by most sailing ships and south of the well-nigh abandoned sailing-ship routes for Java Head. Officers and men took turns standing atop the burning coal, digging deeper and deeper toward the fire sending aloft thick smoke and acrid fumes, and struggling to pour seawater into its heart. As in the incident that prompted Conrad's story "Youth," the coal fire proved impossible to extinguish and spread through bulkheads into the cargo. Like the second officer in Conrad's tale, Making answered the call to abandon ship. He and the rest of the men— and the ship's dog and cat—boarded two lifeboats and plunged decades backward into the age of sail.

By immense good fortune, *Parisiana* burned only about twenty-six miles from St. Paul Island, a tiny speck, volcanic in origin, that offered the only hope of refuge. The islet at 38°43′ south and 77°30′ east offered only a hope, however, for its two-mile diameter and 862–foot elevation made it very easy to miss, and as the lifeboats stood just off the burning freighter in the dark, the wind rose and fog swept across the rising sea. Making's diary traces the shock he felt in going from a modern steel freighter to a lifeboat in which each wave seemed an individual enemy. The already exhausted men proved themselves unable to row well, and one man lost his oar overboard; the captain ordered sea anchors to be streamed and the men rested. The next morning Making struggled to use his sextant in the bobbing boat, and in the early afternoon the two boats began moving under oars toward St. Paul, directly to windward. The men rowed until four the following morning, when the sea anchor was let out again because the men could no longer row. The next day they managed to sail, but so close-hauled they barely kept their course. "For the time being at all events we had beaten the wind and the sea," Making recounted. "We had captured and harnessed our enemy, the prevailing westerly wind, and were making it do the work for us."[28]

But exhaustion led to trouble. Making passed out, and the seaman who grabbed the tiller from him, accustomed to steering *Parisiana* from a wheel with a compass mounted beyond it, mistook the significance of the lifeboat compass mounted behind himself and so swung the boat about and steered exactly away from the island. When Making woke and realized what had happened, he ordered the lugsail reset and then swung the boat slowly. No one aboard, not even the exhausted master, knew enough about boats to realize he had silently corrected the mistake in the hope of avoiding a panic. "The fact that none of them realized what the veriest land-lubber could not help but notice is an indication of their condition," Making recalled.[29] The next day a gale struck.

In the storm the lifeboat almost capsized, and the men threw overboard life belts, canvas, rope, and a lot of other equipment to lighten it so that bailing might save them. The next day Making calculated that their boat now stood forty-six miles from St. Paul, and in the following days he "gained a practical working knowledge of the life of a galley slave." For a long time, the elderly master held the tiller and maintained hope and discipline, even after his mind began to wander. Making remembered marveling at the old man's "sea sense," but finally, with the master, chief engineer, and third engineer all delirious or unconscious, Making found himself in command.

A sober understanding of seamanship emanates from the lines of Making's narrative, one reinforced by the diary entries. The lifeboat commanded by the first mate and carrying the bosun arrived thirty-six hours before the one carrying Making, simply as a result of not having sailed two hours in the opposite direction. The fourth engineer in the first mate's boat succumbed to delirium and illness and died shortly after the second boat reached the islet, confirming Making's suspicion that engine-room officers and men were even less prepared for an open-boat passage than the deck hands, who themselves proved of limited use in the lifeboat. Only old-time seamen, experienced masters-in-sail, and experts like the *Parisiana* bosun did well once lifeboats inched away from davit lines.

Moreover, the skills of a steamship mariner proved only slightly useful in castaway circumstances. Not until January 11 did a steamer rescue Making and the other castaways. After nearly a month in a lifeboat and slogging about a volcanic islet trying to catch fish, club seals, snare rabbits, and gather limpets from rocks exposed at low tide, Making was more than ready to be taken off. But he knew how lucky the *Parisiana* castaways had been. On St. Paul they had found a tiny fishing schooner from Réunion, too small to carry the castaways but able enough to go for help, and a survival cache of biscuit and tinned meat left in 1892 by the French naval vessel *Labourdonnais*. St. Paul Island might be an insignificant speck in the South Indian Ocean, but the French navy knew its importance to castaways unlikely to make either Africa or Australia, and in the steamship age less and less likely to be found by a sailing ship following shifting winds. *Parisiana* burned in summer, and the lack of clothing and shelter on St. Paul did not begin to matter until January, when the first autumn chill reminded Making not only that the islet offered few of the delights of storybook islands but that few of his steam-engined men had the survival skills of Robinson Crusoe.[30] In *In Sail and Steam* Making analyzed important issues about the survival ability of castaways from ships like *Waratah,* issues that preoccupied many thoughtful people in the years when steam supplanted sail, ocean liners went missing, and even master mariners wondered what sailing-ship training meant aboard steamships.

A change had swept across the sea by 1910. Some of the *Parisiana* castaways found it difficult to reach an island thirty miles away, and a few days in an open boat killed one officer and incapacitated several others. Once ashore, many of the castaways lacked the knowledge necessary to survive for long. Bereft of modern technology and lacking traditional skills, the *Parisiana* castaways eked out an existence that makes Dixon's remark about the domestic cat wandering one of the Crozet Islands worthy of more than passing notice. The *Parisiana* castaways got their dog and cat safely to St. Paul, but so anxious were they to be taken off that they forgot the cat they had worked to save. Only the broken-in food store, some graves, a rough-cut inscription, and the cat remained on the island as evidence of transient occupation by a new breed of mariner.

Dixon knew how to read the signs left behind by castaways, and by great good luck his interest in winds and out-of-the-way places connected him directly with one of the more incisive inquirers into turn-of-the-century castaway behavior, H. De Vere Stacpoole, author of *The Blue Lagoon* and *The Beach of Dreams.*

In his constant search for wind, Dixon once took his ship eighteen hundred miles off the regular sea-lanes in the Pacific and found himself off Pukapuka, an atoll about two hundred miles northeast of the Low Archipelago, a place rarely visited owing to its isolation and its peculiar place in the trade winds. Ordinarily, ships passed it hundreds of miles to the east, and reaching it meant beating against the trade winds. Dixon recalled how inviting the atoll seemed, and since his ship stood becalmed, he ordered out his gig and went exploring. After describing in minute detail several techniques that experienced mariners use to brings boats across reefs and through surf, he explained that a sense of adventure made him and his men try to land. They succeeded, and found themselves almost in paradise. "As it grew deeper, farther out, it shaded into blue till, at its deepest, it was a deep but brilliant blue," he recalled of the most beautiful lagoon he had ever seen. "No painter ever put such wonderful coloring on canvas." He explored a bit, found huts, cooking utensils made from coconuts, piles of oyster shells, and an oar. "We concluded that castaways had lived on the lonely lagoon, but the mute evidence of their occupation did not take on very much romantic significance until about a year later, when I was in London."[31] Dixon regained his ship and wound up making the fastest passage of the year between Portland, Oregon, and England. Relaxed and happy, strolling about London, he passed a theater advertising the stage production of Stacpoole's novel *The Blue Lagoon.*

Five minutes after the curtain rose, Dixon saw that Basil Deane, the producer, had created a stage set that nearly duplicated the lagoon on Pukapuka, and Dixon wound up writing to Stacpoole, who wrote back, delighted to have discovered the mariner view of things.

The Blue Lagoon remains a racy, sexy novel a century after its 1908 publication, and only a few years ago Hollywood made it into a box-office success. It and similar early twentieth-century novels get no more notice from literary scholars today than MacLean's *South by Java Head,* and discovering how London received the book takes effort. Against the backdrop of real-life adventures like those of Dixon, the plot is not particularly far-fetched, and certainly Dixon found it realistic. A very young boy and girl, passengers aboard a sailing ship suddenly afire in the South Pacific, are tossed into the ship's dinghy in the care of an elderly, simpleminded sailor. Separated immediately from the other boats, the three eventually take refuge aboard a derelict ship, then leave that refuge in the well-supplied dinghy for an uninhabited atoll of breathtaking beauty. In a short time the aged seaman dies, and the children grow up as gentle savages, naked, reveling in eternal summer. Eventually they become lovers, then parents, and then by the simplest of accidents, losing one of the oars of their boat, they become castaways again until found by someone still searching for them many years later. One book reviewer praised the novel as "a successful attempt to do a rather difficult and unusual thing," but pointed out, too, that other authors had explored the same theme.[32] *The Blue Lagoon* endures not only as a small window into the open-mindedness of the Edwardian era—and as an example of naturalism, a minor literary genre defined by its focus on human behavior away from the restraints of civilization—but also as a study of post-shipwreck human behavior.

Nine years after it appeared, Stacpoole embraced the castaway situation again, this time putting a wealthy young woman passenger named Cleo in a ship's boat crewed by two distinctly unsavory seamen. They reach the South Indian Ocean island of Kerguelen, a barren, forbidding place sailing ships routinely checked for signs of shipwreck and survivors.[33] Stacpoole made it the scene of *The Beach of Dreams*, a novel focused on what many readers still consider nightmare.

Flung into a boat by the terrified crew of a steam yacht run down in the night by a sailing ship, Cleo, along with the two crewmen, alone survives the collision to reach Kerguelen and a special sort of hell. Both seamen quickly become apathetic and fatalistic, and Stacpoole uses their apathy to introduce his harsh themes of castaway life. The seamen cannot adapt to suddenly changed circumstances, he writes of their lack of interest in inventorying the boat locker Cleo opens. "Men are like that, especially men of the people, and when you read of Crusoes and their wonderful doings on desert islands you read Romance."[34] In a few paragraphs, Stacpoole introduces the new breed of steamship seaman.

Cleo soon begins to understand her two companions and realize that she alone has intent, direction, and self-discipline, despite being flung from civilization into fearsome wilderness. "The idea of a Providence to such a person is like brandy," Stacpoole philosophizes, arguing that any person accustomed to living with a roof overhead and then "condemned to live in the open" suffers nothing for the first few hours, but "then there gradually comes upon him a weariness and distress almost unimaginable to those who have not experienced it," coupled with a craving for a roof and walls "to protect him from the great open spaces that seem sucking away his individuality."[35]

After the more useful of the two seamen drowns while collecting birds' eggs, the other tries to rape Cleo as she sleeps. Having suspected his intention and armed herself with a knife, she kills him and begins to survive alone. The novel focuses as much on the mental state necessary to survival as on the usefulness of the food and equipment the woman finds, deposited by a warship for the use of castaways. "This power to live in the moment is the form of strength that brings men through battles and women through adversity," Stacpoole writes. "On Kerguelen it is salvation."[36]

But more than attitude and activity are needed to save her from fever. She is found and nursed back to health by another castaway, a seaman of the sailing ship that rammed the steam yacht and itself sank a short while after. Together the pair make their perilous way to a harbor on the opposite side of the island, where they encounter Chinese sealers, who attack the seaman when he asks for help. The old-style seaman and Cleo together stab to death four of the sealers and send the remaining eleven scurrying to safety. "For a moment she had lived in the Stone Age, she had fought like a savage animal and with the fury of the female, more terrific than the rage of the male," Stacpoole muses after the bloodiest of knife-fight descriptions. "It is a long journey to the Stone Age and back and the man or woman who makes it is never quite, quite the same again."[37] The two eventually reach safety. Cleo is forever altered, but the windjammer seaman seems remarkably unaffected; he is, after all, a sailing-ship man.

Making, Dixon, and a host of other early twentieth-century seamen boldly entered two raging controversies, one concerning missing ships, the other the training of all merchant-service apprentice officers in sailing ships. "Of course I missed the thrill and romance of sail," Pryce Mitchell remembered in his 1933 autobiography, *Deep Water,* of his first passage as cadet officer aboard a steamship touching at Singapore, Penang, and Ceylon en route from Shanghai to London. He busied himself studying navigation and forced himself not to think of sailing vessels, but in time he realized the usefulness of sailing-ship-era skills in modern emergencies.

Of all the master mariners to tell tales of missing and foundered steamships, lifeboats and lifeboat passage making, and the training in sail of cadets bound for careers aboard steamships, perhaps Joseph Conrad proved the most eloquent. His 1923 "Memorandum on the Scheme for Fitting Out a Sailing Ship" argues "that there can be very few sailing-ship officers left now who have had the experience of the care of upwards of a hundred people on board a 1300-ton ship," and who remember what he himself knew from having drilled his own lifeboat crews every morning until his men could prepare the boats for launching in seven minutes. Conrad served as master of a sailing ship that carried not only the usual ship's boats but also two roomy lifeboats stored just abaft the fore rigging and always ready to be freed, raised three inches, and swung out. Few young men sailed in such vessels in 1923. Too many apprentice officers were at sea under the tutelage of master mariners who knew little or nothing of sail and who emphasized only the scheduled, exact-navigation passage making that Making, Mitchell, and others discovered to be so different from what they had learned in sailing ships. "An insight into the old conditions cannot do any harm and may be found useful on occasion," Conrad argued in suggesting that cadets be sent to sea in small sailing ships and taught to handle themselves in the rarest of emergencies, perhaps the foundering of a liner.[38]

As the *Waratah* episode clearly illustrated, steamships did badly in extreme circumstances, and the open-air, deep-water merchant seamen able to snatch safety from extremity had largely disappeared. "I fear that the oar, as a working implement, shall become presently as obsolete as the sail," Conrad mused in *The North Sea on the Edge of War*, his 1919 memoir, of an engine-driven, electrically lit pilot boat. "More and more is mankind reducing its physical activities to pulling levers and twirling little wheels. Progress! Yet the older methods of meeting natural forces demanded intelligence too; an equally fine readiness of wits. And readiness of wits working in combination with the strength of muscles made a more complete man."[39]

So *Waratah* went missing and stays missing. However it foundered, perhaps its officers and crew lacked the experience of sailing-ship men who might have brought its passengers to St. Paul, to Kerguelen, to safety somewhere.[40] Amid all the nostalgia, all the boasting, all the scrupulously accurate quoting from letters and logs, the grumbling deep-water ancient mariners of the 1920s and '30s make it clear that something more than *Waratah* had gone missing. The era of the ship's boat crew had ended, and the era of crewmen who twirled little wheels, often far, far below deck, far beneath the open sky, had triumphed.[41]

Debate on the subject continued into the 1930s, then revived in World War II, and flared occasionally in the years following. In 1949 Fred W. Ellis, a retired British merchant navy master mariner, recalled his turn-of-the-century apprenticeship in a sailing ship. On one occasion his ship was just north of Cape Horn when a Pacific Steam Navigation Company liner overhauled it. All the liner officers waved their hats, for they had trained in sail and valued its lessons. "Lady passengers waved their handkerchiefs, men whistled and cheered; we had our number and ensign hoisted; then she forged ahead."[42] In that moment Ellis realized the startling discrepancy between the steamship officers seeing something they respected and passengers seeing merely a romantic anachronism. He understood why his cantankerous, steamship-hating master hoisted a gracious greeting: the man enjoyed superiority and knew its responsibilities.

The superiority of sailing-ship men perhaps explains more than the disappearance of *Waratah*.[43] Remarkably few early twentieth-century authors examined the modern steamship wholly dependent on engines. Here and there the contemporary scholar finds a book like *Ocean Steamships: A Popular Account of Their Construction, Development, Management, and Appliances*, a handsomely illustrated, multiauthored volume published in 1891, but library shelves mostly prove curiously empty. Between about 1885 and 1910, even the most fervent technology-embracing authors looked away from the modern steamship as they rhapsodized about skyscrapers and luxury trains, dynamos and steel mills. Too many things about the steamship seemed vaguely disturbing and somehow unresolved. The six authors of *Ocean Steamships* ignored the entire issue of lifeboats and scuttled past all questions of large-ship behavior in great storms. In the quarter century before *Waratah* went missing, the literature of modern seafaring contracted, almost in presentiment of disaster.

All too often, U-boat crews closed on lifeboats to discover castaways not only disoriented and shocked but utterly unaware of the nearest land. (Courtesy of the Mariners' Museum)

CRUMBS

ON A PLEASANT SUNDAY MORNING IN JUNE 1943, THE FREIGHTER *Sebastian Cermeno,* bound from Mombasa in Kenya to South America by way of Durban, South Africa, sank 342 miles southeast of Madagascar, ten minutes after being hit by two torpedoes fired from a German U-boat. The submarine surfaced among the five lifeboats, its captain demanded the name of the ship and a summary of its cargo, then it submerged, leaving behind only boats amid wreckage. After searching for additional survivors and then transferring men to balance the loads in each boat, merchant seaman Henry Patenall and fifteen other men began steering northwest toward South Africa.

In the first night they lost contact with the other boats, suffered seasickness, began rationing food and water, and learned how cold an open boat can be in equatorial waters. Nine days later they sighted land, conned their twenty-two-foot lifeboat through reefs and surf, and beached on a jungle-fringed strand unknown to them, and seemingly uninhabited. About a year later, Patenall published his account, "Madagascar via Lifeboat," in *Travel,* contributing his part to the burgeoning genre of World War II lifeboat narratives. Along with other authors, he avoided all mention of why his ship came to be where an enemy found it.

Why should *any* ship bound from Mombasa to South America via Durban be 342 miles *southeast* of Madagascar?

By 1944, of course, readers of *Travel* knew that war had severed traditional sea-lanes and forced merchant ships onto roundabout, awkward, and supposedly secret routes through seas where British and American naval forces provided no convoy protection. *Travel* readers believed in the certainty of Allied victory, and already their memories of the darkest days of the war had begun to fade. Within a decade of Patenall's matter-of-fact account, Cold War anxiety had buried the previously shocking realization that for years German raiders and U-boats and Japanese warships had roamed almost at will around the world. Few historians wrote of the near panic that gripped Americans in early 1942.

A year before the sinking of *Sebastian Cermeno,* the premier American boating magazine, *The Rudder,* warned its yachtsmen readers to watch for Japanese warships approaching United States coasts. In a two-page spread it reproduced silhouettes of fourteen classes of medium- and small-size Imperial Navy warships, but admitted that the United States had no images of the giant 50,000-ton battleships thought to be everywhere in the Pacific. While most of the visual information implies a direct communication from United States Naval Headquarters, the tone borders on panic. "For instance, if you happen to be fishing off Catalina Island, Key West or Montauk Point, you might see a naval vessel coming up over the horizon," the piece begins, offering examples of how yachtsmen must be able to identify the silhouettes instantly, not wait to see national flags. Warfare moved too fast in 1942 to allow for flag identification, perhaps too fast to identify silhouettes. "When you are *absolutely positive* and not before, you pick up your radio phone and call the Coast Guard to make your report. Just one such correct report may be the means of saving a great many lives and tremendous property damage." In the aftermath of the Pearl Harbor attack, a navy marshaling so many vessels capable of thirty-two- or even thirty-five-knot speeds seemed utterly capable of invading Hawaii or attacking San Francisco or San Diego or even Galveston.[1] If its admirals stationed oilers at appropriate places, it could send a battle fleet into the Atlantic, north from Cape Horn, west from Dakar, northwest from St. Paul's Rocks, from nowhere, exactly as *Life* and other national news magazines had been warning since the days immediately following the British evacuation of Dunkirk.

"This Is How the United States May Be Invaded," trumpeted *Life* on June 24, 1940, as British and French troops, reeling from the German *blitzkrieg* that had driven them across The Netherlands and Belgium to the beaches of the English Channel, waded into the sea at Dunkirk toward anything that would float.[2] No longer did the Royal Navy seem strong, let alone invincible, as hundreds of thousands of troops scrambled aboard a ragtag fleet of small craft; no longer did South America seem secure from attack by a combined German, Italian, and—since the Hitler-Stalin pact—Soviet force. Another post-Dunkirk *Life* article warned shocked Americans about thousands of German sympathizers south of the Rio Grande. "Nazi Fifth Column and Communist Allies Are Active in Mexico," a hair-raising article, told of Western Hemisphere fruits of the Hitler-Stalin pact ripening not only in Mexico but in Brazil, Argentina, and elsewhere. Invasion might come from the south, by sea and air combined.[3] After the fall of Dunkirk and the discovery of the Vichy fleet in the Caribbean, warnings about the Axis war machine spewed fast and furious from the United States news media, until by early 1942 Americans feared that German, Italian, and Japanese forces might be anywhere.

The Imperial Japanese Navy might well join forces with the *Kriegsmarine* ships wreaking havoc just off the United States Atlantic and Gulf coasts. In the spring of 1942, as U-boats operated almost at will throughout the Caribbean and the Gulf of Mexico, even torpedoing vessels in the mouth of the Mississippi River, the United States Navy had come to rely on irregular assistance indeed. "Elderly ladies with binoculars were particularly adept at spotting submarines in shoal waters from their front porches," according to the official history of the Caribbean campaigns. Civilian pilots, especially those flying for Pan American Airways, glimpsed German submarines seemingly everywhere, and Gulf Coast newspaper editors quickly surmised that the U-boats refueled at secret bases in the Louisiana bayous. War Department officials shared their concern and in July 1942 organized the Coast Guard Beach Patrol Unit, which everyone along the coast quickly dubbed "the Cajun Coast Guard." It enrolled Cajun men from age eighteen to sixty-five and set them to patrolling bayous, inspecting shrimp boats, and rescuing downed airmen. All along the Atlantic and Gulf coasts, beachcombers began finding grisly and disturbing evidence of inshore combat, although not many parents had the experience of one man in Galveston, who found an oil-covered lifeboat oar from *William C. McTarnahan,* a tanker captained by his son.[4] By then, however, no one in Galveston doubted the reach of *Kriegsmarine* power. Day and night, civilian contractors labored under army direction to fortify the city against invasion from the sea or from secret Mexican bases or from islands and rendezvous points of which almost no one in the United States had ever heard.

What had been perhaps willful ignorance of the so-called European war before the Pearl Harbor attack became by February 1942 a near panic. The popular memory of World War I censorship engendered skepticism and hard feeling, especially along the United States coasts, and confusion in the War Department only angered alongshore civilians suddenly caught up in the war. Not for months did the navy extinguish offshore buoy lights used by U-boat commanders to silhouette merchant ships, and not until the summer of 1942 did the War Department realize that the bright lights of coastal cities also backlit ships.

World news brought reports of Axis victories at places remote, widespread, and almost impossible to locate in atlases. "The tentacles of the Japanese octopus spread over the whole vast area of Malaysia and the islands of the South-west Pacific," remembered Stanley Rogers in 1944, using the sea monster metaphor so many correspondents favored. "First the Philippines fell, then, reaching south, the tentacles embraced the Malay Peninsula, Sarawak, Dutch Borneo, Celebes, New Guinea, and Java, with hundreds of smaller islands. Java, the long island which is divided from Sumatra by the narrow Sunda Strait, fell simultaneously with Sumatra."[5] Along with a handful of other wartime writers, Rogers connected American prewar geographical ignorance to Allied perception of Axis might. Until June 1942, he argued, most Americans had no understanding of the strategic importance of Midway Island. Peacetime complacency had made geography an unimportant grade school subject.

Prewar ignorance alone did not explain the darkening wartime mood. Rogers argued that official secrecy and censorship led thoughtful newspaper readers—and gleaners of alongshore wreckage—to mistrust a government that had lied to its citizenry in World War I.[6] Whatever the morale of the British and American people, public *understanding* of the war at sea was also influenced by prewar writers like Homer Lea, whose 1909 scenario of American-Japanese war (hastily reissued in 1942), *The Valor of Ignorance,* offered a nonfiction projection every bit as chilling as William McFee's 1930 novel *North of Suez,* about Anglo-German naval conflict. Such scenarios, along with wartime books like Oswald Gilmour's taut, gritty *Singapore to Freedom,* produced a framework onto which civilians projected all sorts of troubling events.[7]

A confusion of official pronouncements, oil-smothered lifeboat oars washed up on beaches, accounts sent home by naval personnel and merchant seamen, prewar information published by steamship companies, and propaganda, both Allied and Axis, combined to worry Americans, in spite of morale-boosting films like the 1942 Hollywood spectacular *Wake Island.* Far too many Americans along the coasts raised blackout curtains cautiously every morning, expecting to see U-boats, German landing craft, and coal-scuttle-helmeted soldiers racing along beaches.[8]

Journalists and editors routinely voiced doubts about the quality of official information. In an August 5, 1942, article entitled "U.S. Reckons Up Its Losses," the editors of *Life* published a map and chart showing spatially and numerically what they knew of merchant-ship sinkings in the western Atlantic, including the Gulf of Mexico, through June of that year. "The chart below bluntly contradicts assurances from responsible officials that things are getting better," the editors concluded somberly. "All through 1942, things have gotten worse." Moreover, so far as the editors could learn, worse was only getting worse. "Sinkings have reached the highest level of the war, and acknowledged losses by July 23 broke through the 400 mark."[9] The word "acknowledged" implies that at least some sinkings went officially unmentioned, perhaps even unknown in the confusion ruling the seas.

Even upbeat books published with the wholehearted support of the United States Navy offered glimpses into spreading disaster. Robert Trumbull's 1942 bestseller *The Raft* focused on the extraordinary survival of three navy fliers downed in the Pacific. But any astute reader quickly noted the cartographic vagueness, then the slightest of hints that the castaways worried about finding Imperial Japanese Navy forces far further south than the War Department had implied. The fliers drifted a thousand miles away from their rendezvous with a carrier force probing the fringes of Japanese might, crossed sailing-ship sea-lanes long abandoned, and washed ashore on an island censors forbade naming. When the castaways waded ashore they fully expected to be gunned down by Japanese marines.[10] Axis forces might be anywhere, including far behind Allied lines.

War tangled and twisted well-established steamship routes and shattered the easygoing confidence of most steamship masters, officers, and crewmen. As it deflected ships away from established routes and known ports into seas and ports unfrequented since sailing-ship days, it challenged mariners in novel, sometimes terrible ways, perhaps chiefly by eliminating almost any chance of rescue in the event of trouble. It deflected *Sebastian Cermeno* a thousand miles away from the Mozambique Channel route into seas crossed by no peacetime steamship, not even vessels operating between Cape Town and Freemantle, and it threw the men of that freighter entirely upon their own resources. Even when Allied ships followed peacetime routes, they often sailed into an oblivion mariners thought technology had vanquished.

On February 19, 1942, U-129 sank the Norwegian tanker *Nordvangen* just off the north coast of Trinidad, near Galera Point. The master of the tanker, weary of the United States Navy's inability to protect his ship or even issue advice, had determined to sail alone rather than steam north in an unofficial, unescorted convoy that he thought made an obvious target. Two torpedoes destroyed the ship instantly. No SOS message went out, no wreckage drifted ashore, no one survived. "The result was as if the *Nordvangen* had sailed off the edge of the world," concluded one naval-warfare historian in 1988. "At the end of the war she was not even listed as having been sunk, although the spot where she perished was only two miles from the Point."[11] U-boats subsequently sank so many ships off Galera Point that the location came to be known as Torpedo Junction, for it lay at the end of the

Africa-to-South America wartime route followed by ships like *Sebastian Cermeno* skirting Madagascar and bound past the hazards English-speaking mariners called St. Paul's Rocks.

Penedos de São Pedro e São Paulo, St. Paul's Rocks, had slipped into obscurity decades before 1942. About 55 miles north of the equator and roughly 525 miles east of Brazil, it had long been a dangerous seamark for sailing vessels bound to Asian waters from Europe and North America. In order to catch currents and prevailing winds, especially the edge of the southeast trade winds, shipmasters of East Indiamen outward bound from Europe sailed southwest to the Rocks, sometimes meeting there ships bound from North America, then swung southeast, beginning the long "easting down" that carried them to Canton or Australia. The mass of rocks, barely seventy feet above water, bereft of vegetation and smothered in guano, claimed many sailing ships that steered too close. When steam power made the old follow-the-wind routes unnecessary, what had been an important seamark as late as 1890 became by the 1920s something confused with St. Paul Island and mentioned by aging seamen but never sighted from liners, warships, or even tramp steamers. It rapidly vanished too from top-quality twentieth-century world maps and globes. No longer did Penedos de São Pedro e São Paulo lie athwart the customary route to Java Head.

Yet by 1941 *Kriegsmarine* commanders knew the usefulness of St. Paul's Rocks as a place to shelter rendezvousing U-boats far from the overtasked British and American navies but suddenly smack on steamship routes.

U-boats, not Japanese submarines but the long-distance U-cruisers the German High Command nicknamed *Monsunbooten* after the first sailed from Germany into the Arabian Sea in the 1942 monsoon season, wreaked havoc among Allied shipping not just in the South Atlantic but across the Indian Ocean. While often supplied from *milchkuh*-class U-boats—huge submarines that carried fuel oil, torpedoes and other ammunition, medical teams, and food (each "milk cow" could bake 800 loaves of bread a day when necessary) and that rendezvoused in unfrequented locations—U-boats also worked from their base in Penang in the Malacca Strait, half a world away from St. Paul's Rocks. From Penang sailed U-862, the only U-boat to operate in the Pacific Ocean, which penetrated Sydney Harbor on Christmas Eve 1944 and sank an American freighter. Supplied first from Europe by Italian submarines of the *Merkator* class, then by specially designed, non-combatant supply U-boats, then by Japanese forces, Penang-based U-boats demonstrated the long reach of the *Kriegsmarine*. The first wave of monsoon U-boats refueled from the *Brake*, a surface tanker that rendezvoused with them about 450 miles south of Mauritius in early September 1942, almost exactly where *Sebastian Cermeno* sank a year later.

Like the U-boats refueling from U-tankers in the equatorial Atlantic, the Indian Ocean flotilla feared little from the near-vanquished British, Dutch, and American navies. After all, a Japanese carrier fleet had steamed with impunity across much of the Indian Ocean, shelled Ceylon, then swept the India and Malaya coasts clear of Allied shipping. The German cruiser *Komet* had arrived in the Pacific after steaming along the entire Arctic coast of the Soviet Union, then swinging south through the Bering Strait. British naval units had all but retired to Cape Town and could scarcely safeguard merchant shipping bypassing the Mediterranean around the Cape of Good Hope, making for Australia not by the usual route past Penang but steering far into the empty southern latitudes, far south even of the old sailing-ship goal of Java Head. St. Paul's Rocks and Penang alike belonged to the *Kriegsmarine* and the Imperial Japanese Navy, as prewar sea-lanes shifted, unraveled, and became a joke to the masters of Allied merchant vessels who expected Axis surface raiders and submarines everywhere and all too frequently found them.

Midway into 1941 Britain almost lost the war. Nowadays only a handful of military historians understand the knife-edge on which Britain balanced after June 1940. Few recount the desperation that drove the British Empire to exchange bases for fifty aged United States destroyers in the so-called Lend-Lease experiment. Few analyze the rising terror in United States War Department circles that drove Roosevelt to announce an American "neutrality zone" and to order United States Navy ships to escort British convoys partway across the Atlantic. Almost none probe the paralyzed fascination with which the American General Staff viewed the carrier-equipped Vichy French fleet poised at Martinique, within easy striking distance of the Panama Canal, the Venezuelan oil refineries, even Galveston, New Orleans, and Mobile. Now and then a perceptive undergraduate wonders at the Woody Guthrie song honoring the destroyer *Reuben James,* or asks about the contemporary ownership of Ascension Island, but most of the time military history and wartime cartography arouse little interest.

The grim days of 1939 to 1943 generally receive little formal attention even in university history departments, and almost no one knows enough geography nowadays to realize the accomplishments of Axis naval forces. World war forces everyone caught up in it to reexamine maps and charts and globes, to locate well-known or obscure places or regions suddenly become critically important, to know the approaches to places, even to places like home. Wartime censorship makes thoughtful citizens sift through public and private data, and perhaps wonder at an inspiring *Travel* story that nonetheless suggests that U-boats and Japanese submarines had closed Mozambique Channel to Allied merchantmen. But peace reduces prior interest to a scarcely stifled yawn, then to a confident assurance that newspaper maps provide all the geographical knowledge any educated reader needs. Peace makes everyone forget that not so long ago lifeboats set off for land thousands of miles distant, their occupants knowing that for decades no one had sailed north or south or east or west from the nowhere where ships sank almost instantly after torpedoing.

Minutes and seconds mark time, and mark space too, especially the vast nowhere of untraveled seas. Just as every hour divides into minutes, so every degree of latitude and longitude divides into minutes and seconds. The recent historical past, the past of written memoranda which sometimes takes form as *minutes* of meetings, flows fitfully in dates, eras, long periods of time, short bursts of minutes like those few in which the battleship *Arizona* sank in Pearl Harbor, all more or less explained, or at least noticed, by historians and history teachers. But the schoolroom or public library globe or map or chart nowadays contains no minutes, no seconds, only the crosshatch of longitude and latitude lines, and usually not even all of them, but only those divisible by five. Much of the schoolroom atlas or globe today, like much of the atlas or globe in 1939, is empty, offering inquirers only vast swaths of sea unrelieved by anything except an ornate compass rose or publisher's trademark, or ruled into a giant grid of longitude-latitude lines. The modern-day globe no longer depicts the sea-lanes traveled by sailing ships following prevailing winds; neither does it trace the great circle routes of steamships so prominently lined out on globes produced in the 1920s. Often contemporary atlas sheets do not mesh perfectly, leaving inquirers wondering where the South Atlantic merges into the Indian Ocean, and few globes or atlases prove detailed enough to locate St. Paul's Rocks or Java Head, let alone the spot where *Parisiana* burned.

At 10°13′ north latitude, 109°10′ west longitude lies Clipperton Island, a coral atoll most easily found on a superb globe by tracing a line west into the Pacific from Managua, on the west coast of Nicaragua. Finding Clipperton Island detailed in any encyclopedia proves much more difficult, as difficult as it did in 1939. Clipperton Island has never mattered much, even to France, which claims it still. But in 1942 the existence of Clipperton Island made yachtsmen take *The Rudder*'s warnings about Japanese warships seriously. If an Imperial Japanese Navy battle fleet appeared off Catalina Island en route to lay waste the San Diego naval base, it might well have refueled from oilers sheltered at Clipperton.

In 1940 the novelist John Steinbeck, then working as a marine biologist, sailed from San Diego aboard a fishing boat chartered for a collecting expedition in the Gulf of California. "The planes roared over in formation and the submarines were quiet and ominous," he wrote of a harbor suffused with tension. While Steinbeck and his colleagues bought bread and eggs and fresh meat, stowed away ice, got their last haircuts, and endured questions from curious dockside loafers, the military transshipped munitions and matériel and Steinbeck wondered at the military mind-set. "The men who directed this mechanism were true realists. They knew an enemy would emerge, and when one did, they had explosives to deposit on him," he wrote later in *The Log from the Sea of Cortez*.[12] Steinbeck glimpsed the likelihood of having explosives dropped on him and on San Diego, and he challenged a naval gunnery officer to explain the impact of a heavy shell on a city street. The officer refused, making Steinbeck wonder how much genuine thought about unrestricted warfare lay behind the hurried, disciplined training he saw all around him. In 1940 every major navy had battleships capable of firing immense explosive shells *twenty-five miles.* San Diego might well be attacked by the Japanese, by surprise, and a yachtsman or marine biologist might well be the first to sight the attackers twenty miles offshore and steaming so fast that no berthed warship could confront them.

As the fishing boat *Western Flyer* proceeded slowly along the coasts of Baja California, Steinbeck grew more and more concerned about paralleling not just a dangerous and little-settled littoral but the coast of a neutral nation no more concerned about the "European war" than about maintaining its lighthouses and buoys. Gradually he and his colleagues, and even the happy-go-lucky master of the little fishing boat, took refuge in reading the latest edition of the *Coast Pilot,* a publication of the United States Navy Hydrographic Office, and especially its companion *Sailing Directions for the West Coasts of Mexico and Central America, 1937, Corrections to January 1940.* Without its instructions, Steinbeck knew, navigation shifted from risky to disastrous, at least for strangers, but the book offered an insight into attitude too. "In the first place, the compilers of this book are cynical men," he began, trying to analyze *Coast Pilot* thinking. "They try to write calmly and objectively, but now and then a little bitterness creeps in." Their bitterness came from the willingness of mariners to approach treacherous inlets unmarked by buoys, to follow contorted channels filled with shifting sandbars, to get into shallow water *at night.* It came too from the unwillingness of the Mexican government to mark its coast as well as the federal government marked the coast a few hundred miles north. Intermittently, Steinbeck found himself relying more on the writings of Francesco Clavigero, an eighteenth-century Jesuit with a good eye for coastwise navigation, than on modern guides.[13] Too few ships ventured into the half-charted Baja California shallow waters to justify much in the way of navigational aids or accurate sailing directions. Too few people lived along the Pacific or Gulf of California coasts to notice much of anything afloat, say an Imperial Japanese Navy fleet stealing north from Clipperton Island, hugging the desolate shore of a neutral nation.

Clipperton Island lies as far from La Paz in Baja California as La Paz lies from San Diego. Clipperton Island, had Steinbeck thought to notice, lies within easy *fishing-trawler* distance from a naval base he saw making up for lost time while looking over its shoulder.

Steinbeck missed the irony in the navy's use of the term "sailing directions." A reliable diesel engine powered his chartered fishing vessel, and long before his venture steam-powered ships had replaced schooners in the gulf. But the United States Navy Hydrographic Office retained "sailing" partly from inertia, but also from common sense. Many of the coasts and channels and inlets that it and the British Admiralty Hydrographic Office described had been explored by men in sailing vessels and by 1940 had never produced enough trade to justify steamship service. Until the 1880s, in fact, Lloyd's *Register of Shipping* recorded mostly sailing vessels, and even three decades later many coastal trade steamers carried vestigial masts and sails useful for conserving fuel and in emergencies. Away from western Europe, Canada, and the United States, small sailing vessels, usually schooners, called at the tiny ports coast-pilot authors knew rarely attracted larger, deep-draft vessels. Sailing directions existed chiefly for the masters of small sailing craft, men who needed to know about places deep-water mariners never saw but scorned as "primitive ports."

Now and then in the decades before World War II, a rare author stumbled upon the deeper significance of sailing directions and coast-pilot advice, and the fast-deepening ignorance about places like Clipperton Island. One of the most experienced, hardheaded tramp-steamer voyagers of the first half of the twentieth century, H. M. Tomlinson, used Admiralty sailing directions in both his nonfiction books and his novels, always emphasizing that as liners and freighters grew larger by the year, small ports became less and less visited and, in their own way, more and more "romantic" to European and American travelers intent on exploring out-of-the-way places. In his 1924 *Tide Marks: Being Some Records of a Journey to the Beaches of the Moluccas and the Forest of Malaya in 1923,* Tomlinson described how a tramp-steamer master handed him a copy of *The Red Sea and Gulf of Aden Pilot* as an example of a genre intellectuals ought to know. "This is probably the first reference to it by a critic," he wrote of a work filled with admissions of incomplete information and data useless to masters of large steamships. Tomlinson quoted freely from the sailing directions, noting that one stretch of coast is described as "mainly inhabited by the Jemeba tribe, who generally have a bad character," that certain channels are not to be attempted "unless the sun is astern of the ship and a good lookout is kept," that mariners landing in lifeboats on one coast can expect no aid, for the strand "was formerly inhabited, and remains of dwellings are still to be seen," but it is now deserted. Using the island of Socotra as his chief example, he asserted that geographical fuzziness produced confusion not only about history and present-day trade but about cultural clashes in the offing. "We know the South Pole better than that island, although from prehistoric time every maker of specifics has depended on Socotran aloes," he concluded.[14]

Two years later, in *Gifts of Fortune and Hints for Those About to Travel,* Tomlinson quoted at length from Indian Ocean sailing directions, making explicit the vast untidiness of the islands west of Aden. Berri Berri Road in Rau Strait is treacherous because it has never been charted, and "the inhabitants are quarrelsome and warlike" on Suruake in the Goram Islands.[15] But only in *Gallion's Reach: A Romance,* his masterful novel of 1927, did Tomlinson extrapolate from the sailing-direction genre a meaning growing less obvious every year.

East of African coasts of "radiant bronze and brass," of coasts "burnt out" and bereft of human habitation, steaming into a heavy, windless sea lit by a metallic sunset, his *Altair* crosses from the Red Sea into the Indian Ocean, a crossing marked only by an island left judiciously far abeam. "Off the island of Socotra they found some air, and their ship began to sway," Tomlinson wrote of this ordinary tramp steamer making an ordinary passage at the very end of the monsoon season. "Socotra, a serrated confusion on the horizon, was far to starboard," and a newcomer to the waters sees only "an obscurity of clouds, sea glint, and shadow." But nobody aboard *Altair* has ever landed on Socotra. "I've never met a man who has," remarks the captain. "But there it always is, somehow. If you want to know about some of the things here, you'll get it from Sinbad."[16] Of course, Tomlinson knew what he did because he had read the relevant coast pilots and sailing directions, and then made a subtle intellectual leap from travel narrative to scenario envisioning.

Gallion's Reach concerns, among other stark and sudden alterations in situation, shipwreck. East of Socotra, in a typical monsoon gale, *Altair* strikes the submerged hulk of a steamship floating bottom up and loses its rudder. Deprived of its steering, the steamer swings beam-on to the sea, and while competent, stubborn efforts seem at first to offer success, its master, officers, and crew discover damage to the forward part of the hull producing an implacable leak. Within minutes *Altair* sinks, thrusting what remains of its people into lifeboats.

Sink the steamer, put crew and passengers in lifeboats, and time turns back upon itself, slamming castaways into the legends of Sinbad, the precise observations of Clavigero, the minutiae of Hydrographic Office advisories. "I would sooner read any volume of *Directions for Pilots* than the Latin poets," Tomlinson asserted in *Gifts of Fortune*. "And I should like to ask whether Ceram Laut has not been sighted since 1898."[17] By then he knew at least some of the Far East tales of castaways in ship's lifeboats setting out from sinking or burning vessels toward whatever havens sailing directions recommended. "But by the time we had the boats provisioned and away all was quiet again, except for the flames. We made the Pelews. Anyhow, my boat did. I never heard what became of the other two."[18] So ends one nonfiction fire-at-sea tale Tomlinson recounted in *Tide Marks*, the sort of tale that made him grapple with the events he imagined in *Gallion's Reach* and understand what most mariners knew but what Steinbeck and almost every other sea traveler misunderstood until 1939.

Once any steamship, even any diesel-powered trawler, left established sea-lanes it ventured into waters less known every year after the end of sailing-ship days. And once any steamship, even in established sea-lanes, foundered, its lifeboats sailed, not according to the thrust of steam or diesel engines, but according to the conditions sailing directions still enumerated, often to places no modern mariner ever touched.

Steinbeck had observed and tried to explain the phenomenon by using the name Sea of Cortez instead of Gulf of California. "We had been drifting in some kind of dual world—a parallel realistic world; and the preoccupations of the world we came from, which are considered realistic, were to us filled with mental mirage," he wrote simply.[19] He had not come far in airplane miles from San Diego, but *Western Flyer* had gotten eerily far from ordinary sea-lanes. Out of radio contact, but constantly in touch with the shore as he and his colleagues collected specimens in the shallows and visited with the locals, Steinbeck found himself in another sort of time in which the writings of the eighteenth-century Jesuit Clavigero often made more sense than the vague directions of the local coast pilot. Despite the sailing directions and charts and compass and at least a few buoys, *Western*

Flyer all too frequently motored just above uncharted ledges and rocks unknown to the masters of engine-equipped vessels—who kept well clear of entire reaches of hazard—but once intimately familiar to the masters of Spanish sailing craft working close inshore. Steinbeck found parallel worlds not only when he compared the quietude of the Sea of Cortez with the bustle of San Diego Harbor but also when he compared modern sailing directions with the antique local knowledge of Clavigero.

Sometime between 1880 and 1920, then, technological change forever altered ocean navigation and mariner perception of ocean travel, and transformed the geographical understanding of educated landsmen. Steamships routinely proceeded along a handful of shortest-distance routes, and much navigational knowledge slipped quickly into obscurity. St. Paul's Rocks became legend and Java Head only a magic name.

Sailing ships followed the prevailing winds, their masters taking advantage of regular and seasonal winds to make passages short in time but often immensely long in miles. The Atlantic trade winds hurried United States–bound ships west from London, but only after masters had sailed far south from the English Channel and the Bay of Biscay. Ships sailing to the Far East from Britain and New York proceeded west and south, through the Doldrums, finding at St. Paul's Rocks the trades that led to the "Roaring Forties" of the South Atlantic, which swept East Indiamen and then clippers to Chinese ports and to Australia, past the Crozet Islands and Kerguelen Island. The continual gales that slowed Magellan and his successors trying to double Cape Horn from the east sped the sailing ships carrying grain eastward from Australia to London and Hamburg. Finding and keeping the wind preoccupied mariners before the steamship era, and by the early nineteenth century, governments, shipmasters' associations, and casual conversation combined to give masters more or less accurate information about global wind patterns, and many sailing ships plied regular routes determined almost wholly by winds expected and found. But storms blew sailing ships into half-known regions, masters regularly experimented with weather hunches and fragmentary information that seduced them off regular routes, and agents frequently chartered small, shallow-draft sailing vessels to ports visited rarely, if at all, by long-distance ships. For all that late nineteenth-century sailing-ship masters followed prevailing-wind routes, they frequently found themselves in unfamiliar regions known only through sailing directions.

In an age of uncertain navigation, masters-in-sail knew that following well-known routes meant encountering expected and useful winds most of the time, but also the certainty of meeting other ships and the likelihood of raising seamarks without difficulty. Java Head proved hard to miss, most of the time.

At the Indian Ocean entrance to Sunda Strait, Java Head announced the beginning of the Far East, the outermost portal to the South China Sea. From the sixteenth century on, European sailing vessels followed the route of the earliest Spanish and Dutch adventurers, who learned that much of the year prevailing winds practically deposited square-rigged ships at Java Head, and that Sunda Strait provided the only direct passage through which a sailing ship might be navigated toward Borneo, the Philippines, and the whole coast of China. Through the centuries, European and American merchant vessels and warships made Java Head their first landfall, although sometimes they noticed Cocos a few hundred miles offshore. But other vessels made for Java Head too, especially whaling ships.

"With a fair, fresh wind, the *Pequod* was now drawing nigh to these straits," writes Herman Melville in *Moby-Dick*. Throughout his novel, the former whaleman writes precisely and eloquently about wind. "And yet, 'tis a noble and heroic thing, the wind! Who ever conquered it? In every fight it has the last bitterest blow," Ahab muses before voicing the opposite view. "And yet, I say again, and swear it now, that there's something all glorious and gracious in the wind." Masters of whaling ships, as Melville shows, are students of winds, prevailing and otherwise: they must steer their ships anywhere whales may be found, "off the Persian Gulf, or in the Bengal Bay, or China Seas," and so must know "Monsoons, Pampas, Nor-Westers, Harmattans, Trades." And it is the trades that bring *Pequod*, chapter by chapter, to Java Head.[20]

Gradually, the ship approaches one of the most famous seamarks in the mid-nineteenth-century marine universe. "The *Pequod* still held on her way north-eastward toward the island of Java; a gentle air impelling her keel, so that in the surrounding serenity her three tall tapering masts mildly waved to that languid breeze, as three palms on a plain." In his eighty-seventh chapter Melville waxes geographic, detailing how "the long and narrow peninsula of Malacca, extending south-eastward from the territories of Birmah, forms the most southerly point of all Asia. In a continuous line from that peninsula stretch the long islands of Sumatra, Java, Bally, and Timor." Mariners watch for one opening especially. "This rampart is pierced by several sally-ports for the convenience of ships and whales; conspicuous among which are the straits of Sunda and Malacca. By the straits of Sunda, chiefly, vessels bound to China from the west, emerge into the China seas," passing as they do "that vast rampart of islands, buttressed by that bold green promontory, known to seamen as Java Head."[21]

Although still troubled by Malay pirates, the grand procession of ships moves by day and by night too, since the broad, deep strait is known for its regular, seasonal winds. Java Head stands in *Moby-Dick*, as it did in the minds of so many seamen in the age of sail, as the landfall marking the opening of the Far East, the point of departure for the Indian Ocean, Cape of Good Hope, and the Atlantic Ocean washing home.[22] In 1851 Java Head stood in the first rank of world seamarks, a mark Melville had to touch.

A half century later, Java Head had become a place of memory, a seamark touched by the last tired sailing vessels, like the bark *Judea* Joseph Conrad describes in his 1898 short story "Youth." Making a passage from Liverpool to Bangkok in a worn-out vessel is wonderful to Marlow, its young second officer, for it is his first trip to the East. "Then we entered the Indian Ocean and steered northerly for Java Head," he reminisces about his first passage. But the men discover their cargo of coal is afire, and 190 miles south of Java Head a routine passage turns terrifying.[23]

In roughly the same position as Melville's *Pequod, Judea* founders, and Marlow remarks, "And then I knew that I would see the East first as commander of a small boat." In a fourteen-foot boat, with a jury rig made from an oar and a boat awning, Marlow and two seamen keep company with the other two castaway boats. His excitement so great that he refuses to report the sails of a ship hull down on the horizon for fear it will carry him and the men back to Britain, Marlow steers toward the magical Far East. "I was steering for Java—another blessed name—like Bangkok, you know."[24]

Eventually the three boats make a tiny harbor they discover by means of its fitful red light, and the men make fast against a jetty. A short while later their peace is interrupted by the arrival of a small steamer commanded by a man livid with rage that the red light failed as his ship turned into the channel. He learns of the castaways and promises to take them to Singapore.[25] Marlow ends his reminiscences abruptly, suddenly understanding that in his youth he saw the end of sail and the overwhelming changes wrought by steam, including the expectation of buoys and other navigational aids like red lights. But he is deeply satisfied to have arrived in the East under sail and to have made his first landfall at Java Head.

"On July 5th we passed Java Head at the entrance to the Straits of Sunda," wrote a sixteen-year-old British merchant-service cadet assigned to the sailing ship *Aristomene* in 1903. "Java Head is a high hill, covered with a thick forest. At the foot of it is a white lighthouse on a square rock, called Friar's Rock."[26] Young Victor Making knew the significance of Java Head as the traditional portal opening on Asian waters, and he knew that he passed it only because British custom and law in 1903 still mandated five years of training in sailing vessels for all merchant-service cadets. Had he gone directly into a steamship, he would have avoided Java Head, and so missed the approach to a place in some ways unchanged for centuries, the place described again and again in sailing directions emphasizing winds.

"In the North-west Monsoon, i.e., from the middle of October to the middle of March, but especially in December and January, the southern tropic should be crossed several degrees to the westward of meridian of Java Head, when a direct course can be steered for Sunda Strait, or to make Engano Island, or the land about Flat Point, the southern extreme of Suma-tra," warned the Lords Commissioners of the Admiralty in their 1867 *China Sea Directory*. "Great care must be taken during this monsoon not to fall to leeward of Java Head, for the westerly winds blow with great violence along the south coast of Java, and their strength, united with the strong current setting to the eastward, make it impracticable to beat up along this coast; a vessel may thus have to beat to the southward, and re-enter the S. E. Trade, in order to make sufficient westing to fetch Flat Point." Thus Java Head is not only a portal through a wall of islands but a portal to which, and through which, winds blow regularly but differently with the seasons. While the Southeast Monsoon, "the period of fine season," is essentially an extension of the southeast trade winds, the Northwest Monsoon is irregular, with oc-casional heavy gales, rain, and electrical storms, but it too can be located pre-cisely, between latitudes 10° to 12° south and 2° to 3° north. Winds, not dis-tance, govern the route from the Cape of Good Hope to Java Head, and wise shipmasters "on leaving the Cape steer boldly to the southward so as to run down the easting in lat. 39° or 40° S., where the wind blows almost con-stantly from some western point, and seldom with more strength than will admit of carrying sail." Between the middle of April and the middle of Sep-tember, vessels sailing east should pass the island of St. Paul, "not edge away too quickly to the northward," and then proceed east to the meridian of Java Head, "crossing the southern tropic in about 102° E."[27]

However well known the winds and their seasonal changes, as late as 1867 too much of the seabed remained uncharted for shipmasters to follow sailing directions carelessly. In 1865 HMS *Serpent* went looking for Glendin-ning Shoal, reputed to be in latitude 9°45′ south, longitude 98°45′ east, ex-actly on the November-to-January best-wind Indian Ocean route to Java Head. The explorers found no bottom thirty miles in any direction from the position, but rumors of uncharted shoals continued to worry cautious shipmasters.[28]

Even headlands as familiar as Java Head tricked the unwary. "When making the head in hazy weather, the appearance of the land eastward of

Tanjung Sangian Sira, between it and Tanjung Sodong, bears much resemblance to that of the west point of Java," warned the United States *East Indies Pilot* in 1924, "with the adjacent hills on Princes Island, and the low land in such circumstances not being distinguishable at a distance, the position of it has been mistaken for the entrance to Princes Channel." In 1903 Making watched the Dutch lighthouse keeper dip his national flag and the master of *Aristomene* return the salute, perhaps without realizing how much the light had improved navigation, especially at night. "A flashing white light, 165 feet (50.3 m) above high water, visible 18 miles, is exhibited from a white iron framework, 65 feet (19.8 m) in height on First Point," the *Pilot* explained of the lighthouse atop the rock mariners understood as the first of many points they passed in the straits.[29] Yet only ten years later, lighthouse notwithstanding, the United States Navy reissued its sailing directions with a new warning of offshore hazards and a sketch of the Java Head to make the seamark more recognizable to officers aboard the few fast-moving steamships that neared it.[30] Glendinning Shoal might or might not worry steamship masters in 1934, but making a landfall at Java Head did, because by then Java Head had become an out-of-the-way place rarely passed by steamships.

Of course, the 1934 sailing directions make almost no mention of winds, prevailing or otherwise, except to note that vessels approaching Sunda Strait from the north during the East Monsoon will find sharply defined tide rips. The Admiralty's 1867 *China Sea Directory* existed chiefly for masters of sailing ships, but thereafter British and United States sailing directions become less focused on winds and more fixated on compass-run approaches to specific landfalls. After about 1900, too, sailing directions begin to suggest that shipmasters pick up pilots near particular landfalls, and the 1935 *Sailing Directions for Sunda Strait* emphasizes that "there are European and native pilots at Merak on the Java shore northeast of Merak Island, and at Old Aujer" who would take steamships safely to Batavia.[31]

Dusty volumes of sailing directions for any sea and any coast demonstrate that steam replaced sail extraordinarily quickly, and that being master of an oceangoing vessel increasingly meant a responsibility for long-distance passages only, not blundering about archipelagos and entering primitive ports—or even smaller harbors like Penang or Batavia. Sailing directions increasingly emphasized deep-water ports and merely repeated decades-old information about small harbors served only by local vessels. Now and then even the most comprehensive, up-to-date sailing directions omitted modern aids to navigation cited in directions published only a few years earlier. In 1934 a shipmaster new to the East and relying only on United

States sailing directions might not learn of the fine Dutch lighthouse at Java Head.

Nevertheless, Java Head remained a prominent touchstone in the education of sailing-ship men and in the literature of the sea. A man who reminisced about St. Paul's Rocks or Java Head advertised his traditional authority as a man who had trained in sail and who knew things steamship men would almost certainly never learn.

Conrad wrote from sailing-ship experience as well salted as Making's. His first berth as second mate provided the details of "Youth," and his later career provided the perspective that enriches that story and so much of his other writing as well. His ship *Palestine* was a very old vessel, and it never delivered its cargo of coal to Bangkok in 1881. Instead it burned and had to be abandoned in Banka Straits, a shipwreck that thrust Conrad into a small boat. Perhaps Conrad became a writer because he wanted to be a writer, but no one can deny that when he was nearly forty, an experienced master of sailing ships and one known for choosing chancy but fast routes across the Coral Sea and through the Torres Straits, positions for men like him suddenly became very scarce, and he thought himself too old to begin again as a mate in a steamship. Steamships put Conrad out of work. Steamships made a mystery of Java Head. Steamships put Singapore on the map.

Singapore is a creation of the steamship era, a modern city in a way few port cities are modern. It is a creation almost wholly of the steamship and the Suez Canal, and rooted in the prescience of Sir Stamford Raffles, who hoisted the British flag at a fishing village in 1819. Raffles convinced the Honorable East India Company that locally owned trading vessels from Siam, Cambodia, Champa, Cochin China, and even China would find their way to a British duty-free port and that if enough "country ships" arrived, the Dutch East India Company would lose its monopoly on East Asian trade. Bit by bit Singapore trade grew, but most of the place remained a mangrove swamp, simply because sailing vessels entered the East past Java Head, to avoid terrific headwinds that made reaching the new outpost well-nigh impossible.[32]

With great regularity, the Northeast Monsoons defeated shipmasters sailing north into the China Sea. Other master mariners reminisced about the Northeast Monsoons as John Herries McCulloch did in his 1933 memoir *A Million Miles in Sail*. "It was often attempted, but it invariably meant months of unavailing struggle," he wrote of clipper-ship masters struggling north of Sunda Strait toward the vicinity of Singapore. "In the monsoon season, the wise skipper of a windjammer ran his easting down until he passed Australia, and then headed straight north through the Pacific to the latitude of port." The regularity and significance of the monsoons under-pin the writings of Melville, Conrad, Making, and hundreds of other au-thors concerned with Java Head and the East in sailing-ship days. The mon-soons dictated that sailing ships ran the Eastern Passage from New York to China, the south-of-Australia-then-due-north "Great Circle" route twenty thousand miles long. Fluctuating winds often carried ships south of the Roaring Forties into the frigid seas of 50° to 55° south latitude, and some-times even to 60° south, the southern limit of the albatross, which to wise masters meant the beginning of dangerous pack ice. By the 1920s the bird McCulloch estimated ranged over twenty-seven million miles of sea rarely encountered a steamship en route to Singapore or anywhere else.[33] In the first years of the twentieth century, steamships leaving western Europe passed through the Suez Canal, then the Red Sea and Gulf of Aden of Tom-linson's *Gifts of Fortune* and *Gallion's Reach;* midway across the Indian Ocean they joined the sea-lanes of ships proceeding east from North America to make a landfall at Pulo Weh, rather than Java Head.

Running east after Pulo Weh, the steamship passenger found Sumatra to starboard as the vessel entered the searing heat and thunderstorms of the Malacca Strait, a place so dangerous to sailing ships that few veteran pas-sengers entered it until steamship days. Ahead lay Singapore, by the end of the 1930s the seventh city in the world in shipped tonnage, a bustling duty-free port heavily fortified by the British Admiralty, a crossroads of Asiatic and world shipping. To port, sometimes worthy of a brief pause even by great liners, lay Penang on the coast of Malaya, another city born of steamship service.

When William Brown saw Penang at the turn of the century as a junior mate on a British India Steam Navigation Company steamship, the tiny Eu-ropean settlement had settled into a decline, shipping little except *nappi,* rotten fish buried in cloth bags for months or years, the chief constituent of a Burmese gastronomic delicacy called *balatchong.* The smell sickened Brown, but he noticed not only that Penang boasted a handful of spotless, white-painted bungalows inhabited by a tight-knit community of Scots, but that its "native quarters" were clean and prosperous, in marked contrast to what he considered the starving filthiness of every other city on the run of his small coasting steamer.

Unlike a great many European steamship officers, Brown treated all lo-cals—Malays, Chinese, Burmese, and others—with respect. He soon had a place as first officer of *Amboyna,* a steamship of the Koe Guan Steamship Company, a firm owned entirely by a Chinese family itself quite newly ar-rived on the west coast of Malaya. Gradually Brown learned to navigate the straits, learning from the master of *Amboyna* what that Englishman had learned from Ah Chong, the founder of the company, who had learned to navigate a sailing junk in those treacherous waters. What few charts existed, say of the Mergui Archipelago, proved wholly inaccurate, but in time Brown knew the route as well as Ah Chong, who occasionally shipped as passenger aboard the little steamer. When the great tin-mining boom hit Penang, Brown found his navigational knowledge in great demand, and in 1906, at age thirty-two, he became its harbor pilot. In this capacity he guided freighters and eventually passenger liners into a very difficult anchorage opening on a new, wealthy city, one of the few in the British Empire in which different races mixed easily. That, at least, was the case until the late 1920s, when so many Europeans descended upon Penang that the idiosyncratic harmony of Brown's early decades gave way to a racism that helped drive him home to Scotland.

Penang offers an intriguing window on the steamship era. An insignificant port served chiefly by sailing vessels and later by small, oceangoing coastal steamers, it boomed first with the discovery of tin, then the perfecting of rubber planting, then the building of a railroad link to the interior. As late as 1908, Penang had no wharves at all: steamships anchored in the harbor as had sailing ships, and loaded cargo from lighters.

The tin boom overwhelmed the port. Its first wharves could accommodate only four oceangoing ships at a time, but for months a dozen ships arrived daily in a port that for decades boasted not a single tugboat. Liners up to eighteen thousand tons berthed under their own power, Brown ordering their helmsmen to take advantage of tricky winds and currents; and fast mail steamers, pausing only to unload high-priced package cargo, merely anchored in the Ka Channel for a few hours. Colliers arriving from Britain anchored for lightering as Brown directed, or else ran up the Prai River under his guidance. As Penang pilot, Brown took ships to the Province Wellesley shore across the narrow strait, conning steamships in ways that would have astounded anyone in London or New York. When he took a steamer away from the Prai River wharves, he let the incoming tide sweep the ship's stern upriver, away from the wharf, then unmoored the bow. As the stern of the steamship crashed into the jungle growth across the river, he ordered engines full ahead to swing the bow past the wharf end and downstream, toward the strait. Away from the tugs of London and New York, in places like Penang, pilots like Brown used techniques mastered in days of sail and borrowed from local old-timers like Ah Chong. Steamship masters, especially younger steamship masters, gratefully surrendered their responsibilities to pilots who combined sailing-ship experience with local knowledge.

"I am not suggesting that there was anything more romantic or dangerous in the work than lies in, say, plumbing," Brown recalled in 1940. "It requires experience and confidence, no more."[34] But Brown was modest about more than his skill at hurriedly piloting liners on the London–Tokyo run. Penang pilotage required endurance and bravery too, for all that he says so little about it, and while pilotage was not compulsory at Penang, the United States Hydrographic Office's 1933 *Sailing Directions for Malacca Strait and Sumatra* emphasized twice that prudent shipmasters picked up pilots before trying to enter port.[35] All a deep-water master had to do was find the approach buoy; Brown or a fellow pilot would arrive in a small boat and pilot in the steamer.

Brown often picked up his inbound ships fifteen miles from the anchorage, at the north- or south-channel approach buoys, leaving his small launch to clamber up the rope ladder flung down from the deck of a steamship rolling in the swell. In fair weather the wait proved pleasant, and since in the years before radio the lighthouse keepers reported inbound ships, Brown often had only short waits. In bad weather, the little launch endured terrible danger, as it did one Sunday when Brown took his wife for a picnic before piloting a Peninsular & Oriental intermediate liner up the south channel. From a bluff, Brown spied funnel smoke on the horizon, and they went back to the launch, noticing that the wind had risen. As he and his wife and crewmen ran the launch seaward, the sky darkened and a squall hit, obscuring the approach buoy.

Brown knew that the liner master would not near the approach buoy but would expect to find the Penang pilot farther offshore. Away from the shelter of Penang, Brown watched his wife flung from one side of the cabin to the other and saw his tough Malay boatmen begin to worry. Afraid to put the launch about lest it capsize in the trough of the waves, Brown soon realized that in all his years he had never seen such a sudden storm, and began speculating where the bodies would wash up. In the driving gloom he spotted the liner heading away from land and danger, but when the steamship master saw the struggling launch, he swung about and made a lee that Brown used to near the ship. Brown raced up the ladder, took charge, and asked permission to enter the south strait more slowly than usual, saying only that the launch needed shelter. "A dozen times I called out to ask how the launch was behaving, and each time apologized for the steamer's reduced speed," Brown remembered. Finally the steamship master asked about the concern, and when Brown told him his wife was aboard, prostrate with seasickness, the man roared, "Do you mean to say your wife is down there on the launch? Get her up here at once."[36] But Brown knew that in such seas no line could winch up his wife safely, and he guided the liner and its launch into port despite a raging argument with the steamship master. After anchoring, the master, a friend of Brown's for many years, left the bridge without a word. He expected Brown and his crew to run small-boat risks to safeguard shipping, but not to risk the lives of passengers, especially women. Pilots like Brown existed to reduce risk to everyone, except themselves.

In all major ports, pilots relieved masters from following the intricacies of sailing directions, but as steamships grew larger by the decade, deep-sea masters encountered fewer pilots with the old-fashioned sailing-ship flair. Instead of Brown's masterful use of wind and tide, pilots now used tugboats, sometimes many tugboats, to overcome natural forces. By 1939 many masters of large freighters, tankers, and passenger liners served in ships running little more than back-and-forth bus routes along established sea-lanes with one or two intermediate stops, each stop studded with pilots. Masters seldom needed to consult the volumes of sailing directions shelved on chartroom bulkheads.

Along established sea-lanes linking ports by the shortest possible routes—not the routes of favorable winds and currents—steamships operated on schedules so nearly predictable that passengers bet, not on the hour land would appear at the end of the passage, but on the minute. Not for nothing did so many trans-Atlantic passengers speak glibly of the "Atlantic ferry," or so many passengers putting out from San Francisco expect Honolulu, Tokyo, and Singapore to materialize right on time.

Penang became the western world's gateway to Singapore and the rest of the East, and the Strait of Malacca replaced Sunda Strait and its suddenly quaint seamark, Java Head. Seemingly insignificant mileage differentials made steamships avoid longtime seamarks. Steamships proceeding from Cape Town to Singapore past Java Head logged a distance of some 5,624 miles. On the new, preferred route, from Cape Town through the Malacca Strait, past Penang, then to Singapore, they ran just a bit less, 5,560 miles.[37] The thirty-six-mile difference totally altered Indian Ocean sea-lane traffic. Although just north of Sumatra eastward-steaming ships encountered the hundreds steaming east from the Suez Canal and Arabian Sea and the hundreds more steaming west toward Europe and the Atlantic coast of North America, only a handful of sailing and steam vessels moved through Sunda Strait by 1935, and of those most were local steamships on the Singapore–Freemantle route. Java Head became as St. Paul's Rocks, a place almost as obscure to young men whose steamships never passed it.

Sometime after 1900, St. Paul's Rocks slipped from sailing directions and mariner knowledge into oblivion, becoming essentially a navigational hazard well marked on charts as a place to be avoided. Even in sailing-ship days, St. Paul's Rocks had nothing to recommend a visit. In its 1873 sailing directions entitled *The Coast of Brazil,* the United States Hydrographic Office wrote that no one had carefully examined the rocks until 1799, when a curious shipmaster studied their massing and channels and ascertained that the group consisted of three islets and ten rocks rising from great depths. The highest point on the group is only sixty or seventy feet above sea level, and various warships and survey vessels tested how far the rocks might be seen. One warship, USS *Portsmouth,* discovered that a masthead lookout could see them twelve miles off, and that from the deck, about twenty feet above the sea, men could see the highest point when about eight miles off. Guano splattered over the taller rocks made them somewhat more visible than they might have been without it, but the sailing directions conclude that "in a dark night they would hardly be seen more than 2 miles off, owing to the whiteness of their summits."[38] Depending on the direction of approach, the group of islets and rocks changes appearance rapidly, but no survey by anyone found the remotest reason for approaching very closely. Seabirds swarmed over the place, and while *Portsmouth* found a small cove on the middle islet—its opening protected from surf by the northern islet—where its boats landed without difficulty, its officers located nothing of value except the eggs of sea birds and a surfeit of fish. In two hours, the boats brought in enough fish, mostly groupers, to feed one hundred men.

St. Paul's Rocks lack fresh water, and *Portsmouth* found "some remains of a wreck, such as pieces of copper and timber; also places cut out of the rock, evidently intended for collecting rain-water." Barring its warm climate, a more barren, less hospitable menace to navigation could not be imagined, and no mariner thought of the group as a refuge from shipwreck, even as a destination for survivors in lifeboats. Castaways did not live long on St. Paul's Rocks. Masters of sailing ships used the group as a seamark to check their longitude reckoning, but prudent mariners avoided any close contact. As steam replaced sail, St. Paul's Rocks became a treacherous hazard which steamship men heard about in old tales but never saw.

Only the *Kriegsmarine* recognized in St. Paul's Rocks a useful shelter that broke the Atlantic swells, a few tiny crumbs on an immense chart where U-boats could resupply in safety, one of many such places blundered into during wartime by hapless merchant vessels far beyond naval protection. War forced steamships into seas empty since sailing-ship days, and in those seas steamships encountered enemy forces.

Ignorant of what lay ahead, then, *Sebastian Cermeno* steamed into a secret *Kriegsmarine* rendezvous area at 30° south, approximately due south of Mauritius and southeast of Madagascar. Six U-boats, each maintaining strict radio silence, had gathered around the supply tanker *Charlotte Schliemann* for fuel, provisions, and ammunition. But none of the six U-boats torpedoed the American merchantman. A seventh U-boat, U-511, making a passage from France to Japan and a day to westward of the rendezvous zone, made the kill. Not until long after the war ended did naval historians begin to piece together the torpedoing and subsequent events. Two *Sebastian Cermeno* castaways somehow arrived at Fort Dauphin in South Africa after two weeks at sea, but the arrival of Patenall's lifeboat in Madagascar remains "off the charts."[39] The progress of at least one Italian submarine en route from Bordeaux to Japan through the rendezvous area remains a mystery as well. But above all, the mere fact that late June 1943 saw *seven* U-boats southeast of Madagascar reminds anyone looking backward at World War II of the immense spread of Axis sea power—and the reason why lifeboats from ships like *Sebastian Cermeno* sailed in lonely seas indeed. Axis naval power first scattered Allied merchant shipping worldwide and then methodically destroyed it, thrusting castaways into lifeboats in regions sailing directions scarcely mentioned.

BLIGH

"I WAS THINKING OF THE *Bounty*'s LAUNCH AS OUR LITTLE SQUARESAIL bellied out to the following wind and Timoe was dropping away astern," James Norman Hall related to *Atlantic Monthly* readers in 1934. "What a trifle our voyage was compared with that of Bligh and his men! And yet, twenty-five miles of sea can be as rough on a given day as thirty-six hundred miles of it, and a small heavily laden boat, whether bound for Mangareva or Timor, can take only so much water over the gunwales."[1] Fitted with a jury-rigged mast and a sail made from a deck awning, a rig that Conrad knew when *Palestine* foundered, Hall's whaleboat headed away from the uninhabited coral atoll on which the trading schooner *Pro Patria* had struck days before. Hall had every reason to think of Captain Bligh and his castaway crew, for he had taken passage in the ninety-ton, Chinese-owned, French-flag trading schooner solely to visit Pitcairn Island, the isolated spot settled by the mutineers of HMS *Bounty*. Deeply ensnared by the puzzling contradictions of the 1789 adventure that he and Charles Nordhoff had described in their 1932 *Mutiny on the Bounty,* Hall knew that only firsthand knowledge of Pitcairn would enable him and his coauthor to complete the third volume of their *Bounty*-focused work. The second volume of the trilogy, *Men against the Sea,* appeared almost simultaneously with *The Tale of a Shipwreck,* the book-format version of Hall's magazine article, a juicy firsthand account of shipwreck and small-boat passage.

However devastating the shipwreck, Timoe offered plenty of fish, lobster, and coconuts, and so many pools of fresh water that at first the castaways scarcely noticed the well made by long-vanished residents. "We sat down to meals even better than those we had enjoyed on shipboard," Hall recounts of days spent off-loading the schooner, erecting tents, and piling up great stores of supplies, then strolling beaches while gathering driftwood for the cooking fire.[2] So blissful did he find the atoll—except for the hermit crabs—that for a short while he considered spending six months alone there, as an "experiment in solitude," but longing for his wife and children caused him to take a place in the whaleboat setting out for the nearest inhabited island.

Hall rescued his books first from the wreck of *Pro Patria*, bringing them ashore at night wrapped in a piece of oilcloth he cut from the cabin table, but on Timoe he found he had no need of them. "We were completely cut off from the world, for the time being at least," he mused, knowing that the schooner lacked wireless equipment to report its foundering. "Christian and the *Bounty* people had not been more isolated." But even after the salvage work ended and he had hours of empty time, Hall felt no need to read.[3]

Books, he decided, belong in the densest of cities, because books offer individuals the aloneness that nurtures individual thought. Cast up on Timoe, wandering its immense beaches, thinking about leaving it in a jury-rigged whaleboat, Hall realized not only something of the late eighteenth-century microcosm of the *Bounty* launch but something more complex than his earlier dismay at American culture.

Hall had fled the United States for Tahiti in 1920, disgusted by transformations caused by applied science, big business, and commercialization, especially advertising. Home from his Great War experiences as a fighter pilot and a wounded prisoner of war, Hall found what "seemed to be a kind of hysteria everywhere, a natural reaction, perhaps, to the sudden release from the strain of war," which drove him to the South Seas. In the next twelve years, he reported that he "had time and to spare for the pleasures of reading and reflection, and to make a belated start at raising a family," and he and Nordhoff—another refugee from fast-paced urban life—became intrigued with South Seas history, especially with the tales told by reflective mariners like the aging seamen who populate books by Conrad and Tomlinson in which time and space flow as waves and currents.[4]

On Timoe and in the whaleboat—but never aboard *Pro Patria*—Hall struggled to imagine the circumstances of the *Bounty* launch and the thoughts and emotions of its occupants. But he knew the fallacy of such imaginings. For all that he could study the five inches of freeboard in the whaleboat, see how every wave threatened to swamp the boat as it left the shallows, think about how cold he felt in the downpours of tropical rain, he knew his mind was of the twentieth century, as modern as the schooner *Pro Patria,* so unlike the eighteenth-century *Bounty*. In 1933 the year 1789 seemed almost unreachable even from Timoe, and suddenly Hall began to wonder how readers would accept his books. Would they understand the subdued dignity with which Bligh and his men arrived in Koupang Bay?

"Thus, often, are the great, the truly heroic achievements in human history brought quietly and obscurely to their conclusion. At least, so it was in the eighteenth century; but one can imagine the ballyhoo that would be made— that is made—in our day over achievements far less remarkable."[5] Perhaps Hall referred to Gertrude Ederle swimming the English Channel or Lindbergh flying the Atlantic, but certainly he stewed about the twentieth-century mind-set as he confronted a tale from long before the steamship era.

Like *Men against the Sea*, his *Tale of a Shipwreck* originated in his loathing of the too rapid modernization he saw everywhere after the Armistice. "I love change only in its aspect of slow and cautious advancement and slow and imperceptible decay," he wrote self-analytically. "And I dislike change in manners, customs, and habits of thought as much as I do in material aspects." Hall flew fighter planes in combat, but he so hated automobiles that he refused to learn to drive one. Automobiles, he argued fiercely, symbolized the postwar fragmentation of small-town Iowa life and the destruction of the American rural landscape, for they made possible the real-estate speculations on which big business thrived. In his view, the automobile destroyed the railroad-corridor-balanced landscape of cities, small towns, suburbs, and rural regions and replaced it with one of endemic commercial disorder and sprawl.[6] Once convinced that even his Iowa hometown had succumbed to the crassest commercialism, he joined his like-minded flying-corps friend Nordhoff in a 1920 sojourn in the South Seas.

In Tahiti and in the more remote islands of the South Pacific, the two young men found portals into other centuries. Even after Hall's whaleboat made Mangareva, the castaways had only marginally improved their condition, for only once every four or five months did a schooner make the nine-hundred-mile passage from Tahiti. "There was no wireless station, so the loss of the *Pro Patria* could not be reported," Hall wrote. "Convenient though a wireless station would have been for us then, I was glad that none existed." Without a wireless station, his family and friends would wonder at a schooner marked "overdue," but the wireless station that would relieve their concern would certainly ruin Mangareva, a place blissfully isolated from daily news bulletins. Hall imagined transporting Rikitea village and its beach to Manhattan, between Thirty-second and Forty-second Streets, as a time- and space-travel experiment, but could not imagine how New Yorkers might evaluate the insertion. He thought that its very spaciousness might put off New Yorkers more than the lack of technology.[7] Nowhere else does Hall so closely approach the reason behind his move to the South Seas, his writing *The Tale of a Shipwreck,* and his coauthoring a book about an eighteenth-century open-boat passage.

In any ship's-boat passage Hall glimpsed something of the premodern era, before steamships, wireless telegraphy, and a social disorder masquerading as progress. Whatever changes might come to Tahiti eventually, the *Pro Patria* wreck confirmed Hall's thinking that a ship's-boat passage remains essentially timeless, the best portal into the premodern past and the postmodern future he envisioned when he imagined Rikitea dropped into midtown Manhattan.

A deepening interest in ship's-boat passages marked the years immediately following World War I, when Conrad wrote about the wheels and levers and electric lights destroying seamanship. While only a small part of the interest can be traced to the memoirs of elderly masters-in-sail contemptuous of steamships and steamship seamen, that part proves important in any contemporary analysis. Beginning in the 1860s, narratives of ship's-boat passages not only recounted harrowing marine experiences but offered complex and subtle moral, social, and cultural perspectives of immediate use in any evaluation of technology-driven modernization. After World War I, the narratives spoke to young men like Nordhoff and Hall to whom the Great War seemed an abyss into which questions of heroism, endurance, manhood, and even humanity had plummeted.[8]

Whatever the original reasons for publishing them—and the reasons range from defense against charges of incompetence to correcting newspaper misstatements to chronicling great events for posterity—in the 1920s boat-passage narratives gathered new meaning as a way of understanding a prewar era not so much long distant as on the other side of a great chasm. In the 1920s only a handful of sailing ships remained, and only a handful of ship's boats remained at sea, lashed upside down to cabin tops or carried in stern or quarter davits. When Hall and his contemporaries thought of ship's boats, they thought of boats from the 1860s perhaps, or the 1880s: boats still in service alongshore everywhere, but boats wearing out except in regions like the South Seas.[9] Hall and Nordhoff accepted the everyday presence of launches and gigs and jolly boats and whaleboats that clipper-ship masters like Josiah Mitchell would have understood in 1866, the year his *Hornet* burned.

Mitchell sailed his twenty-one-foot ship's longboat four thousand miles, carrying himself and fourteen other men to safety in Hawaii, a spectacular feat that aroused worldwide admiration. Mark Twain himself dispatched an account of it by schooner from Hawaii to his newspaper, the *Sacramento Union*, thereby scooping several other reporters linked to wire services.[10] Several months later, having read journals kept by three of the castaways and having interviewed some of the men again, Twain published "Forty-three

This September 29, 1866, *Harper's Weekly* illustration of the *Hornet* disaster focuses more on the crowded ship's boats than on the burning hulk.

Days in an Open Boat," in *Harper's Magazine,* a much expanded version of his initial dispatch. Three decades later, he recalled not only the extraordinary longboat passage but the way his reporting of it catapulted him to fame. In an 1899 *Century Magazine* essay entitled "My Debut as a Literary Person," he recounted filing the original newspaper story and writing the *Harper's* article. He also reproduced portions of the diaries kept by Captain Mitchell and his two passengers, the Ferguson brothers; at the time of the sinking one brother was a recent graduate of Trinity College in Connecticut, the other a student there. When he wrote his *Century* piece, Twain was living in Hartford, and he and Henry Ferguson, by then a professor at Trinity, had become friends and had discussed the significance of the longboat passage. "The interest of this story is unquenchable; it is of the sort that time cannot decay," he concluded at the close of his *Century* article. "I have not looked at the diaries for thirty-two years, but I find that they have lost nothing in that time."[11] What had been a sensational story in the early fall of 1866—*Harper's Weekly* ran a full-page illustration of the boat leaving the

Replicas of seventeenth-century ship's boats show few differences from
the *Bounty* launch or even from late nineteenth-century ship's boats.
Always the boat exists to freight anchors and other heavy objects.
(Photo by author)

burning clipper, along with a lengthy story—endured as a great tale of survival, one Twain's 1899 article only reemphasized.

Unlike so many ship's-boat passages, that of the *Hornet* longboat stands
recorded in three substantial diaries, each much more than a log. Students
of the passage, including Alexander Crosby Brown, whose 1974 *Longboat to
Hawaii* reproduces all three diaries and Twain's remarks about them, enjoy
not only detailed accounts of what happened but the subsequent thoughts

of three articulate men about the events. The journals open as *Hornet* departs New York and do not close until its longboat reaches Hawaii, so readers learn how the fire and subsequent boat passage figured in the larger views of the three men. Moreover, the daily entries juxtapose the remarks of an experienced mariner against those of two astute landsmen.

In outline at least, the story is straightforward. Against standing orders, the mate went belowdecks at night with a lantern to draw some varnish from a barrel, and the resulting explosion and fire leaped up through the booby hatch and ignited the sails of the clipper ghosting through a near dead calm. Mitchell reacted at once, launching three boats (damaging two in the process) but having time to load no more than ten days' provisions, embarking his crew and two passengers, and keeping the boats together for a day in the hopes that some other ship might be drawn by the glow and, later, by the immense pillar of smoke. Several days later, after Mitchell divided the supplies equally and took an additional man into the longboat, the boats separated and began feeling their way north from a point about a thousand miles west of the Galapagos Islands toward several vaguely charted archipelagos. After wasting a good deal of time sailing over the supposed locations of the islands, Mitchell headed for the Hawaiian chain. The crewmen behaved well if not superbly, only a few incidents of insubordination and disobedience marring a passage marked by storms and other hardships. The longboat made land in Hawaii, its men exhausted and suffering from thirst, starvation, and exposure, and within days sailed into a fame vastly more definite than the islands Mitchell tried to find.

Mysterious islands link the *Hornet* longboat passage to earlier narratives, especially Bligh's, and indeed give the *Hornet* journals an eerie, antique tone. "If our chronometer sights are to be depended on in the least it is hopeless to think of reaching any of the Revillagegedo Islands on the coast and our only chance is being picked up or the bare possibility of making Henderson Island, latitude 24° longitude 128°," Henry Ferguson confided to his journal on May 27, twenty-four days after abandoning ship. "The wind has headed us more still West Northwest."[12] Unfortunately cartographers had mistakenly platted Henderson Island at 24° *north* latitude, when in reality the island lies at 24° *south* latitude. And by May 23 the prevailing winds had driven the longboat ever farther from the Revilla Gigedo Islands off Mexico, a barren archipelago just north of Clipperton Island, toward which Mitchell had first tried to sail before being defeated by prevailing winds.

Caught up in the perplexities of islands far to windward, other islands mischarted, and several islands discovered by whalemen but marked "doubtful" on his charts, Mitchell plowed slowly on, aiming for American Group, an archipelago listed as "doubtful" in the twenty-ninth edition of Bowditch's *New American Practical Navigator*. "It seems like a forlorn hope but it is our best chance as our path runs right through the path of both upward and downward ships, and if we can live so long and they are really there, we have tolerable certainty of fetching them, as they are to leeward and we can calculate with confidence on the Trades," wrote Henry Ferguson on May 30, a day before Mitchell began his entry with the laconic statement, "American Group, gave up Henderson."

Day by day, the entries reveal the extraordinary faith of Mitchell, a Congregationalist, and the Fergusons, both Episcopalians. While all three implicitly entertained concepts of distinctly personal salvation, all three saw their predicament too in the theological frames of two Christian denominations famous for doctrinal precision. "Please God we are saved will all have an awful lesson against carelessness," Henry Ferguson wrote on May 12.

Several weeks later, Samuel Ferguson noted that he "could not make out to read the full service this A.M., but did the Prayers, Psalms, (both) Epistle and Gospel this P.M. and feel comforted by them." Until he lost his glasses overboard, Mitchell read from a book of Congregational prayers, and even after that struggled to read a little every day. On June 7 Henry Ferguson recorded that he "read the service and heartily prayed that the time is not far off when we can return thanks to God in His Holy Church," then quoted from the *Book of Common Prayer*, "From plague, pestilence, and famine, from battle and murder, and from sudden death, Good Lord deliver us." On the same day Mitchell wrote, "Another beautiful June Sabbath finds us still alive in this boat. Oh may the day be blessed and sanctified unto us to the salvation of our souls as nearer approach to heaven. How much we think of dear ones at home this sacred day and their privileges. We are fast starving to death. God have mercy on our souls." Stripped of almost everything, with time for thought and prayer, the three men perceived God all around them.

In his 1866 *Harper's Magazine* article Mark Twain dealt at length with the religious faith of the three writers, including their understanding of divine intervention in a passage among uncharted and nonexistent islands. "This strange voyage, in its entirety, is an eloquent witness of the watchful presence of an all-powerful Providence," he maintained. On approaching Hawaii, for example, the crew lowered their sail to pass through a reef, thought better of it when they saw the surf, but then found themselves too weak to raise the sail again and bear off. Drifting helplessly, they saw an opening in the reef and felt their boat swept through it and onto a tiny beach interrupting miles of cliffs, something Twain took as the hand of Providence.[13] Twain's view only strengthened when he, Mitchell, and the Fergusons took passage in the bark *Smyrniote* from Hawaii to California, and the men talked together for days about the boat passage.

Yet thirty-three years later, Twain made no mention of Providence. Perhaps he assumed that reproducing portions of Samuel Ferguson's diary would more than convince readers that faith in God supported the three diarists, but it is also possible that something had changed in American culture. "It is an amazing adventure. There is nothing of its sort in history that surpasses it in impossibilities made possible. In one extraordinary detail—the survival of *every person* in the boat—it probably stands alone in the history of adventures of its kind," he wrote in summarizing the longboat passage. No longer did divine Providence succor the castaways or direct the drifting boat through the reef. He characterized Mitchell as "a bright, simple-hearted, unassuming, plucky, and most companionable man," consigned the mental breakdown of the seamen to a category in psychology, and focused on the physical privations, especially the lack of food, above all else.[14]

To be sure, Twain himself had changed. A year after the *Century* piece appeared, he published "The Man That Corrupted Hadleyburg," one of several stories focused on hypocrisy, religious and otherwise, and this new perspective seems to have reshaped his view of the *Hornet* narratives. But then again, perhaps Twain intuited that turn-of-the-century readers were a different breed from those of 1866, that they no longer lived the spiritual life—at least on dry land—as directly as had the *Hornet* diarists three decades earlier, and that they no longer thought of ships as full-rigged and delicate and in need of protection from God.

In the intervening period a great change had swept through American culture. Well before the shipwreck, Ferguson saw islands as hazards only, something the speeding *Hornet* must avoid. "There are some nasty little islands just sticking up somewhere about here according to some charts," he wrote in his diary.[15] Until 1880, when the authors of *Pilot Book for the West Coast of Mexico* exploded the fiction of islands lying off California about 140° west longitude, mariners feared a string of islands likely to snare fast-moving vessels, and clipper ships kept well clear of their rumored locale. By the turn of the century, however, oceangoing passengers expected charts of well-traveled routes to be accurate, to show no dubious hazards, and to guide steamships running on schedules most travelers supposed immune to all but the worst gales. No longer did passengers worry about winds and storms routinely bringing them into badly charted waters. For passengers, and for many steamship officers and crewmen, by 1900 the old fear of wreck and

God had gone, vanquished by technology that was soon to produce the unsinkable liner *Titanic*.

Changes in attitudes came so rapidly with the steamship that in 1918 Conrad casually noted that the whole ocean had been well charted.[16] Yet as late as 1880, when D. J. Munro shipped as an apprentice aboard a sailing ship bound from Britain to the South Seas, he encountered a captain so locked in navigational and religious tradition that he owned only a Bible and a set of sailing directions. The master issued every crewman a Bible and demanded it be read cover to cover as sailing directions to the next world. He refused to allow clergymen passengers to usurp his right to hold divine service, and while he did not coerce passengers into attending, "his regulations as to smoking, walking on deck, etc., became drastic or otherwise according to their church attendance." Written half a century after the author had served as an officer aboard steamships and in Royal Navy warships, Munro's account is untinctured by condescension. "The Terror," as cadets called their tradition-minded master mariner, had navigated a sailing ship among a host of grave dangers and had had to accept the continuous likelihood of disaster and death. Munro understood shipboard theology—especially its linkage of sailing directions with divine writ and religious faith with faith in master and mates—to be common everywhere in sailing ships. And after his wartime experiences, he knew the worth of it. Even great steamships sometimes founder and faith proved important in lifeboats.[17] But most passengers accepted steamships as the safe transport of a modern world moving beyond traditional theology and old-fashioned hazards.

The post–Great War changes remarked by Nordhoff and Hall actually began much earlier, perhaps as early as the Civil War, and could be observed anywhere that had been touched by railroads, telephone lines, daily newspapers, and other accoutrements of modernization, including the ocean-going steamship. Long before Hall and Nordhoff fled to the South Seas, Robert Louis Stevenson had moved to Samoa to enjoy a sort of quiet already lost elsewhere. "The sound of wheels or the din of machinery was hardly known in the island," wrote one visitor who heard only the distant beat of the surf during his 1892 visit.[18] By 1920, as steamships supplanted sailing vessels and reduced many ports and islands to casual, suddenly old-fashioned maritime service, more than one traditional mariner made the connection between the rise of sophisticated, modern technology and the frenzy it engendered and the erosion of traditional religious faith. In analyzing the loss of the liner *Empress of Ireland*, Conrad mused in 1914, "We have been accustoming ourselves to put our trust in material, technical skill, invention, and scientific contrivances to such an extent that we have come

Period photographs of grounded steamships, such as this photograph of *Cartago*, often reveal ship's boats—not lifeboats—at work in the shallows. (Courtesy of the Mariners' Museum)

at last to believe that with these things we can overcome the immortal gods themselves."[19]

Into the increasingly remote world of half-charted islands and old-time religious faith stumbled first Stevenson, then Hall and Nordhoff. The two Americans' itinerant lives in the South Seas thrust them into maritime adventures their twentieth-century readers knew through old books like *Two Years before the Mast* and into an oral tradition of adventures known far less well. Many of these conflicting tales focused on Captain William Bligh and the *Bounty* mutiny.

Bligh had published his one-sided view of the mutiny almost at once, followed in 1834 by the more or less official version, John Barrow's *The Mutiny and Piratical Seizure of HMS Bounty*. As early as 1921, when they published *Faery Lands of the South Seas*, Nordhoff and Hall had begun to appreciate what Bligh's writings, especially his *Voyage to the South Seas for the*

Purpose of Conveying the Bread-fruit Tree to the West Indies, in His Majesty's Ship the Bounty, had to offer. "Don't think I am cynical in saying this," wrote Hall in *Faery Lands* after examining South Seas modernization by missionary, advertising, and canned food. "I respect and envy men who possess real faith; they are the ones by whom every great task is accomplished."[20] After a decade in the archipelagos, Nordhoff and Hall understood that the passage of the *Bounty* launch rewarded renewed attention in a secular age uncomfortable with older ideas of faith.

Men against the Sea is historical fiction based on the log Bligh kept in the launch. Nordhoff and Hall cast the story as a narrative written by Thomas Ledward, acting surgeon aboard *Bounty,* while he was recuperating in the Koupang seaman's hospital. Ledward became their inquirer into small-boat seamanship of an earlier era.

Overcrowded and poorly provisioned, sailing largely in uncharted waters, the *Bounty* launch performed brilliantly thanks in large part to its commander and his faith in God and the boat itself. Nordhoff and Hall present a Bligh markedly different in the launch than aboard HMS *Bounty,* a man somehow changed for the better by an immense challenge the authors characterize as a test. "But God expected us to play our part. We should not have had his help, otherwise," their chief character asserts at the close of the book, speaking of God while staring lovingly at the launch he commanded.[21]

Always the twenty-three-foot-long launch is the nineteenth character. Nordhoff and Hall make it an extension of Bligh and an instrument of God. It needs and receives continuous care, and gradually the reader learns that a most capable hand holds its tiller. Late in the story, Bligh hurriedly strips and dives into the sea, carrying the grapnel line below to the grapnel from which it has parted. He surfaces, dives again, makes fast the line, and is dragged into the launch. "I was none too eager to go down; but I'll ask no one to do what I fear to do myself," Bligh says, before Nordhoff and Hall begin a general discussion of great sharks and marine crocodiles.[22] Superb leadership ranks second only to superb seamanship, and Bligh uses both to preserve the launch and make the passage to Timor.

Nordhoff and Hall benefit from their firsthand knowledge of ship's-boat work in emphasizing Bligh's masterful handling of the launch. A long description of his conning the boat through the Great Barrier Reef reflects their experiences in shooting reef-channels. The Almighty, the launch itself, maybe Bligh, perhaps some combination of them all, saves the men and sweeps them into the placid blue lagoon through a passage modern charts name the Bligh Boat Entrance.

Men against the Sea also results from the extensive research Nordhoff and Hall conducted long-distance in the archives of the British Admiralty. The authors had access to the original drawings of the launch and knew that the open boat was twenty-three feet in length, six feet and nine inches in beam, and two feet and nine inches deep. Unlike many similar small craft, the *Bounty* launch carried two masts, one in the bows, the other amidships, each setting a dipping lugsail. *Bounty* carried two other boats, a jolly boat and a cutter, both substantially smaller than the launch and perhaps used more frequently. The Admiralty intended that the launch do the heavy work necessitated by working a sailing ship in uncharted places, especially carrying out anchors when the ship had to be moved on windless days—or had gone aground.[23] While built to heavier military specifications, the launch resembled the longboat of the *Hornet* castaways.

Nordhoff and Hall succeeded in their effort to introduce an eighteenth-century feat to twentieth-century readers, but they only scratched the surface of an adventure that still attracts inquirers. Forty years after their three novels appeared, Richard Hough published his *Captain Bligh and Mr. Christian: The Men and the Mutiny*, an effort to reexamine the mutiny and subsequent events in the light of newly discovered materials. Like Nordhoff and Hall, Hough took the precaution of venturing along Bligh's route in a boat not much larger than the launch. However, Hough wrote without sustained personal experience in small boats like the launch, and his book, along with others, lacks the authentic tone of *Men against the Sea*.

Anyone who follows accounts of Bligh's accomplishment through the nineteenth century will discover that it grew more and more mythic as decades passed. Amasa Delano, whose *Narrative of Voyages and Travels . . . in the Pacific and Oriental Islands* of 1817 includes the first account of the Bligh passage published in the United States, summarized the adventure in a long paragraph that described it more as a voyage of discovery than one of superhuman seamanship, and implied that Bligh's experience in navigating unknown waters mattered more than anything else; subsequent Admiralty efforts in the South Seas support Delano's view. Delano shows Bligh to have had very special skills even among eighteenth-century British naval officers. What he accomplished in his longboat derived not only from his religious faith, self-confidence, and seamanship but from his experience in surveying coasts unknown to him. Very few British navy officers knew how to manage small craft among coral reefs struck by massive surf, and perhaps only a few enlisted men had the boyhood experience necessary for making shore. What Nordhoff and Hall perceived as an old skill fast disappearing,

Delano determined to have been a rare one always, the mastery of which made Bligh an exceptional seaman.

In 1791 Captain Edward Edwards, charged with capturing the mutineers, arrived in Tahiti in HMS *Pandora* and quickly sent his launch out in pursuit of a schooner built by those *Bounty* mutineers who had not fled to Pitcairn. Sailors in the launch and in the *Pandora* pinnace captured the schooner, which Edwards then outfitted as an auxiliary search vessel before departing for other islands. Immediately after this victory, however, immense surf and uncharted hazards wreaked havoc with Edwards's expedition. Off Palmerston Island he lost a midshipman and four seamen in a cutter, and off Oahtooah the schooner sailed out of sight forever. Then Edwards misjudged an opening in a reef found by another of his small craft, and ran his ship into and over the coral. In the midst of efforts to slide sails under the hull and across the holes, *Pandora* sank so fast in the lagoon that not everyone had time to jump overboard. Within a few days Edward and his tiny flotilla of four ship's boats had begun following Bligh's course to Timor, even recognizing islands Bligh had charted.

George Hamilton, surgeon aboard *Pandora,* kept a journal during this passage that unwittingly showcases the excellence of Bligh's navigation. The *Pandora* flotilla blundered from one reef to another and escaped wreck more by luck than skill, at least in Hamilton's view. Certainly the presence of prisoners in the boats created problems unknown to Bligh and almost impossible for Edwards to solve, but somehow from the beginning the *Pandora* men fared worse than the *Bounty* castaways. They suffered sooner from sunstroke, thirst, and hunger. Morale worsened rapidly, many of the men began drinking their own urine, and the long open-sea passage led only to subsequent battles with surf as bad as any along the Great Barrier Reef.

Beset by thirst, hunger, sunlight, heat, and local inhabitants anxious to kill them, but perhaps above all by their own inexperience in navigating half-charted waters, Edwards, Hamilton, and the rest of the *Pandora* castaways eventually reached Koupang and enjoyed a gracious welcome from Dutch officials.[24] What the Dutch governor-general and his officers thought about another party of castaway Royal Navy personnel arriving in the same port no one recorded, but the *Pandora* disaster made subsequent mariners especially cautious.

Exploration proceeded very slowly in the great region Bligh had traversed in the *Bounty* launch. No word reached Britain about Christian and his fellows for another two decades, perhaps because the high costs of searching uncharted, reef-studded seas forestalled Admiralty effort. Not until late in the nineteenth century did mariners know something about the special conditions the *Bounty* and *Pandora* castaways endured. In the late 1880s Robert Louis Stevenson explored much of the southern Pacific as a passenger aboard the small trading schooner *Casco* and learned that the archipelagos of islands remained extremely dangerous and usually unvisited, except by local trading vessels. The vast parallelogram Stevenson penetrated, extending "from tropic to tropic, and from perhaps 120 degrees W. to 150 degrees E.," dismayed the *Casco*'s master, and only the continuous entreaties of Stevenson and his friends caused him to enter the region. "The reputation of the place is consequently infamous," Stevenson concluded. "Insurance offices exclude it from their field." In a few words Stevenson summed up not only the vast stretch of ocean out of which Bligh sailed his launch to Timor but also the chief problems with navigating it a century after the *Bounty* mutiny and the *Pandora* wreck. Ordinary knowledge of winds failed in the region, powerful currents swirled seemingly haphazardly, and insurance companies refused to insure vessels navigating it. Whatever the dangers of wind, current, surf, and incorrect charts, the prohibition of insurance firms mandated that the region remain infrequently visited, especially by deep-water sailing ships, and later even by steamships that could breast the bizarre currents that swept Stevenson toward islands he hoped to avoid. "I began to be sorry for cartographers, but my captain was more sorry for himself to be afloat in such a labyrinth," he concluded soberly.[25]

Stevenson, and later Nordhoff and Hall, recognized the all-powerful importance of the "local boat," the small craft—designed for particular waters and pulled by local men intimately familiar with reefs and currents—that pulled out to trading schooners anchored just beyond the reef. In the South Seas, the local boat frequently meant an outrigger-equipped canoe, the craft developed over millennia, but by the 1880s it often meant a sort of whaleboat, a very special ship's boat introduced into the Pacific by whalers.[26] Quickly accepted by locals as a short-distance vessel good for carrying passengers and very light cargo through surf, the whaleboat became the mainstay of "lightering," the small business of transporting passengers and freight from tiny schooners hovering beyond reefs and surf.

By 1920, when Nordhoff and Hall arrived in Tahiti, the local whaleboat had become the usual means of transportation at any island. The schooner's whaleboat served as ship's boat in all sorts of conditions, and as lifeboat if the schooner foundered; its presence became a hallmark of South Seas waters decades after the old-style whaling industry collapsed. But in Tahiti, masters, crew, and passengers alike still expected local-boat service by locals trained in the immense, traditional surf-racing canoes, which astounded visitors like Paul Gauguin in 1900.[27]

Elsewhere on the oceans, however, ships and schooners carried boats unlike the double-ended whaleboats, more like the boats of the ship *Serica.*

In 1868 a great hurricane destroyed *Serica* in the Indian Ocean, dropping its spars onto all three of its boats, then dismasting it, and then slamming the drifting masts and other spars through its hull. Thomas Cubbins, master of *Serica,* wrote a narrative of a boat passage most unlike that of the *Hornet* castaways. With one boat completely useless, he appears to have put his wife and two young children in the damaged second-largest boat, having decided not to risk them in the less damaged smallest boat, which he thought unfitted for storms. Neither boat performed well in the seas around the derelict, and after *Serica* sank, they separated, making first for Mauritius, then turning back toward Madagascar.

Cubbins and his family endured travails ranging from thirst to seamen stealing the last of the provisions, but these paled next to their attempt to land through surf. Despite his best efforts, the boat capsized in relatively shallow water, and while trying to save his life belt–equipped wife, Cubbins felt the boat roll atop him. With broken ribs but a clear mind, he ordered his wife and children to hold hands across the keel of the overturned boat, but just as they did so the boat rolled again, drowning his children.[28] Fighting the undertow, then flung up on the sand, the remaining passengers of the boat reached the safety inshore of the surf.

Twined through narratives like those of the *Hornet, Pandora,* and *Serica* boat passages runs the thread that Hall and Nordhoff discovered and followed, the thread that makes Bligh's passage so important. Most ships carried several boats, always one large one like the *Hornet* longboat or the *Bounty* launch kept especially for anchor work. In calms, or before storms hit anchorages, ship masters often lowered full-sized anchors into cradles slung beneath their launches or longboats, then ordered men to row to particular spots, towing heavy cables astern, and drop the anchors. By alternately dropping and retrieving anchors, a sailing ship might be moved across a becalmed harbor or swung about to approach a wharf. This brutal and slow work depended on a large ship's boat capable of carrying a massive and unwieldy anchor, and sometimes a windlass as well to help the men lower the anchor without damaging the boat or hurting themselves. But launches and longboats had an additional feature, one critically important in any analysis of passages like Bligh's or that of the *Serica* castaways.

Usually the largest ship's boat had a nearly square stern. The broad stern better supported the massive anchor slung beneath it, of course, but also increased the general freight capacity of the boat. Although seamen tried to avoid loading anything heavy into a beached boat or one floating in shallows, a big ship's boat could carry anything from a cannon to several tons of freight. As Dana recounted in *Two Years before the Mast*, loading hides or anything else into a ship's boat away from a wharf usually involved men wading at least waist-deep out to the boat, then dumping in their burdens. While the square stern made possible the transport of filled water casks from beaches, that operation so taxed the seamanship of merchant mariners and naval men alike that many chose to tow filled casks out to anchored ships rather than attempt to move them aboard boats barely afloat in the shallows.[29] In navies, the square stern meant that a launch might carry not only a very light cannon in the bow but also a squad or more of marines along with the sailors manning its oars. Moreover, the square stern made a roomy area called the stern sheets, a quite comfortable place for officers and passengers to sit. Further forward, a launch or longboat might have room for eight or even more men to row at a time, and most had room for two small masts. Such boats performed slowly but well in harbors, and they often worked well offshore, as when the *Pandora* launch chased and captured the mutineers' schooner. But the square stern presented terrible difficulties in storms and surf.

Against the flat transom the sea and surf rammed with implacable force. While the sharp or bluff bow of the launch or longboat simultaneously divided and lifted above the waves, the square stern lifted only slightly—and it never broke a following sea crashing against it. In any conditions other than flat calm, the square stern demanded constant care by anyone steering the boat. In storms, especially those approaching from astern, the vulnerable square stern often required changing course, and if reefs or islands made such changes impossible, then the occupants bailed continuously. But as the master of *Serica* learned, in surf the square stern proved disastrous, for as the water shoaled, the waves built up against the transom, making the boat almost impossible to steer, either with rudder or with steering oar.

Everywhere in the world, apprentices in sailing ships learned that the launch or longboat ought not sail in high seas or in surf, but everywhere in the world they learned too that when sailing ships foundered, masters tended to choose the biggest ship's boat for themselves, their families, and their passengers. A big boat was often the most seaworthy in many conditions, and it carried lots of provisions. Only at the end of the passage, when it had to land through surf, did its square stern become a liability to the tired, weak, and desperate people it carried.

Not surprisingly, then, many epic ship's-boat passages involved boats designed for heavy-weather and inshore work, boats unlike the *Bounty* launch but resembling whaleboats and often called quarter boats or gigs. Small ship's boats, double-ended like whaleboats but usually a bit shorter, more heavily built, and much deeper in hull shape, hung from davits on the quarters of many naval and merchant ships. The quarters, where the aft sides of the ship meet the transom, offered protection against all but great storms, and lay under the immediate eye of whatever officer held the deck. Should a crewman fall from the rigging, the cry of "Man overboard!" could be followed instantly by the officer flinging over the life buoy, ordering whatever maneuvers he saw fit to stop the ship, and sending away the quarter boat in search of the man.

The small, double-ended boats served other purposes too. In the merchant service, as so many old seamen recounted after about 1900, they carried masters through surf on business, and they figure prominently in all sorts of narratives from Dana's *Two Years before the Mast* to Making's *In Sail and Steam.* In such boats, masters often trained their apprentices in pulling, sailing, and other boat-handling skills, and many masters also used their quarter boats for pleasure cruising. In the world's navies, they carried messages and sometimes small armed forces through almost any kind of seas and into all but the worst surf. Into the twentieth century, explorers preferred them to any other sort of small boat. In the right hands, the double-ended quarter boat or gig did superb work.

When USS *Saginaw* foundered on Ocean Island in 1870, its captain sent five volunteers in a gig toward Hawaii in search of help. The twenty-foot-long, double-ended gig, slightly modified by having a crude deck built over it with holes for the men rowing and steering, carried its crew through heavy weather. Suffering from diarrhea for weeks, the men soon proved almost too weak to row or manage the sails in foul weather, and lost three sea anchors in five separate gales. But the gig performed flawlessly despite the men's illness and the loss of the sea anchors, at one time sailing up and over an immense floating log; it finally failed only when the men inadvertently sailed it into a strange current setting into the Hawaiian surf. The gig capsized, and all but one man drowned, a crewman who stayed with the boat as it crashed through the breakers, then drifted through quiet water inshore of the reef.[30] While almost too small to do well on the long passage east from Ocean Island, the gig practically landed itself and might well have brought its occupants safely on shore had they been more aware of the hazards before them.

An even smaller naval gig, captured by Confederate forces early in the Civil War, performed superbly in a small-boat passage, this time outward from a collapsing would-be nation. In a tale that resembles those of escapees from Singapore in early 1942, John Taylor Wood recounted the flight of a group of men from the ashes of the Confederacy in May 1865. As secretary of war for the Confederates, Wood expected to be tried for treason by the victorious Union side, and he used his military knowledge to reach the St. Johns River in Florida, where he remembered a planter had earlier surprised and captured USS *Columbine* and kept one of its gigs.

Wood knew such a boat needed little description, even in 1894 when he wrote his memoirs, and he called it merely "a man-of-war's small four-oared gig," scarcely able to carry six men with their improvised supplies. Camping at night, plagued by heavy rain by day, always wary of alligators, the men made their way south to the headwaters of the river, eventually reaching Lake Harney. From there they portaged the boat on an oxcart to the headwaters of Indian River, taking two days to cover eighteen miles through woods swarming with sand flies and mosquitoes. After "a little caulking and pitching," the gig descended the Indian River to Cape Canaveral and entered what Wood knew as the "lagoon," the long sound inshore of the barrier islands. Day after day the men sailed south, sometimes dragging the gig through the shallows, then traversing Juniper Narrows. After dragging the gig a half mile over sand dunes, the six men passed through the shallows and into Florida Channel, heading for the Bahamas.

Wood had little real knowledge of the capabilities of the gig and no more experience at sea than his comrades, but he was perfectly willing to enter deep water. The wind blew from dead ahead, however, forcing the boat south along the Florida coast and eventually near a Federal warship. Despite their pulling in to the beach and tipping the gig upside down on the sand, the Federal cruiser put one of its ten-oared cutters in the water under the command of a very young midshipman. Wood and his comrades admitted to being former "rebels" but successfully argued that they were only simple farmers digging turtle eggs. "As they shoved off," Wood remembered, "the coxswain said to the youngster, 'That looks like a man-of-war's gig, sir,' but he paid no attention to him." A few days later, after trading gunpowder for food with some Seminoles, the men pursued and captured a tiny sloop crewed by three Confederate deserters. Within moments, Wood and his five friends exchanged boats, and the deserters had the gig that had served so well in its southward passage.[31]

The *Columbine* gig appears to have been fitted with a wineglass transom, so called because its outline resembles the cross-section of a wineglass. Under Wood's command it proved well enough suited to all sorts of estuarine and other shallow-water encounters, and to being dragged through swamps and sand dunes too. But above all, it repeatedly had to enter light surf, for its occupants chose to camp each night on the beach, in part so they could search for food and avoid cruising warships, but also because with no compass they feared nighttime navigation. However much Wood worried about their ability to sail the gig to Bermuda, he had little trepidation about landing or launching it through light surf, for the wineglass transom parted at least some of the waves. Within hours after seizing the sloop, however, he and his friends knew that they had acquired a sort of prison, for it could land only at harbors, and the few harbors along the Florida coast had become home to deserters, pirates, and other villains. Taylor's group was soon recaptured by Federal forces.

The marriage of whaleboat and quarter boat or gig produced a variety of other small, double-ended ship's boats, most built for specialty purposes, but a few used in long passages. In 1916, after their ship *Endurance* had been crushed by ice, some of the men of Ernest Shackleton's Antarctic exploration expedition sailed from Elephant Island via the South Shetland Group to South Georgia Island in a 22½-foot-long double-ended boat built especially for polar exploration. F. A. Worsley, master of the *Endurance,* described the boat a few years after his frigid ordeal as "more lightly built than is required by the Board of Trade." Light construction meant much to Arctic explorers, who had to drag a boat over ice, but it worried them once they had to make an unanticipated ocean passage in the vessel. Despite their scanty resources, the castaways essentially refitted the boat for the lengthy passage, and they did so from long experience. "Shackleton and several of us had been trained in square-rigged ships," Worsley noted laconically, and throughout his book the voice of tradition speaks loudly.[32]

Worsley may have learned about gigs during his training aboard square-rigged ships, or perhaps his New Zealand boyhood made him aware of the whaleboats and other local craft so prevalent in the South and West Pacific. In any event, he had ordered custom-built ship's boats for the Antarctic expedition that were nothing like the ordinary launches and longboats carried aboard warships and seagoing merchant vessels. Instead he opted for something derived from inshore boats like the Maine peapod and from the old-style whaleboat, a craft that strikingly resembled the island-built boats Nordhoff and Hall knew as whaleboats.

Men against the Sea endures as powerful historical fiction, but what Delano and others knew as Bligh's great achievement vanished with the end of the square-rigged ships. It became somehow mythic, as did the Shackleton passage. When Nordhoff and Hall published their book in 1934, few readers understood any longer the nuances of ship's-boat design and construction, let alone the sailing qualities that distinguished launches and longboats from whaleboats, quarter boats, and gigs. Moreover, by the 1930s few merchant officers and seamen expected to land in unknown harbors, let alone through surf crashing on barrier beaches, in their own ship's boats. Steamship men anticipated shore boats every bit as much as they expected experienced local pilots, accurate charts, and solid wharves. Writers of historical fiction no longer assumed that readers cared about boats carried aboard sailing vessels, and book illustrators cared even less. Only rarely did a novelist understand the nuances of ship's boats, and alongshore boats too, as well as W. Clark Russell did in his 1892 story *Alone on a Wide, Wide Sea,* the story of a vacationing woman who is blown far out to sea after her boatman dies of heart failure, then rescued after a long bout of unconsciousness and amnesia. Bit by bit Russell makes the shore boat the key to more than one mystery, at one point demonstrating that it is every bit as indicative of origins as the woman's attire. But by the 1930s, however, all but the most dedicated of historical novelists accepted vague terms like *rowboat.*

In 1912 Conrad mused on the change sweeping the Thames estuary. "Only six weeks ago I was on the river in an ancient, rough, ship's boat, fitted with a two-cylinder motor-engine of 7½ h.p.," he wrote in an essay entitled "The *Titanic* Inquiry." "Just a common ship's boat, which the man who owns her uses for taking the workmen and stevedores to and from the ships loading at the buoys off Greenhithe. She would have carried some thirty people." About his estimate of its capacity, he added, "I know what I am talking about."[33]

In the Thames estuary after about 1890, many had discovered that the square-stern ship's launch or longboat intended to support heavy anchors supported heavy, slow-turning engines equally well. As engines grew more powerful by the year, the square-shaped stern balanced the squatting caused by ever more powerful propellers. In sheltered waters, the ship's launch or longboat evolved by 1950 into the isosceles-triangle-shaped powerboat that by 1960 traditional mariners and Coast Guard personnel knew to be unstable in following seas, likely to broach when running inlets, and utterly useless in surf. But after the 1890s, the quarter boat or whaleboat or gig evolved into something just as specific but far less common: the ship's lifeboat. Tales of long passages in ship's boats became prologue to tales of long passages in the double-ended small craft everyone casually called lifeboats. Designed and equipped to navigate uncharted seas, withstand terrific storms, and reach land through shallows and surf, lifeboats lacked engines. In time, their oars, masts, and sails became their defining elements, and only mariners appreciated their double-ended hulls.

The square-stern navy boat alongside *Carter Braxton* is designed for harbor work only, not deep-sea passage making. (Courtesy of the Mariners' Museum, from the A. Aubrey Bodine Collection)

4

BOATS

"IT REQUIRED SOME CARE TO AVOID HAVING OUR FRAIL LIFEBOAT staved in by the floating timbers, but we managed to clear this menace successfully," wrote Captain N. P. Benson two years after his four-masted lumber schooner foundered 2,700 miles off the Chilean coast in 1913.[1] Benson understated his accomplishment.

He and his men successfully launched a lifeboat in the teeth of a hurricane. *The Log of the El Dorado* reads differently than most castaway accounts. Written by a wry and self-deprecating master mariner, it emphasizes details most narratives omit, focusing on the initial joy of a greenhorn crew unaware of the enormity of the lifeboat passage ahead of it and making clear the role of good luck. As the storm overwhelmed the schooner, Benson managed to grab the medicine chest but forgot the chronometer while collecting some canned goods, and only at the last moment did he climb down a line into the lifeboat floating alongside. Under his leadership, his ten-man crew— eight of them landsmen never at sea before—embarked on a perilous passage.

Winter forced Benson to steer away from Pitcairn Island, only about 560 miles to the west, and instead make for Easter Island, some 700 miles northeast by east, a decision that quickly demoralized his greenhorns. With the help of his competent mates and his many years of experience in sailing small boats, Benson commanded his craft through a succession of serious gales and maintained crew morale, slipping owing to saltwater sores, cramps, and overcrowding. Eventually, they raised Easter Island. "I think that was the pleasantest Sunday morning of my life, and if we had been at all a religious lot the day would certainly have suggested prayer," he recorded of the general relief. "I am not even aware that we took the trouble to thank the Almighty for His care of us, but let no man doubt that we did not feel grateful, for we did."[2]

How *El Dorado* came to be equipped with a lifeboat remains unclear, but Benson emphasized that the lifeboat made salvation possible. About the boat he said remarkably little. "She was of whaleboat pattern, a good seaboat and fast," but sometimes too fast, swamping several times in one gale because he could not keep its pointed stern angled toward following seas. Without a sea anchor to hold its bow to the waves, Benson tried making a drogue from six blankets and trailing it astern, discovering that it slowed the boat just enough to make it manageable, although intermittently fifteen or twenty gallons of ocean came over the gunwales.[3] As a ship's boat it had proved awkward enough to be only rarely used. But as a lifeboat it did well, although it scarcely resembled the boats generally known in 1913 as lifeboats.

For decades after about 1850, lifeboat connoted the shore-based rescue vessel used by the British Royal National Lifeboat Institution (RNLI). As a cursory inquiry into editions of the *Encyclopaedia Britannica* reveals, the institution was an object of moral, religious, and technological interest. The eleventh edition (1911) offers a window into an organization that had become what the editors called "perhaps the grandest of England's charitable societies." It began in 1824 as the Royal National Institution for the Preservation of Life from Shipwreck, largely through the work of several upper-class men distressed by the large-scale loss of life along Britain's coasts. While concern for alongshore rescue had prompted one designer to build an "unimmergible" boat as early as 1785 (under the patronage of the Prince of Wales, later King George IV), a shipwreck in 1789 spurred widespread interest. After thousands watched the ship *Adventure* break up in heavy surf and its crew drown, the people of South Shields offered a reward to anyone producing a boat capable of rescue work. Following much effort and considerable wrangling, Henry Greathead saw the first of his boats used, and by 1803 had built eighteen boats for England, five for Scotland, and eight for other nations. Each boat was set up as a local charity, manned by volunteers and supported by local resources, often those of a resident nobleman whose lands touched the sea. One benefactor, having saved 305 lives along the coast of the Isle of Man, determined on larger-scale effort. In 1823 Sir William Hillary began agitating for a nationwide system, and within a year Parliament, the king, and the archbishop of Canterbury combined to support the newly chartered national charity.

A rash of shipwrecks in the late 1840s caused a reorganization, and thereafter only royalty—starting with the prince consort, then the duke of Northumberland—headed the charity, with Queen Victoria an active and generous contributor. As the decades passed, the institution grew stronger, sponsoring competitions for improved rescue boats and other equipment, offering medals for heroism, even providing barometers to impoverished

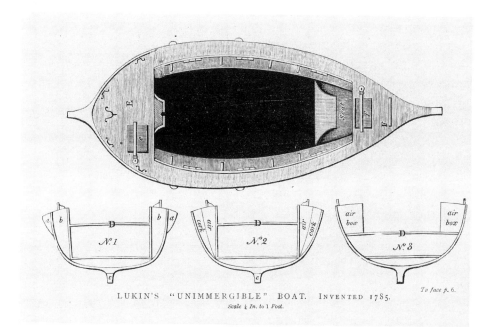

LUKIN'S "UNIMMERGIBLE" BOAT. INVENTED 1785.
Scale ¼ In. to 1 Foot.

In his *History of the Life-Boat and Its Work,* Richard Lewis provided illustrations of the first unsinkable, shore-based lifeboat.

fishermen. After the death of Victoria, it underwent reorganization again to emerge an even more powerful force headed by King Edward VII. Despite disasters like the sinking of two lifeboats off the coast of Lancashire in December 1886, when twenty-seven of the twenty-nine crewmen drowned, the renamed charity led efforts to remove half-sunken wreckage that endangered its rescue crews, build a communications network along the coasts, and create a pension system for its elderly retired volunteers. Between 1829 and 1908 the RNLI saved 47,983 lives, its efforts supported wholly by charitable giving. It drew praise from a succession of government committees that saw it as not only financially prudent but technologically innovative.

While daring rescues helped keep the institution in the public eye, fund-raising activities ensured continuous public attention. In coastal towns, including seaside resorts popular with urban vacationers, fund-raising often consisted of a stately perambulation of the lifeboat, pulled by its crew on its wheeled carriage through commercial and residential streets. On a sunny summer day, thousands of adults and children might see the boat, touch it, and contribute to its maintenance. Had the institution been tax supported, such a boat would have stayed indoors in its shed, perhaps as well maintained and well respected, but probably not nearly so well beloved.

In this nineteenth-century print, found at a New England church rummage sale, an RNLI lifeboat begins transferring passengers and crew from a stricken vessel whose own boats are already smashed. (Author's collection)

In the last years of the nineteenth century, Britons developed a fierce and permanent love of the RNLI. Not surprisingly, a charity that routinely provided hairbreadth escapes and rescues, reassured Britons of their maritime courage and skill, and prospered above the disputes of civil government attracted not only transgenerational allegiance and contributions ranging from the schoolchild's penny to the millionaire's bequest but also a host of cultural corollaries, including the Lifeboat Lager Ale so often held aloft in British pubs to toast the latest success of the local lifeboat.

Even to the present day, the RNLI remains a thoroughly British institution. Only once has the institution "gone foreign" in British alongshore slang, but that single venture recast its image forever. In World War II RNLI lifeboats helped evacuate some 338,000 retreating British and French soldiers from Dunkirk. The blue-painted motor lifeboats, some manned by their own crews, some by Royal Navy parties, sailed from British ports into the legend of the great Dunkirk triumph. Into the wildest confusion of modern warfare came the barges and fishing boats and other small craft—the last nowadays remembered as "the little boats" commandeered by the Admiralty. Some yachts arrived in the Dunkirk shallows crewed by their owners—C. H. Lightoller, the senior surviving officer of *Titanic,* appeared in his sixty-foot-long *Sundowner* and brought home 127 men—and many smaller vessels arrived under tow. Some of these were crewed by civilian volunteers, some by one or two Royal Navy sailors ordered to work close to the sands and extricate soldiers standing chest-deep.

Eighteen RNLI lifeboats worked the beaches of Dunkirk, and another worked the English Channel itself, rescuing men from the ships and boats sunk en route by German aircraft. Two of the lifeboats at Dunkirk carried their own crews. Royal Navy personnel manned the others after the regular crews protested that the lifeboats could not perform as ordered. Soldiers saw only the boats they had known since childhood, however, the blue-painted rescue craft carrying out a wholly new sort of rescue.

Buried in the mass of reports about the two-week-long evacuation are many accounts of the lifeboats. One night, after the Ramsgate lifeboat—manned by its regular crew—delivered 800 men 160 at a time from the beaches to a ship waiting in the night made darker by the smoke from burning oil tanks ashore, a naval officer shouted down, "I cannot see who you are; are you a naval party?" and the coxswain answered, "No, sir, we are members of the crew of the Ramsgate lifeboat." Such remarks are everywhere in printed sources, and sometimes in contemporaneous reports. "On behalf of every officer and man on this ship I should like to express to you our unbounded admiration of the magnificent behaviour of the crew of the lifeboat *Lord Southborough* during the recent evacuation from Dunkirk," reads one letter of commendation from HMS *Icarus.*[4]

Propaganda contributed to the Dunkirk legend, particularly in respect to the part played by RNLI lifeboats. Despite the unexpected success of the operation—the Admiralty had hoped to evacuate at best 45,000 troops—and Winston Churchill's warning to Parliament that "wars are not won by evacuations," British officials emphasized triumph, not only to restore morale at home but to pique the admiration of the United States. Churchill's "Dunkirk speech" and the attendant whirlwind of British propaganda convinced even *Life* magazine and the *New York Times* that a massive defeat had become a victory.[5] British sleight-of-hand shifted British and American public attention from disaster by land to success by sea, success involving the rescue of hundreds of thousands of soldiers.

Within months authors like E. Keble Chatterton published books arguing that Dunkirk epitomized maritime daring, that *at sea* Britain remained supreme. In *The Epic of Dunkirk,* Chatterton wrote of the RNLI lifeboats that "nobody had ever visualized employment of these special craft collectively and for the same purpose. It is regrettable that no records exist of the fine work which these craft performed, but activities were more essential than reports at that time."[6] Similar vagueness characterizes John Masefield's early 1941 *The Nine Days Wonder: The Operation Dynamo,* although his slim book at least tries to give a day-by-day account of the evacuation. In contemporaneous accounts, great men in little boats snatch victory from defeat and bring home a cheerful army. No mention was made in print of what one Scotland Yard inspector reported along the Dover–London railroad line one night: the flurry of sparks as demoralized soldiers flung their rifles from the speeding coaches.[7]

By May 1940 the RNLI had elsewhere rescued some two thousand people, including men from downed German bombers, since the beginning of hostilities. In its annual fund-raising booklet—its title now changed from *The Story of the Life-Boat* to *The Lifeboat Service in Peace and War*—which appeared in the early spring of 1940, the RNLI reported that for 116 years its boats rescued an average of eleven people per week, except during the years of World War I, when the average went up to twenty-one a week. But in the first six months of World War II, its boats rescued an average of sixty-eight people a week, an increase demanding much-increased public support. "Who would not feel that it was part of his national duty in war time to do his share in helping the Life-Boat Service to rescue life?" the booklet concludes.

The lifeboat described in the pamphlet is a specialized vessel, one having five particular characteristics: great strength, self-bailing cockpits, an engine that remains running when the hull goes underwater, the ability to continue working when damaged, and great buoyancy. Strength comes from a double-thick hull made from "woods of half the Empire": Canadian rock elm and red cedar, Burmese teak, Honduran mahogany. Flotation originates in self-bailing cockpits with scuppers so precisely designed that they empty the cockpits in twelve seconds while admitting no water from surging seas. Engines have double sets of spark plugs and double magnetos, and air intake and exhaust systems stretch high above the vessel so that the lifeboat can operate when almost fully submerged. And compartmentalization keeps the vessel working after being holed, something more than likely to happen.

An RNLI lifeboat might well be one of the so-called cabin classes, boats forty-six or fifty-one feet long, closed to the sea along almost their entire lengths and scarcely able to operate in shoal water. The institution operated smaller boats, but even the "surf motor life-boat" stretched thirty-two feet, and the institution had begun experiments in which submergible, bulldozer-like tractors forced such boats and their carriages across beaches, through shallows, and into surf. The volunteer crewmen of all the motor lifeboats, and especially of the larger classes, were well aware of the limitations of vessels intended to launch from harbors and perform in either deep water or storm-tossed shoal, but not to reach into beaches except to discharge castaways. Although even a big motor lifeboat might nudge into shallow water, especially along a shallow beach, so that wounded passengers could be carried off and able victims jump overboard and wade ashore, thus freeing the lifeboat to make another trip, the RNLI intended that passengers disembark at wharves whenever possible. Most RNLI lifeboats could not well approach shallows, even on a calm day, and pick up passengers in any number. Not only was the typical lifeboat too high-sided to make such pickups easy, but as people clambered into it and it settled lower into the water, it often went aground no matter how well handled.

The sands of the Dunkirk beaches, some RNLI personnel pointed out to the Royal Navy, extended from three-fourths of a mile to one-and-a-half miles seaward at low tide, making the use of RNLI lifeboats as embarking craft rather ridiculous. However unwittingly, the RNLI crews shattered the fragile deep-sea authority of the Royal Navy officers, who themselves knew almost nothing about how the Dunkirk evacuation might work. In a manner characteristic throughout the war when faced with questions from merchant-service men and fishermen, the Royal Navy officers lost patience at once and dismissed most of the RNLI crews. Only much later, when military historians pieced together reports of grounded RNLI lifeboats manned by Royal Navy seamen, did the expert questions of the RNLI men make sense. What stands out so markedly from the confrontation at Dover is that the Royal Navy never asked the RNLI men to suggest what might be done. In the end, both merchant sailors and naval seamen had to learn what the handful of RNLI men aboard two motor lifeboats off Dunkirk taught by example when they refused to enter shallow water. Given the massive size of motor lifeboats, evacuation proceeded in ways earlier generations of RNLI lifeboatmen knew so well: by oared boat.

A half century after the founding of the RNLI, its secretary, Richard Lewis, published a detailed history not only of the charity itself but of the development of its classic thirty-three-foot rowing-and-sailing lifeboat. Hampered by a paucity of records from its first years, and by a long-simmering dispute concerning first inventors and first definitions, he focused on the 1850 design competition from which emerged an improved lifeboat combining the best features of older ones. This new design was additionally self-righting, self-bailing, lighter (for easier launching from beaches), and less expensive to build. The competition judges insisted on a fast-rowing boat that could move through high surf, and hoped to identify a fast-sailing

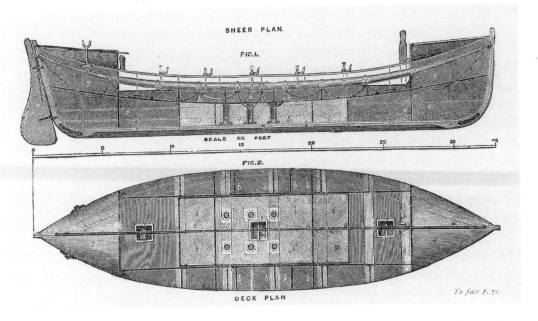

SHEER PLAN.

FIC.I.

SCALE OF FEET

FIC.2.

DECK PLAN

To face p. 72.

Richard Lewis detailed the beach-launched boat that in time epitomized "lifeboat" everywhere in Great Britain, a boat that only ten men rowed. (From Lewis, *History of the Life-Boat*)

design as well. Out of the competition came both a design and a definition, the latter shaping lexicography long after new designs supplanted the thirty-three-foot one. "Although the word *life-boat* has not in itself any definite meaning, it is generally understood as signifying a boat specially constructed for saving life in storms and heavy seas, when ordinary open boats could not do so except at the imminent peril or certain death to those within them," Lewis concluded of the shore-based rescue vessel.[8] Much of his 1874 *History of the Life-Boat and Its Work* emphasizes the boat itself and distinguishes the lifeboat from other open boats.

Yet even Lewis felt compelled to address what by the time had become a standard question: Why not propel lifeboats by steam? In his view, questioners simply misjudged the conditions in which lifeboats worked. No one could imagine a way of keeping the coal fire alight while thousands of gallons of seawater crashed into the boat over and over again.[9] Other issues, such as maintaining a good draft in the stack, seemed equally insurmountable. Not until 1888 did the RNLI commission the first steam-powered lifeboat, and that vessel operated with suction-pump drive rather than with a screw propeller, since everyone feared that striking a rock would twist the propeller shaft and render the boat helpless. The so-called hydr tem required a larger engine that consumed more coal than ordi boats' engines, and quickly the RNLI found itself pushed towar larger and larger hulls solely to accommodate bulky machine bunkers.[10] But by 1923, when A. J. Dawson published *Britain The Story of a Century of Heroic Service*, the efforts at designing and op- ing steam-powered lifeboats had become merely a technical footnote: gaso- line engines had replaced steam, and the RNLI had embarked on a long effort to make its large "motor-lifeboats" as reliable as rowing ones.

More than tradition explains the public devotion to oared lifeboats. Along many stretches of coast, oared boats worked into the twentieth cen- tury, and as late as 1912 they attracted sustained attention as paradigms of successful design. In *The Life-Boat and Its Story,* Noel T. Methley argued that the RNLI oared lifeboat was the only true lifeboat, because only it could work successfully in broken water, the defining environment of coastal ship- wreck. Alongshore the RNLI crewmen confronted every hazard from es- tuary tidal currents to breakers, usually intensified and usually in combina- tion. The lifeboat must survive through breaking crests of rollers reaching the sand, maneuver in cross seas, turn in deep, shattered troughs, and right itself after capsizing. Unlike Lewis, whom he quoted approvingly, Methley devoted much attention to the self-righting attribute. What distinguishes Methley's book from others, especially later ones, is his detailed analysis of the RNLI lifeboat working close inshore, in heavy, breaking surf. Methley understood the reason for the high bow and stern of Viking and other very early small craft, mentioning the Nile boats described by Herodotus and the oared lighters of the Moorish ports of Safi and Sallee: the hull shape not only helps prevent capsizing but either rights the boat or at least aids in its righting later.

Between 1920 and the nine days of the Dunkirk evacuation in 1940, the British understanding of the word *lifeboat* bifurcated. For one generation, the RNLI stood for alongshore rescue, short-distance effort in shallow, often storm-wracked seas to assist vessels sinking or stranded close inshore among rocks and sandbars. These activities often gathered throngs of onlookers. But as the engine-driven, fifty-foot-long lifeboats went farther to sea, their crews encountered situations in which steamships and other long-distance craft had arrived to help the distressed vessels targeted by RNLI efforts. The events seemed real enough when reported by the BBC, but they happened offshore and out of sight. Somehow the century-old image of volunteers manning oared boats and pushing out through the surf endured in spite of

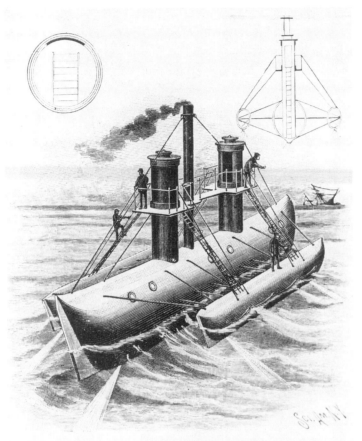

THE WILLIAM F. JAMES LIFEBOAT

One inventor after another suggested steam-powered, shore-based lifeboats, but experienced mariners believed the inventors would need rescuing themselves. (Courtesy of the Mariners' Museum)

technological change, perhaps because the volunteer crews of the older oared boats figured prominently in print media ranging from ale advertisements to novels like Thomas Stanley Treanor's 1892 *Heroes of the Goodwin Sands.*

The image endured in the United States too, despite similar change in rescue craft. By 1940 the Coast Guard had motorized lifeboats no more able than the RNLI vessels to snatch soldiers from the shallows, but its rowing boats remained firmly placed in the popular imagination. Combined from older services in 1915, by the beginning of World War II the Coast Guard had begun to lose its expertise with beach-launched oared lifeboats, partly because its men stayed in service fewer years than RNLI crews, but chiefly because the engine-driven, thirty-six-foot-long, seven-ton lifeboat had largely supplanted older oared boats. Unlike the volunteer RNLI crews, whose close knowledge of waters near individual lifeboat stations enabled them to work wonders in storms but made them uneasy working in strange places, Coast Guard men shifted from one station to another and so learned to adapt to novel situations.[11] Four years after Dunkirk, Coast Guard men piloted landing craft through the surf on D-Day, for example, accomplishing the sort of shallow-water success that deep-water navy men grudgingly admire, knowing it originates in a long tradition of surf rescue.

The Coast Guard had its origins in 1807, when the Massachusetts Humane Society erected a boathouse and acquired what it called a "surfboat" for the beach at Cohasset. By the time Congress appropriated funds for lifeboat stations in 1847, the society had sixteen lifeboats in service along the Massachusetts coast alone, but the rest of the Atlantic coast had almost none.[12] Not until the 1870s, after mistakes in maintaining boathouses and boats and experiments with volunteer crews, did the Life-Saving Service emerge as a competent branch of government. Yet until after the turn of the century it relied on local men—often employed full-time only in winter— to launch locally designed boats through surf.[13]

When the German ship *Elizabeth* struck on the coast of Virginia in January 1887, five members of the Life-Saving Service drowned in an attempt to rescue its crew. An account of the incident published in *Harper's Weekly* paid no special attention to their boat, which proved no more able to return through the surf than the one launched from *Elizabeth* itself. Indeed, the title of the story, "The Wreck of the Life-Boats," emphasizes that two lifeboats foundered, one launched from shore by the service, the other launched from the wrecked ship. To the readers of *Harper's Weekly,* and to the inspectors of the Life-Saving Service, "lifeboat" designated a variety of open, oared boat, some belonging to the service and some belonging to ships. Only locals, including the men working full- or part-time for the service, referred to certain service-owned rescue boats as "surfboats." Launched through the surf from wheeled carriages, the boats often had to return through the surf with their crews and passengers, a feat that by 1940 the Coast Guard preferred not to chance, even though the boats had improved mightily over the decades.

Yet at the beginning of World War II the Coast Guard still used a few beach-launched, oared surfboats, the standard one being twenty-five feet long and weighing 2,200 pounds. While self-bailing, the boat did not self-right: once capsized, it had to be heaved back on its keel by its crew. Equipped with air tanks and a water-ballast system operated by a hand pump, the standard surfboat was a holdover from an earlier era. Indeed, it carried sails. Local conditions and tradition made it unacceptable in New England waters, where the Coast Guard still maintained what it called "Monomoy surfboats," heavy, whaleboatlike oar-and-sail vessels neither self-bailing nor self-righting but nevertheless remarkably suited to winter surf like that the Cape Cod writer Joseph C. Lincoln described in his 1924 novel of the old Life-Saving Service, *Rugged Water*. "The boat leaped forward into the breakers," the novel begins, describing something Cape Codders had watched dry-mouthed for decades. "Then another mighty pull, and a rise that lifted them up and up. Flying foam, a deluge of icy water, a series of strokes, and then a coast. They were over the first breaker."[14]

In its efforts to stop rumrunners, the Coast Guard emphasized power-boats, and as steamships replaced first full-rigged ships and then even schooners and other small coastal craft, the Coast Guard recorded fewer and fewer surfboat launchings each year. As 1920s and '30s newspaper accounts emphasized Coast Guard police activities, all but the most daring rescues vanished into the limbo of annual government reports.[15] In Britain, mean-while, free from the foolishness of Prohibition and dependent on the RNLI for alongshore lifesaving, every rescue worked its way into local newspa-pers and national fund-raising activity alike, and the word *lifeboat* contin-ued to connote for most people a small, double-ended, oared boat launched into the surf.

What postwar military historians identify as "the legend of Dunkirk" not only endures but prospers. Christian Brann's 1990 *The Little Ships of Dunkirk* opens with a letter from Prince Philip remembering how "cockle fishermen, lifeboatmen, yachtsmen and members of the Royal Navy put to sea in anything that would float," and contains a chapter on the surviving RNLI lifeboats, now converted to pleasure craft. "The lifeboatmen had a very personal attachment to their boats and did not take kindly to the idea of beaching them at La Panne," writes Brann, who remarks too that the lifeboatmen considered the naval reserve officers less competent seamen than themselves. Brann also notes that the Royal Navy officers who replaced the RNLI crews kept no detailed logs, so that no historian can reconstruct what individual lifeboats accomplished. But then he falls into the wartime propaganda paean, summarizing the daring of motorized lifeboats under fire. Unwittingly emphasizing that the boats did not take the ground in the shallows to load troops as the navy first proposed to their crews, and avoid-ing the reasons why one vanished without trace, he offers no analysis of the Royal Navy decision and the moral issues implicit in using civilian rescue craft in a military operation. Prince Philip's less-than-candid remark about lifeboatmen and Brann's repeating of wartime vague praise combine to further the legend.[16] But any summer afternoon yachtsman knows that the engine-driven RNLI lifeboats could scarcely lift soldiers directly from shallow-water beaches: the boats drew too much water. Moreover, naval logs assert that smaller engine-driven boats launched from destroyers and other warships fared no better.

"The naval vessels used cutters and whalers," David Divine reports in his *Nine Days of Dunkirk.* But they did not use them as well as many expected, and the coxswain of the RNLI *Prudential* noted that he and his crew found that "naval ratings who manned wherries were not skilled at handling small boats under such conditions" and so exchanged places with them. Amateur yachtsmen and even soldiers expert in shallow-water pleasure boating discovered that naval crews in cutters, whalers, and wherries simply lacked the experience to board men in shoal water without putting their vessels aground and sometimes damaging propellers and rudders. "The procedure was to tow the whaler and cutter to the beach and swing them round and cast off tow in about 3 feet 6 inches (my draught being 2 feet 6 inches)," reported the owner of the yacht *Constant Nymph,* one of the first pleasure boats to cross to Dunkirk. "The cutter then dropped her grapnel and went in as close as she dared without grounding the whaler, and troops waded out to board them."[17] In the Dunkirk shallows—but not at Dover—Royal Navy officers readily turned over small-craft operations to civilians who took charge and thereafter issued orders to sailors and soldiers alike. Royal Navy watch-keeping systems partly explain the willingness: aboard destroyers and other smaller warships, gun crews and boat crews consisted of the same group of men. Harassed commanders repelling bombers and fighters knew that sending in their own boat crews meant pulling men from guns. But early failures of whalers and cutters, launches and wherries manned by sailors also convinced them that only shallow-water expertise might save the day.

A century earlier, an anonymous writer in *Nautical Magazine* had foreseen the uselessness of naval boats in a disaster like Dunkirk. He observed, "I have often thought it strange that our men-of-war are not furnished, either with a whale boat or coble; our ships of war have not a boat *fit to take a beach in any surf whatsoever.*" He went on to argue that the Royal Navy ought to look to indigenous alongshore craft from Africa, India, and the South Seas, and then to boats ranging from Thames River wherries to French navy boats, all of which offered bits and pieces of good design that might improve the alongshore fishing boats of Great Britain in ways the Royal Navy might someday need. After all, how could the Royal Navy impress "the savages of the South Seas Islands" if it had no boats capable of working in surf and shallows and instead asked for transport in local canoes in a way that gave "barbarous people a somewhat mean opinion of our boasted superiority?"[18]

But such suggestions fell on deaf ears, and in the end most naval authorities understood boat work as deep-sea ship-to-ship transfer or ship-to-wharf transport demanding large boats. Then too, few admirals envisioned wholesale evacuations from shallows-fringed beaches, but instead imagined lifting troops from wharves. By 1940, when RNLI lifeboats had grown massive compared with ones built even a generation earlier, both Royal Navy and United States Navy boats had become almost too heavy to run into shallows. Ever larger warships meant ever larger boats, of course, but engines forced naval architects to redesign naval small craft, and the designers faced the same problems that confronted the RNLI.

The 1920 *Boat Book of the United States Navy* points out that engine power had radically altered boat design. Steam-powered ship's boats came in 50-, 40-, and 30-foot lengths, the larger ones being carried aboard battleships and other large warships. But by 1920 the navy owned all sorts of gasoline-powered boats too: motor sailing launches 50, 40, 36, 33, 30, and 24 feet long; motor barges 40 feet long; motor dories 21 feet long, and on and on, all adaptations of older hull types to modern motive power. "These boats are especially designed for ships' heavy work, such as carrying stores or large liberty parties or landing forces, carrying out anchors, weighing kedge anchors," the *Boat Book* reports of boats used in a time of transition, when the navy expected motor sailing launches to sail long distances and work under engine power only intermittently. Bligh might have recognized even the largest of the launches, for its hull still took its form from the task of ferrying anchors, although anchors had become far larger. The navy's square-stern cutters and double-ended whaleboats would have been familiar to Bligh and to the men of the wrecked *Saginaw*, but no longer did *gig* designate the sort of naval small boat one might use to escape through bayous, to pull across sand dunes, or to sail into the open sea. In 1920 gig meant a motorboat reserved for the use of a ship's commanding officer, and *whaleboat* had been essentially superseded by a new term, *motor whaleboat*. Moreover, the navy had begun phasing out both pulling and sailing boats, preferring the greater speed of gasoline power. By 1920 an oared cutter no longer came equipped with a machine gun in the bow, for example, and higher-speed boats had begun to be termed motorboats.[19]

In the two decades that elapsed between the 1920 *Boat Book* and the 1940 edition of the *Bluejacket's Manual,* the United States Navy, like the Royal Navy, shifted permanently away from rowing and sailing boats and toward engine craft useful for tasks as diverse as "airplane salvage work, gasoline delivery, and survey work." Motorboats now designated boats used to ferry officers and dispatches, and motor launches designated heavy-duty, square-sterned craft useful for carrying stores, liberty parties, and—as always—anchors. But the biggest change came in the whaleboat class. While both navies maintained oared whaleboats fitted with sails, mostly for lifesaving work, both had also shifted emphasis to engine-driven versions which served as lifeboats aboard large warships. "Whaleboats are used as lifeboats, although the open-type motor whaleboat is generally preferred," the *Boat Book* had asserted. No longer did the navy expect much in the way of rowing expertise, although it did provide racing cutters to large warships in order to foster recreational competition. By 1940 British and American warships carried large boats indeed, many larger than any RNLI engine-driven lifeboat. The United States Navy fifty-foot motor launch carried 190 men "seated as closely as possible," and such a launch could scarcely have worked in the shallows of Dunkirk.[20]

Even much smaller naval boats proved awkward, not only in shallow water but at sea too, as a number of naval parties learned bitterly at the beginning of World War II. "Our boat was low in the water, wide and flat-beamed, and not built like a lifeboat by any manner of means," recalled Lieutenant Commander John Morrill in *South from Corregidor* of the thirty-six-foot-long, diesel-powered launch in which he and seventeen other survivors of two sunk minesweepers escaped Bataan and in thirty-one days made a 1,400-mile passage to the north coast of Australia. Between its square stern and low freeboard when loaded with eighteen men and several barrels of diesel fuel—at the beginning of the passage it had only six inches of freeboard—the boat quickly shipped water whenever it encountered a cross sea. But Morrill soon realized worse limitations. The launch drew four feet of water, and its unguarded propeller meant he had to use the utmost care in approaching any shallows, especially those interrupted by rocks or coral. After one of his men made a lead line from some line and an oarlock made otherwise useless by a lack of oars, Morrill learned how to approach shorelines cautiously, "taking great care our propeller didn't graze anything." Like Gilmour fleeing Singapore, he knew he had to hide his craft by day from Japanese aircraft and patrol boats, but getting the launch close enough inshore to camouflage it with chopped-up tree limbs proved continually frustrating, especially near daybreak when haste governed everything. The boat drew far too much water, and it behaved abominably in surf.[21]

"To climb over a reef in a motorized boat is quite a thing," Morrill observed. "The smart thing was to avoid the surf, with our heavily laden boat."[22] The far-off appearance of a Japanese patrol boat convinced him to run south along the margin of the white water rather than risk damage crossing the bar, but by then Morrill knew enough about the limitations of his launch that he preferred open water, even stormy open water.

South from Corregidor skirts the significance of bad weather. To be sure, Morrill and his men encountered what he called "blustery" weather, but never do they seem to have endured full gales. Even in squalls the boat took green water over its sides, and the men bailed frantically, marveling as they bailed at the diesel engine running faithfully even when almost wholly submerged. Six- to eight-foot seas now and then slammed over the gunwales, yet the launch rode well when running before the wind, something Morrill discovered near the end of the passage, when he finally admitted how little he and his crew actually knew about small-boat seamanship. They had heard of boats being "pooped," of waves coming inboard over the stern, but they could not understand why it did not happen to them. The most likely answer lies in the weather. Never once in 1,400 miles did the navy launch ever encounter even a gale, and in every storm but one the wind tended to come from anywhere but aft.[23] Had the men encountered a following sea at the beginning of the passage, when full fuel barrels gave their launch so little freeboard, they might well have foundered.

Over the beginning of Morrill's book hangs a pall of darkness, the gloom of immense oil fires raging on Bataan, Corregidor, Caballo, and especially at Cavite navy yard, and the pall reappears again and again in the narrative. On a clear, sunny day, Morrill recorded, he could not see the beach fifty yards away. "Smoke rolled up as from all the oil tanks in the world on fire, and an inferno of sound dwelt in the somber cloud." *South from Corregidor,* like Gilmour's *Escape from Singapore,* is one of many World War II small-boat escape narratives shadowed by the immense pillars of oil smoke that distinguish World War II narratives from those of the Great War two decades earlier.[24]

For Morrill, the images lingered in part because his launch operated on diesel fuel oil, and to make his escape, he had to carry barrels of diesel fuel where more men might have crouched. Unable to work into the shallows and nudge onto soft, sandy beaches, the motorboats of the United States and Royal navies by 1940 had somehow become one with the high-tech conflagrations ashore. What burned at Singapore, Bataan, Corregidor, and Dunkirk fueled the navy small craft that proved so useless at plucking men from beaches smothered in smoke and flame and from oil blazing atop water.

Over all of Dunkirk and its beaches floated a pall of fuel-oil smoke so vast that shore-dwelling Britons saw it from across the Channel. Dunkirk "had an astounding, terrifying background of giant flames leaping a hundred feet into the air from blazing oil tanks," wrote one British army officer who arrived on the beaches near the end of the evacuation and looked across at the breakwater. Almost every building in Dunkirk seemed alight too, making a "high wall of fire, roaring and darting tongues of flame, with the smoke pouring upwards and disappearing in the blackness of the sky above the rooftops." The lurid red of Dunkirk astonished even the most battle-hardened British soldiers, who fell silent before the "flames, smoke, and the night itself all mingling together to compose a frightful panorama of death and destruction." Into it naval parties had to penetrate, feeling their way across the surf toward flames that might incinerate any boat left aground by the ebbing tide.[25]

Not even the motor whaleboat that most naval commanders expected to use to save life worked well in the shallows. The double-ended motor whaleboat appeared to have the proper shape to venture into surf: after all, its hull made it less likely to take waves inboard from astern. With luck it might navigate surf-roiled shallows, for it could move sternward into deep water through at least low surf, if not the sort described by novelists depicting RNLI and United States Life-Saving Service rescues. But the overgrown motor whalers failed when put to the test, in part because their heavy engines and fuel tanks depressed their sterns to begin with, and in part because under full power the narrow-sterned boats squatted, their sterns pulled perilously low in the water by the thrust of their propellers.

At Dunkirk, Edward Drake Parker, coxswain of the RNLI lifeboat from Margate, understood the shallow-water disasters occurring around him as soon as his big lifeboat grounded under the weight of an uncounted number of soldiers. After high tide lifted off his lifeboat and he transferred the soldiers, he went in again, this time in company with a Royal Navy motor whaleboat fighting a northwest wind making a nasty surf. "The whaler from the destroyer which went in to the shore with us on our last trip was swamped, so was the motor pinnace that was working with the whaler, and so it was all along the sands as far as I could see, both sides of us and there was not a boat left afloat," he reported of modern motorized small craft.[26]

All British and United States navy instructions for launching and landing through surf came from more than a century of RNLI experience. They mesh perfectly with the detailed accounts R. M. Ballantyne provided in his late nineteenth-century novel *The Lifeboat: A Tale of Our Coast Heroes*, of smugglers landing through surf.[27] The smugglers land properly, in a suitable boat, behaving exactly as the RNLI lifeboat crews did, and as did the United

States men J. W. Dalton depicted in his 1923 nonfictional *The Life Savers of Cape Cod*—a book that makes clear the verisimilitude of Joseph C. Lincoln's fictional *Rugged Water* published a year later. The RNLI and the United States Life-Saving Service both knew proper procedures for launching and landing through surf, and while such landings were never safe, the procedures made possible landings onlookers often evaluated as reckless if not suicidal. Above all, the procedures called for an oared boat, not an engine-driven one.

On some types of beaches, the RNLI counseled, the boat should be spun about at the last moment and landed stern-first, something that put its bow to the sea and left the boat aimed for launching. In other situations, the boat should back in the entire way, keeping its bow to the sea and touching land with its stern. Of course, neither method worked with engine boats, since both ensured immediate and permanent damage to propeller, propeller shaft, and rudder, and perhaps to engine too. The RNLI knew and accepted this. It assumed no one would land except in some sort of boat adapted to the task, and in most instances that boat would be a lifeboat, by which the RNLI meant not only its own oared lifeboats and the American surfboat but the ordinary ship's lifeboat possessed of some extraordinary if frequently overlooked capabilities.

An ordinary ship's lifeboat could be coaxed through surf and onto almost any sort of beach, and it might leave any beach through all but the worst surf, the RNLI told the United States Navy.[28] Implicit in navy instructions, lifted verbatim from RNLI publications, is a simple message, breathtaking in its importance. Only double-ended, oared boats should attempt launching and landing through surf, and only they should work in the shallows along beaches. Engine-driven boats, with the situation-by-situation exception of double-ended lifeboats crewed by RNLI and United States Coast Guard personnel, should stay in deeper water or in harbors.

No wonder the RNLI crews so forcefully questioned Royal Navy officers on being ordered to Dunkirk. And no wonder even now that so many accounts of the Dunkirk evacuation confuse two sorts of lifeboats, one the deeper-water, engine-driven RNLI craft that went ashore only by accident, the other the hundreds of merchant-ship lifeboats that made so much of the evacuation—especially the evacuation from the beaches—possible at all.

From the Port of London arrived 915 ship's boats, most of them lifeboats from docked steamships, 34 equipped with engines, 881 oared.[29] Those ship's lifeboats, not one with a name, lifted thousands of soldiers directly from beaches, but scarcely a photograph of their work survives. The devastating German attack on the beaches made much photography impossible, of course. More important, British censors understood that photographing such small boats at close range necessarily meant capturing faces showing exhaustion, malnutrition, demoralization and cowardice, lack of discipline, and near mutiny. In keeping images of such faces from the British and American public, censors unwittingly eliminated ship's lifeboats from the photographic record. No wonder the public confused published accounts using the word "lifeboat."

Douglas Williams, a *Daily Telegraph* war correspondent, waited almost too long to leave Dunkirk. Away from the sands, at the base of a cliff whose top the Germans either commanded or were about to take, the rocky beach was so narrow that British soldiers lay sleeping one atop another at two in the morning. Williams knew the lifeboat offered the very last hope. "It filled and left and came again. Each trip out and back seemed to take an eternity, for the tide was strong and there were only four oars. The boat was overcrowded, and each man had to climb a rope-ladder when he reached the cargo boat lying 300 yards off shore." In the dark, the Germans began firing artillery rounds into the shallows, but all missed the boat. Williams timed the cycle: each round-trip took twenty-five minutes. The arrival of a second ship's lifeboat filled him with hope: now 120 men of the 1,000 packed on the beach might be evacuated each hour. Williams's *Retreat from Dunkirk* appeared a few months after the evacuation. His account of the evacuation is above all a small-boat narrative that emphasizes the two ship's lifeboats moving steadily back and forth, ferrying soldiers to "speedboats, fishing smacks, cargo boats" lying in deeper water, the Royal Marines aboard the cargo boat yelling "Row harder!" or "In, out! In, out!" as machine-gun fire joined the artillery shelling.[30] Without the ship's lifeboats, Williams knew he and the rest of the thousand men in his group could not reach what became the "little ships" of the Dunkirk legend.

Other accounts corroborate Williams's stark testimony. An army officer remembered standing up to his shoulders in the sea, along with hundreds of other soldiers, when "suddenly out of the blackness, rather ghostly, swam a white shape which materialized into a ship's lifeboat, towed by a motor-boat." Pulled by only four men, the lifeboat accepted about forty soldiers, including the officer. "From the instant I landed on my head in the lifeboat a great burden of responsibility seemed to fall from my shoulders." Suddenly at sea, if only in a few feet of water, the soldier gave his responsibility up to seamen.[31] Ordinary lifeboats worked superbly during the nine days' wonder, surprising all but a handful of observers and still confusing military historians. "Ships used their own boats, the heavy, clumsy lifeboats approved by the Board of Trade and designed for deep-sea work—nothing more inappropriate could have been thought of for the shallows of Dunkirk," wrote Divine in 1959.[32] But the military historian misunderstood: the ship's lifeboats proved perfectly appropriate, rowing from the age of sail into the oil-smoke horror of modern warfare.

Nicholas Drew remembered his first view of Dunkirk from a ship's lifeboat towed from London across the Channel: "We saw the cloud from more than ten miles offshore: it lay upon the land like a vast shadow. As we drew nearer we saw the dull red glow at its base. Out of the shadow came flashes of gunfire."[33] In his 1944 *Amateur Sailor*, Drew provides one of the few firsthand accounts of the workings of the 881 oared ship's lifeboats, one of which lifted Williams from the beach. A yachtsman about to enter the Royal Navy, Drew had been plucked from navigation school by an Admiralty request for volunteers to take ship's lifeboats from the Thames River to Dunkirk. His keen eye for detail and his fierce love of small, open boats combine to make his memoirs, written soon afterward while off-watch aboard a warship, especially revealing.

Drew's companion in the lifeboat spoke softly of Dante as they approached the flaming coast through mist. An engine-driven boat towed the lifeboat into the shallows, then Drew and three other men, accompanied by a naval officer, rowed slowly into the acrid murk, the officer sounding the depths with a down-thrust oar. A shell burst immediately wounded one rower with shrapnel. In about three feet of water the men swung the boat around and backed in, being careful not to ground, and one went overboard, waded ashore, and vanished. "We sat in the boat resting on our oars, now and then giving our blades a shove to keep the boat afloat," Drew remembered, remarking too that he then passed around a flask of rum he had brought along, defying all navy regulations.[34] In a short while the man returned with twenty-two soldiers too exhausted to board, and Drew had to step overboard and shove each of them into the boat. Then he and his three companions began rowing for the engine boat, a long and difficult task owing to the tide, the thick darkness, and the shelling. After finding the engine boat and making fast alongside, Drew watched as the soldiers stumbled aboard and his crew passed along a few rifles. Then Drew took his lifeboat back toward shore for more exhausted, often weaponless men.

Unlike contemporaneous morale-boosting accounts, *Amateur Sailor* depicts the ship's lifeboats rescuing exhausted, demoralized men, many of whom lacked weapons. The French soldiers so mobbed the lifeboat on its second trip that it grounded, and as more soldiers jumped aboard despite Drew's shouts—in French—to stand still and await orders, one of Drew's companions grabbed a rifle and began thrusting its butt into the mass of panicked humanity. Drew feared being left aground in daylight on a falling tide, of course—being a sitting target for artillery and machine-gun fire alike—but what he chiefly feared was being captured while in civilian clothing. Surrounded by panicked men, all in uniforms, Drew realized that as a civilian engaged in military operations he might be executed if captured.

A French sergeant restored order and got the soldiers out of the boat, but not for some time could the demoralized men be made to help Drew and his mates push it into water deep enough to float it. Drew remembers almost weeping, "pushing in short, foolish jerks, choking in impotence and fury," knowing that it "seemed such an utterly useless way to lose one's life, to be stuck and stranded in this amateurish way."[35] Eventually the boat floated, everyone climbed aboard in an orderly fashion, and Drew and his three companions again rowed it to a waiting ship, transferred the men, and returned to the shallows to find nearly two hundred soldiers waiting in a long, silent line.

Trip after trip, throughout the night, Drew and his companions rowed the ship's lifeboat into the shallows. Some of the lifeboats in his flotilla wound up away from their assigned area, one went fast aground, its crew signaling by flashlight that it intended to get free on the incoming tide, and two more made fast to a Dutch barge anchored further away. In the oily murk, ship's lifeboats seemed everywhere, and Drew had trouble keeping track of those in his own group. He recounts hearing a naval officer reassure soldiers by shouting of the lifeboats, "There's dozens of the damned things trailing around."[36] But almost none entered the myth.

In the desperate propaganda war that swirled around the Dunkirk evacuation, and that lingered until Allied forces landed in Normandy in 1944, "lifeboat" meant the RNLI motorized lifeboats to which so many ship's lifeboats transferred men ferried from the beaches, which in turn transferred troops to barges, tugboats, and other larger vessels for the homeward passage. Hastily making Dunkirk into a propaganda victory led to many egregious mistakes—Masefield's book calls a tugboat *Nicholas Drew*, an error for which he later apologized to Drew—and many more omissions, perhaps chiefly those regarding the RNLI crewmen who refused to go to Dunkirk beaches without some sensible instruction and the poor morale of so many British and French soldiers. Yet below the errors and omissions lurks an awareness even in the most optimistic books that engine-driven boats cannot manage certain tasks as well as double-ended, oared ship's lifeboats, and that modern conditions—especially petroleum-fueled modern warfare— often defeat modern motorized small craft.

In storms and in shallow water, particularly in surf, the engine-driven boats of the 1920s and later decades—even engine-driven naval craft—performed far worse than the lifeboat Benson sailed away from *El Dorado*. British English and British popular culture reflected the bifurcated meaning of lifeboat after 1920, the word denoting on the one hand any RNLI vessel, on the other a ship's lifeboat, double-ended, open, usually oared. In American English lifeboat has always meant a ship's lifeboat, and terms like surfboat and its descendant, rescue boat, designate the small craft of the Life-Saving Service and Coast Guard. At Dunkirk, Singapore, Bataan, Corregidor, and countless other places, Allied forces and fleeing civilians discovered—and readers later learned—that modern technology failed to rescue troops and civilians from modern warfare. Away from and into the beaches moved the most conservative, traditional, *old-fashioned* small craft of all those still serving at sea. Only a handful of acute observers, including Nicholas Drew, noticed at the time what an even smaller handful, including Alistair MacLean, murmured after 1945. In defeat, disaster, and confusion, the traditional ship's lifeboat—what seamen around the world knew by 1900 as the "Board of Trade lifeboat" and what seamen still know by that name—offers the best hope for escaping not only a sinking ship but a seacoast writhing in war. It offers the best hope of making a passage to safety too, to life itself.

5

BOARD OF TRADE

In 1941 the British vice-consul at São Luis in Brazil performed an experiment to satisfy his own curiosity. Possessed of a ship's lifeboat that had suddenly come into his care as a piece of distressed British property, he began loading the boat with men, trying to squeeze in eighty-two. Geoffrey Leigh Bryan discovered that he could fit only seventy-four aboard, and he learned from the original occupants of the lifeboat that the presence of a German warship alone had caused eighty-two men to cram themselves into a boat intended to hold no more than fifty-four. Bryan considered the arrival of the lifeboat an indication of *Kriegsmarine* might and a remarkable feat of seamanship. On the coast of northern Brazil, in a small harbor about 360 miles south of Pará and the Amazon Delta, he had first-hand evidence of German warships far further south than anyone had expected. He had firsthand evidence too of the seaworthiness of an ordinary ship's lifeboat, which had sailed some 1,535 miles in twenty-three days carrying far more than its assigned number of castaways.[1]

"For nearly twenty years I have endeavored to live down and forget the story told in this book. As you will discover, it is not a pleasant or entertaining story but one of human suffering and death," begins the preface to Frank West's *Lifeboat Number Seven*. West had no need to mention the lifeboat in his first sentences: he had already honored it in his title. In 1959 West had his copy of the onboard diary he kept after the sinking of *Britannia* in 1941 (the original had been requested in 1946 for permanent exhibition at the National Maritime Museum at Greenwich), and he had the notes he made during his hospital stay in São Luis, but above all he had his memories of the passage in the lifeboat Bryan tried to pack with eighty-two men. West had what Nicholas Drew defined as an amateur interest in warfare, and that interest shaped his view of the sinking of *Britannia* and the subsequent lifeboat passage. As the days passed onboard, it came to West that he was witnessing—and was of course part of—two spectacles, one that of the eighty-two men trying to survive, the other that of the lifeboat itself on passage. "This boat was our real world, the only one we knew," he wrote after three weeks at sea. Only thirty-eight men of the original eighty-two survived, but the ordinary ship's lifeboat, despite being holed by shrapnel and machine-gun bullets, survived to reach a desolate stretch of Brazilian coast.[2]

Lifeboat Number Seven is one of a small collection of lifeboat narratives based upon journals mulled over for years by their writers. Unlike wartime morale-boosting stories, West's book is a penetrating, scathing analysis of survival at sea carefully juxtaposing diary entries against sustained recollection and judgment. *Lifeboat Number Seven* is partly a record of self-education. Eighteen years after the sinking, West looked upon his diary not only as a record of shipwreck and a lifeboat passage but as a chronicle of his brutal and lengthy introduction to an artifact of modern seafaring. *Lifeboat Number Seven* resembles Conrad's "Youth" in its depiction of the learning experience, but its true literary genre is not marine autobiography or bildungsroman, but lifeboat-passage narrative. The genre is distinct from small-boat-passage narratives written by eighteenth- and nineteenth-century men, for whom small boats were an important part of their world. West and other authors discovered lifeboats.[3]

"The lifeboat was one of the standard Board of Trade type, clinker built and with a hull form similar to a whaler—pointed at both ends," West wrote of his microcosm-to-be. "She was twenty-eight feet long, with a ten-foot beam and depth of three feet nine inches." Copper buoyancy tanks ran along each side, covered by longitudinal thwarts, and at the same height ran four cross thwarts from which seated men could row. A small forepeak and stern locker held supplies and drinking water, beneath the cross thwarts bread lockers contained food, and in the foremost cross thwart stood the twenty-foot-long mast that carried a lugsail.[4] In two paragraphs West sketched the sort of boat that *Titanic* and *Lusitania* and *Empress of Ireland* launched decades earlier, that Drew took into the Dunkirk shallows, that Morrill wished for during squalls, that tens of thousands of steamship passengers had glanced at during boring lifeboat drills since the 1890s. Sometimes smaller, sometimes larger, by 1940 occasionally built of steel rather than wood, the Board of Trade lifeboat existed around the world as an artifact of experience and regulation.

Anywhere *Britannia* made port in Europe and the Americas during the peaceful years before 1939, alongshore fishermen and other mariners might glance at the lifeboats ranked along its sides and see not Anchor Line adherence to Board of Trade regulations but small-craft cousins to traditional local boats used by tradition-minded local fishermen. In many parts of western Europe and around the British Isles, even landsmen glancing from the rows of canvas-covered lifeboats to the small fishing boats bobbing elsewhere in the harbor might make the connection. During the depression, when many devotees of vanishing sailing ships began looking at sailing-ship-era details, both seamen and landsmen found traditional small fishing boats described in books like R. Thurston Hopkins's 1931 *Small Sailing Craft*.[5] Products of a deepening interest in artifacts that still reflected skills and attitudes that seemed to be disappearing from the high seas along with full-rigged sailing ships, such books emphasized the pointed stern and lugsail rig of so many seaworthy small craft sailed by highly competent men. The lifeboats slung in davits aboard *Britannia* resembled traditional, very small fishing boats intended to work far at sea and arrive home safely laden with fish, and they reflected, too, older attitudes toward seafaring, especially seafaring in heavy weather.

But they did not speak, at least to experienced seamen, of whaling and whaleboats. Despite naval use of the term *whaleboat,* and even though nineteenth-century whaling boats are double-ended, the Board of Trade lifeboat owed almost nothing to the narrow, lightly built, extremely fast boats whaling ships carried for almost a century. The whaleboat existed to carry a harpooner and four or five rowers, harpoons, and tubs of coiled line alongside a swimming whale. No one intended a whaleboat to transport heavy cargo any more than anyone expected it to stay afloat in storms or make long passages. Despite detailed descriptions of whaleboats in all sorts of nineteenth-century books, however, steamship passengers and other landsmen unthinkingly considered the double-ended whaleboat to be the prototype of the double-ended ship's lifeboat secured in chocks and davits aboard steamships. In reality, the prototypes of Board of Trade lifeboats were both far more traditional and far more modern small craft.

Perhaps the most traditional of all British small fishing boats, the sixareen of the Shetland Islands descends from Norse vessels traceable to at least A.D. 300. Hundreds of years later, the larger double-ended, lapstrake Viking boat that evolved from the smaller types terrified anyone outside of what is now Scandinavia, for it signaled the arrival of raiders in craft capable of landing where no harbors existed. Simply built, its planks or strakes overlapping each other lengthwise in a way that improves watertight integrity, adds strength, and traps enough air as the boat moves for the hull to ride on a partial cushion of bubbles, the Viking longboat ranged far under oars and sail. It roamed at will around the whole of the British Isles and along the entire western coast of Europe, penetrating upriver to Paris and other towns thinking themselves secured by shallow water, and probed as far east as present-day Istanbul. Its figurehead, often a dragon, gave rise to folktales of fabulous monsters told in European villages long after the actual memory of Viking raids faded. Sometimes built larger for trading expeditions and other long passages, the Viking longboat had a smaller cousin, one that survives today in the Shetland Islands, an archipelago under Scandinavian rule until 1469 and still remarkably traditional about what sort of small boat best endures subarctic conditions.

A sixareen exists to carry six or seven men, six oars, a mast and sail, food and water for two days, bait, and approximately six and a half miles of heavy tarred fishing line and buoys—and to carry home the fish caught on the 1,200 longline hooks. The hull is designed to be seaworthy and fast enough to sail home before an oncoming gale, but light enough to be pulled up on beaches remote from any sheltered harbors. Between twenty-eight and thirty feet long, about eight and a half feet in beam, and some four feet deep, a sixareen fully loaded—but without fish—displaces some three tons, although the hull, stripped of everything else, weighs only a little more than three-quarters of a ton. Common everywhere around the Shetlands in 1850, but already long obsolescent by 1920 and almost vanished today, the sixareen still functions well because of its ballasting.

Before leaving for the fishing grounds, Shetland Islanders would load the just-afloat sixareen with some three tons of stone ballast. As the crew caught fish, the men would throw stones overboard, until—if they were lucky—they replaced all three tons of ballast with three tons of fish. Given its flaring sides, the sixareen acquired more buoyancy as its crew loaded it down, and on calm days some sixareens arrived home with five tons of fish aboard. If the men had to unload fish just off a beach (which they frequently did, since they dried the flaked fish on pebble beaches), the sixareen drew less and less water as the men carried the fish ashore, and soon the boat itself could be drawn up out of the waves. In emergencies, the ballast or fish might be thrown overboard to allow the boat to reach shore through shallow water and to leave again without fuss. Without its ballast or its load of fish, the sixareen proved unhandy, but almost never did it move without ballast or load.

Over the centuries Shetland Islanders modified the old Norse square sail into one more like a lugsail trapezoid, but they always kept in mind the chief mission of the sailing rig. To reach their fishing grounds, sixareen fishermen rowed offshore into the prevailing southwesterly winds. Once they made their catch, or once they suspected another gale from Greenland was bearing down on them, they raised the sail and headed home.[6]

Not many Britons see a sixareen today, but the type is only one of a number of seaworthy local boats once known along the coasts of the British Isles. Minus its midship mast and square sail, the sixareen looks less like a Viking longship and more like many other alongshore small fishing craft vanquished by steam-engined fishing boats after the middle of the nineteenth century. In many ways it seems a larger version of a Maine peapod, another fishing boat now rarely encountered by anyone except residents of out-of-the-way Maine towns.

Maine fishermen lobstering inshore invented the peapod sometime after 1850, when they grew dissatisfied with traditional fishing dories. In the 1880s, when Federal Fish Commission inspectors studied Maine coast fishing, the peapod type worked everywhere along the "down east coast," but no one remembered its place of origin. In the first half of the twentieth century, researchers traced the hull design to the seagoing canoes used by the Penobscot and Quoddy tribes, which resembled the large, very seaworthy vessels Nordhoff and Hall found in Polynesia. Smaller versions of the canoes proved definitely superior for inshore work to the mass-produced fishing dory ubiquitous by 1850 on Georges Bank and elsewhere offshore.[7] But nothing of Penobscot and Quoddy construction characterized the peapod, which boasted sawed frames, thick keel, cedar planking, and other heavy elements.

Double-ended and lapstrake like the sixareen but rarely more than sixteen or eighteen feet long, peapods are strong, seaworthy boats meant to be rowed to lobstering areas by men standing up at their oars and facing forward to see the rocky ledges around which lobster live. Extremely stable even when empty, they prove even more so as a lobsterman places most of his weight on one gunwale and pulls aboard a heavy lobster pot. A grown man can balance on one gunwale of a peapod and not capsize the boat, so heavily built and widely flared is its hull. Intended to work in rough water and to withstand sudden gales and their attendant waves, the peapod carries a sprit sail useful mostly for sailing home. Along the Maine coast, afternoon ordinarily brings an onshore wind, and almost invariably severe storms come from the sea as well, so the downwind, or in Maine parlance "downhill," sailing ability of the peapod is far more important than an ability to sail close to the wind or to tack easily.

Peapods fascinate experts in antique small craft. As a nineteenth-century invention based on seventeenth-century prototypes, peapods enjoyed a brief popularity before being replaced by engine-powered boats, but their sea-keeping ability has never been surpassed. Writing at length in *Small Boat Journal* in 1980 about a type of boat he remembered being built by an elderly native of Matinicus Island, at the mouth of Penobscot Bay, Walter J. Simmons explored the Maine coast rivalry implicit in the continuing battle over the name of the type. For many old-timers, the term *peapod* designated a new sort of boat invented on North Haven Island, while *double-ender* identified the improved traditional version built on Matinicus. Usage means more to Simmons than antiquarian significance. "Although the old days are gone, there is still much to learn from past ways," he observes, before detailing the variations among the so-called peapod type. He also makes it clear that while peapods are "superior sea boats," they simply fail to attract the eye of contemporary Americans interested in small boats. Out of the water, in boat shops and in museums, they seem heavily built and ungainly. Only in the water do they impress scrutinizers somewhat familiar with small-boat evaluation.

Peapods need to be heavy; plywood construction would result in a boat that looks right but behaves abominably. As Simmons explains, peapods "are designed to move with the seas rather than fight them," and so must be heavily built of traditional materials in order to ride low enough in the water to take advantage of their full hull shape.[8]

Stories abound about peapods, the most common one involving the refusal of two gale-tossed lobstermen to abandon their peapod and seek refuge aboard a schooner come to their aid—and the two men survived the storm fine. The tales tend to come from the last years of the nineteenth century and the first two decades of the twentieth, from the period when Maine coast fishermen abandoned peapods for engine-powered small craft but they still served other alongshore uses. Mailmen used peapods for decades in making their rounds among the Maine islands, and lighthouse keepers preferred them for ferrying supplies to island lighthouses. But what becomes apparent from listening to elderly alongshore informants is the preference of coastal schooner masters to carry—or tow astern—a peapod as a lifeboat.[9] Just before they decided to carry or tow engine-powered yawl boats to work their schooners more efficiently into and out of harbors, especially on windless days, the impoverished schoonermen running between Maine and Boston, New York, Florida, and even the Caribbean chose the peapod as the most able of all small boats. Their decision to carry a specialized lobster boat as a lifeboat parallels the decisions of mariners in other places. Captain Benson of *El Dorado* seems to have carried a lifeboat remarkably similar to a Pacific Coast fishing boat that resembled not only a large peapod but a sixareen as well.

What naval architects identify as the "Columbia River salmon boat" is another mid-nineteenth-century invention based on older notions, since the first seems to have been built in San Francisco for a man researchers know only as "Greek Joe," immortalized in an 1885 Fish Commission report. Greek and Italian fishermen spread around the world in the nineteenth century, often adapting traditional European hull types to local conditions. In California these immigrants built what was known by the 1880s as the "Italian fishing boat," a double-ended vessel modeled on Neapolitan forebears but modified with attributes of small craft from Spain and the south of France. On the west coast of Florida, Greek immigrants developed the "sponging boat," for the sponge-fishing industry.[10] The innovations of Greek Joe and countless other even more obscure Mediterranean fishermen prove difficult to trace now, because like the peapod their creations came into being at the very end of sail, and few adapted well to the weight of inboard steam or gasoline engines and the thrust of screw propellers.

Greek Joe ordered his new sort of boat in 1868 from a San Francisco boatyard, intending to use it salmon fishing off the Columbia River. A strong, heavy rowing and sailing boat, the twenty-two-foot-long, lapstrake, double-ended boat proved a great success but slightly too small for its purpose. Within a few years it had become the prototype for a popular twenty-eight-foot-long boat capable of staying at sea for a few days, its crew camping at night beneath a tent made of its sail, cooking over a tiny oil-burning stove. "The design of the boat was controlled by the need to carry a great load, yet it had to be of such size that it could be rowed," remarked one Smithsonian Institution researcher.[11] The Columbia River salmon boat was not only traditional in the extreme but also modern in a way almost generally ignored by contemporary historians of small-scale naval architecture: Greek Joe pioneered what might be termed a mass-produced boat.

Off the mouth of the Columbia River fished men who did not own their own Columbia River salmon boats. Instead they worked for wages in small boats owned by the salmon canneries, which ordered one type of boat in very large numbers, thus ensuring a standardization of design and reducing the price of each boat. While using conventional techniques and materials in construction, builders built the boats heavily, since the cannery ownership system invited abuse of boats fishermen did not own themselves. Yet the boats survived an average of ten to fourteen years, working in very difficult seas; in time they were adapted to other uses in Alaskan waters into the 1930s.

"The mass-produced stock boat used for commercial fishing has never been recognized by writers for what it really was: a successful design that had been proven by trial and that had achieved sufficient recognition of this to permit repeat sales," wrote Howard Chapelle in *American Small Sailing Craft: Their Design, Development, and Construction* in 1951.[12] Almost all marine historians focusing on small boats study only those constructed by individual builders whose "signature" sprawls across each, or else study the design of so-called class boats—the Concordia yawls, twelve-meter sailboats, and other designs intended for racing. A "one-off" boat interests researchers simply because it is unique. A stock boat always interests researchers less, since no single example of it says much about individual desires and specialties, and often only an expert can distinguish one stock boat from another.

A prime example of another boat ignored by small-craft scholars, the late nineteenth-century dory offers a somewhat better opportunity than does the Columbia River salmon boat to study mass production and its effects. While researchers know that Massachusetts colonists recognized some sort of dory as early as 1726, not until the middle of the nineteenth century did the boat become a stock ship's boat, carried aboard schooners fishing off New England. By 1870 the term *dory* identified a flat-bottomed, beamy open boat, sharp at the bow and with a very narrow transom, usually rowed but capable of sailing, its flared sides and removable thwarts allowing one to nest inside another. Schooners fishing out of Gloucester and other New England ports carried nests of dories, and once anchored over fishing banks dispatched each dory and its one- or two-man crew to set and retrieve a mile-long line carrying hundreds of baited hooks. Available in five standard sizes—twelve, thirteen, fourteen, fifteen, and sixteen feet—the crude but strong dories formed the basis of boat-shop prosperity well into the 1920s. *The Tenth Census of the United States* reported that in 1880 the town of Salisbury, Massachusetts, alone had seven boat shops producing between 200 and 650 dories each year. Within ten years construction had shifted to a few large shops, and while variations on the dory evolved for special purposes, its main type remained the nesting "banks dory," carried aboard schooners fishing the Grand Banks.

While the banks dory figured in everything from Rudyard Kipling's 1897 novel *Captains Courageous* to Winslow Homer's and William Partridge Burpee's paintings to postcard views of the fishing grounds, it seemed so much a boat "built by the mile and cut off by the foot to suit" that experts continue to focus on its variants. While an empty fifteen-foot dory might be rowed by one man, it grew so unmanageable in rough water that two men ordinarily rowed one. As anyone who has ever rowed an empty banks dory learns, the high flared sides and flat bottom cause the boat to skim downwind in ways that exhaust even powerful, experienced oarsmen. Like the

This November 22, 1879, *Harper's Weekly* illustration entitled "Lost in the Fog on the Banks of New England" is the sort of view that emphasized the sea-keeping ability of dories.

sixareen, a banks dory does best when fully loaded with fish, but banks fishermen never set out with anything other than loaded trawl tubs, lunch, and a bottle of water, and they expected their dories to be hard to row until the end of day, when they started home with a good catch. A bad day of fishing meant a brutal, exhausting row, and any kind of bad weather, including fog, meant that the master and cook of the schooner came downwind fast to collect the half-dozen or dozen dorymen pulling their lines into half-empty dories.

Unlike the round-bottomed sixareen, peapod, and Columbia River salmon boat, the flat-bottomed banks dory struck most observers as crudely, if strongly, built. As dory builders explained to the United States government in the months before Pearl Harbor, the dory is quickly and cheaply built, even by inexperienced labor. Throughout World War II, one builder produced scores of slightly modified fifteen-footers for the Coast Guard, although what the Coast Guard did with them remains one more mystery of wartime procurement.

German U-boats attacked the New England fishing fleet soon after the United States declared war in 1917. Ruthlessly efficient modern warships surfaced to destroy some of the most traditional, often almost worn-out sailing vessels moving under the American flag. Too small to justify a torpedo and carrying no radio to alert United States and British warships, the schooners received only the most casual shot across the bows from surfaced U-boats, then a scuttling charge placed near keels. Almost always, U-boat crews not only observed the rules of warfare they sometimes no longer practiced when attacking steamships but also told fishermen of their sadness at destroying beautiful old vessels often owned by their masters and crews. They knew that their need to disrupt United States fisheries at a critical moment in the Allied supply situation cast U-boat men in the role of aggressors against vessels most naval men considered so harmless and unimportant as to enjoy a sort of tacit neutrality. Moreover, they knew that schooners carried no lifeboats, but only dories not particularly well equipped for passage making. Conversation between fishermen and U-boat officers proceeded haltingly sometimes, but at other times easily, as when one U-boat first officer commented to some fishermen that he had owned a summer home in Maine for twenty-five years and knew the local waters intimately. In almost every instance, U-boat crews took care to make sure fishermen left their schooners well provisioned, because the fishermen had only dories to carry them to safety.

When U-156 stopped the Gloucester schooner *Robert and Richard* in late July 1918, its skipper and crew left the halibut-laden vessel immediately, in dories. Almost at once, the twenty-three men distributed among five dories set sail and began making for the Maine coast, some sixty miles away. Despite fairly pleasant weather, the five dories became separated during the following night, but passing ships picked up all but one, which sailed all the way to Kennebunkport. The captain of U-156 subsequently captured a Canadian steam-powered fishing trawler, fitted it with two rapid-fire guns, and trailed it around the Grand Banks. Approaching unsuspecting fishing schooners, the captain warned their crews into dories and then sank the schooners with scuttling charges. By the time U-156 sailed for home, it had sunk thirty-four vessels, including twenty fishing schooners, spreading alarm and sometimes panic from Cape Cod to Nova Scotia, and populating the Grand Banks with scores of fishermen seeking refuge in dories. More than a hundred survivors reached the tiny French fishing port of St. Pierre alone, and while the Canadian and American public reacted in shock and fury, the dory as small-ship lifeboat more than proved itself. Journalists praising the seamanship of castaway fishermen indirectly extolled the passage-making virtues of inexpensive dories most readers thought fit only for fish.[13]

However useful in skilled hands, the typical banks dory proved too small for carrying more than four or five men. After sinking a small British steamship—and in the process accidentally damaging all its lifeboats—the captain of U-156 had to spare a Canadian schooner to carry the castaways to safety. The schooner *Willie G.* carried six dories, but U-156 had aboard some thirty men, some wounded, from the sunk steamship, and the *Willie G.*'s six dories could not carry both crews to shore twenty-five miles west. *Willie G.* made port at St. Pierre at ten in the morning the next day, still afloat because its dories simply could not handle many more men than it ordinarily carried. It arrived at the same time as a fleet of dories rowed in, carrying the castaway crews of four other schooners sunk in the meantime.[14]

Sixareen and peapod, Columbia River salmon boat and dory, all more or less resembled what in 1914 suddenly became so important in the minds of seamen and newspaper readers alike, the Board of Trade lifeboat that had become a fixture aboard British-flag merchant vessels.[15] Yet long before 1914, indeed long before the foundering of *Titanic* in 1912 or *Waratah*'s vanishing in 1909, all British Empire merchant steamships left port equipped with boats designed for saving life, as did most steamships flying the flag of European nations and the United States. After January 1, 1847, no British vessel could be "cleared outwards foreign" without having aboard the number and type of boats prescribed by Parliament; in Liverpool and elsewhere, harbormasters refused clearance to vessels lacking certificates of compliance until the shipping industry accepted lifeboats as routine appliances.[16] But as Joseph Conrad so brilliantly outlined in his 1900 novel *Lord Jim*, worn-out steamships flying the flags of minor potentates often carried nothing more than a ship's boat useful for routine harbor work, not even the lifeboat big American sailing vessels like *El Dorado* carried.

By the outbreak of World War I the ship's lifeboat, modeled on traditional coastal fishing boats and on coastal fishing boats invented in the latter half of the nineteenth century, struck most ocean liner passengers as one more standardized, factory-built, mass-produced item of the twentieth century. The artifact scarcely changed from one decade to another until by the 1930s it had become the most traditional component of ultramodern liners, like the diesel electric-powered *Morro Castle*, which caught fire off New Jersey in 1934.

By this time, the Board of Trade had for decades been administering its Lifeboat Efficiency Examination to seamen anxious for promotion. As the government department charged with ensuring safety at sea throughout the empire, the board divided its activities into devising and testing improved safety equipment, inspecting existing equipment regularly, and mandating continuous training for officers and seamen alike. The board's worries that fewer and fewer British seamen knew much about small-boat

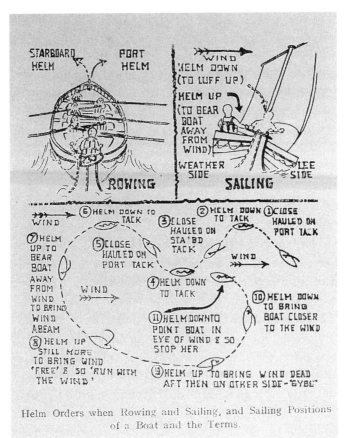

Helm Orders when Rowing and Sailing, and Sailing Positions of a Boat and the Terms.

Managing a lifeboat meant—in the view of Harold S. Blake, the Board of Trade, and every man who passed its examination—making a course under oars and under sail. (From Blake, *The Lifeboat*)

management meant that many steamship passengers watched lifeboat drills for deck crew members—but not engine-room crewmen. Lifeboats became bigger in the first years of the twentieth century—those of *Titanic* easily held sixty people each—but despite the pleas of inventors they remained almost unchanged for decades. Board of Trade examinations, and their United States counterpart, kept alive nineteenth-century notions of seamanship under difficult conditions in lonesome seas rarely traversed by great ships. Fitted securely at the very keel of twentieth-century seafaring, the Lifeboat Efficiency Examination insisted on the importance of traditional skills no longer honored or even remembered in engine rooms and in wheelhouses, let alone in staterooms and saloons.

"A lifeboat is defined by the Board of Trade as being a boat possessing a capacity of not less that 125 cubic feet, pointed at both ends and fitted with some means for giving it increased buoyancy when water-logged," begins Harold S. Blake's 1933 study guide *The Lifeboat: Its Construction, Equipment, and Management: A Guide to the Board of Trade Lifeboat Efficiency Examination.* One of dozens of booklets periodically printed by privately owned waterfront schools for seamen, or by self-described "nautical publishers" aiming at the examination market, *The Lifeboat* opens with the blunt statement that it is written for crewmen lacking any knowledge of seamanship. Unlike other study-guide authors, Blake makes every effort to avoid nautical phraseology and to explain to the post-age-of-sail seamen the reasons underlying Board of Trade requirements; in short, to make his students think in old-fashioned ways about a traditional vessel.[17] From the outset, his little book implies that lifeboats may have to make long passages in storms, and may have to make land on their own.

C. W. T. Layton's *Ship's Lifeboats: A Handbook for the Board of Trade Examination for Certificates in Lifeboat Efficiency* begins with the encouraging assertion that a ship's lifeboat can handle seas that might destroy larger vessels, but then cautions that lifeboatmen must be prepared for long passages. It warns that the Lifeboat Efficiency Examination is a practical examination in which pencil and paper count for very little. Board of Trade examiners demand mental and physical competence together, and they want both *in the boat used in the examination.*

All Board of Trade examination guides begin by painstakingly describing a typical lifeboat, ordinarily a twenty-five-foot-long boat capable of carrying thirty-six adults. The guides acknowledge that Board of Trade regulations permit some variation in lifeboat size and construction; while most boats are lapstrake (sometimes called clinker or clincher), some are carvel-built or smooth-sided, and some of the latter made with a double thickness of wood placed diagonally. But lapstrake or carvel or diagonal, the lifeboat is invariably double-ended and very heavily built, almost overbuilt in the eyes of steamship seamen.

No better introductions to traditional wooden boatbuilding exist than the Board of Trade examination books. Since the board was well aware that lifeboat crews might have to repair damage and leaks while underway, examiners insisted that seamen be able not only to name all the components of a lifeboat but to explain their roles in the larger built whole. A familiarity with both design and the building process could help lifeboatmen understand the behavior of the boat in different conditions and the ways in which the boat could be repaired, and also why some damage might be nearly inconsequential.

At first reading, most examination books seem elementary, but short sentences and single-syllable words—language appropriately geared to the average reader's educational level—mask sophisticated awarenesses. "We will now examine the boat in detail," begins Layton's dissection of the typical lapstrake lifeboat. "Looking at it we notice a strong timber that goes fore and aft underneath the bottom. This is the *Keel.* It is a most important part and is, in fact, the backbone of the boat," he writes, using boldface type to key his paragraph to sketches on a facing page. "Scarphed, or spliced, to the keel are somewhat similar timbers that come up at the fore and after ends. The forward timber is called the *Stem;* the rounded portion, where it turns to meet the keel, is called the *Forefoot;* and the top part that rises a little above the boat is called the *Stem head.*"[18]

Part by part, Layton not only identifies each component but explains how subassemblies work. Only gradually does the larger design of his thinking become discernible. The parts and subassemblies have different uses at different moments: when the lifeboat is suspended from its falls during lowering and recovery, when it is rubbing against the hull of a steamship, when it is slamming to a sea anchor in a storm, and perhaps above all, when it is moving under oars or sailing in all sorts of conditions.

Crewmen studying to be lifeboatmen had to familiarize themselves with the ship's lifeboat both as a traditional, heavily built small boat and as a special-purpose one. In particular, they had to understand the deeper significance of the great hooks bolted to the keel fore and aft. These hooks, connected with the falls that lowered the lifeboat from davits into the sea, identified a ship's lifeboat as fundamentally different from boats it superficially resembled. Yet for all that the lifeboat epitomized Board of Trade thinking about tradition fixed to modernity, the lifeboat sailed under an exceptionally old-fashioned rig.

By 1910 only a handful of traditional British inshore fishing vessels still sailed under lugsails, and not many more vessels elsewhere in Europe or anywhere in North American waters carried what mariners knew as a distinctly outdated or very provincial rig. World War I brought not only a dramatic increase in the demand for fish but a correspondingly rapid development in small gasoline engines. Ten years later, British fishermen worked under power, and most knew little of the lugsail-rigged boats once so common along certain coasts. The small, fast luggers that smuggled lace and brandy across the Channel before and during the Napoleonic Wars had become legend, and only in remote, usually shallow-water ports did their descendants remain. Yet at the start of World War I most British warships still carried lugsail-rigged boats, as did every British merchant ship. By the time *Britannia* foundered in 1941, the Royal Navy had long converted to powerboats, but the Board of Trade remained adamant that merchant vessels carry lifeboats equipped with lugsail rigs, the design of which dated back two centuries.

The lugsail rig long characterized inshore fishing boats along the coasts of Kent, Devon, and Cornwall, and it fitted perfectly into the lifeboat hull design evolving from new inshore fishing boat types invented after the 1860s, especially in North America. As engines replaced sail along the British coast, the lugsail rig became first obsolescent, then an anachronism, but aboard steamship lifeboats it became "the lifeboat rig" or "the Board of Trade rig," something seemingly without origin, known around the world, and enduring through peace and war.

The experiences of Sidmouth fishermen offer a useful guide into the origins of lifeboat rigs. Sidmouth families had fished for centuries along the Devonshire coast, but by the 1850s they were becoming outnumbered by retirees and others discovering the beauty of their village and its low cost of living. Really only a half-mile-long shingle beach across a valley opening with five-hundred-foot-high cliffs on either side, Sidmouth never had much of a harbor. During a misguided harbor improvement scheme late in the nineteenth century, it lost the few rocks that slightly sheltered its fishing boats from southwest gales. Storms frequently smashed ashore, eroding the beach into a series of steplike flats and pulling stones and sand slightly seaward to create a sandbar dangerous to cross in high winds. After long spells of calm weather, the beach might re-form into a gentle incline, but sometimes a series of gales so eroded it that men had to launch and retrieve their boats along a wood incline leading up to the esplanade built for the newcomers. Into the 1920s Sidmouth's impoverished fishing families continued to use lugsail-rigged open boats, thus attracting the attention of age-of-sail enthusiasts.[19]

Sidmouth boats ranged about ten miles on either side of the beach and as far at sea as they could, working at a variety of tasks changing with the seasons. Between May and October the fleet fished for mackerel using drift nets, and from November to March caught herring in nets, in the intervals running occasional trawls for flat fish, working pots for crabs and lobsters, and setting drop lines for cod, whiting, and other near-shore fish. By 1900 many of the boat owners were in the Royal Navy Reserve, receiving a small stipend that enabled them to keep fishing from their hometown in traditional boats roughly twenty to twenty-four feet in length. Usually they left the beach carrying stone ballast in bags that had to be thrown overboard before the vessel returned to shore. Owned by the men who sailed them, sometimes in partnership with local storekeepers, the beamy Sidmouth luggers, locally built of elm planks over wood ribs, each carried a single dipping lugsail that provided tremendous power with the wind behind it but proved tricky to manage inshore.

Both at sea in their own boats and during their regular stints in the naval reserve, Sidmouth fishermen used the lugsail rig. In their fishing boats, the big sail hung suspended from an unstayed mast, the lack of standing rigging giving a great deal of flexibility in setting bulky, stiff nets over the sides of the boats and in retrieving the nets when filled with fish. The lugsails lacked booms along their bottoms, allowing fishermen to move about easily under the so-called loose-footed sail. To be sure, when close to inshore rocks Sidmouth men preferred to row and believed that "two oars be better than two reefs any day" in bad weather.[20] Almost always, in gales or calms, they worked at heavy oars to reach their particularly difficult stretch of home beach. But for their own multiple purposes and for their own stretch of coast, the lugsail rig seemed perfect.

In his 1909 analysis of Sidmouth life, *A Poor Man's House,* Stephen Reynolds observed that the open boats, none longer than twenty-five feet, carried lug sails that brought men home in the worst of storms. No wonder the Royal Navy so immediately enlisted the young men of Sidmouth and then tried so hard to reenlist them into the reserve. At sea in Royal Navy small craft, the fishermen encountered the identical rig, since the navy recognized its simplicity, strength, and power and assumed that any naval party moving in a cutter or other small craft had enough men to row when needed. Sidmouth fishermen now and then glimpsed in civilian life the reason that the Royal Navy—and eighteenth-century smugglers—so favored the lugsail rig: once a twenty-two-foot lugger raced eight miles to seaward in fifty minutes; another time one drove across Folkstone Bay without sails, making harbor to the cheers of onlookers awed by the great gale driving it. Even in the merchant service, an occasional speed-minded master kept his gig lug-rigged. "Whether rowing or sailing, our beautiful gig always won; for sailing she was lugger rigged, a facsimile of the pretty fishing luggers of St. Ives, the Old Man's home town," Fred Ellis recalled in his 1949 memoir of his turn-of-the-century sailing-ship apprenticeship, *Round Cape Horn in Sail.*[21] Most Sidmouth fishermen accepted such exploits easily, although newcomers to the town marveled at sixteen-foot-long luggers landing a thousand mackerel on the beach every morning and larger ones nonchalantly launching through surf.[22] Between hull and lugsail rig, the boats of Sidmouth struck expert observers as both anachronistic and perfectly suited to their tasks—caught on the very verge of the engine-driven modernization transforming both the RNLI and the Royal Navy.

By 1900 modern steamships carried the standing-lug rig around the world, carefully stowed in lifeboats. Though the lugsail rig was fast disappearing even from Sidmouth, the Board of Trade insisted that all steamship officers and any seamen wishing to become lifeboatmen learn to manage a lugsail rig. Mass-produced to rigid standards, the ship's lifeboat of the Board of Trade Lifeboat Efficiency Examination became an international fixture all too frequently taken for granted as its original prototypes faded from memory.

Officers commanding lifeboats during long passages in peace and war marveled at the astonishing ability of the boats. "I do not suppose that it was for a moment anticipated that boats such as these would ever be called upon to undergo a test like the one we were compelled to put them to," asserted Cecil Foster in 1924 of what he called, simply, "Board of Trade boats."[23] His *1700 Miles in Open Boats: The Story of the Loss of the S.S. Trevessa in the Indian Ocean and the Voyage of Her Boats to Safety*, an account of a harrowing passage to Rodrigues Island, details the growing admiration of a steamship master for the twenty-six-foot-long wooden boats he had uncritically accepted as ordinary required safety appliances.

Many lifeboat narratives state or imply that no one aboard lifeboats embarking on thousand-mile passages believed the passages possible, that everyone expected somehow to be rescued in a few days by a passing ship. On another level, merchant-ship officers and crewmen must have felt that when they boarded lifeboats they boarded vessels from an era before the great chasm of modernization, that clambering into a lifeboat meant escaping the present for what seemed at first glance the distant seafaring past when seamen were seamen. This was, of course, an illusion: the Board of Trade lifeboat, for all its lug rig, carried modern equipment.

Early on, buoyancy tanks distinguished ship's lifeboats from all other alongshore or seagoing craft except coast-based lifeboats, and demonstrate a direct link between Board of Trade thinking and the technology evolved by the RNLI. But the technology proved awkward to evaluate. As early as 1848, David Rough, harbormaster at Auckland in New Zealand, cautioned readers of *Nautical Magazine* that he had tested several "so-called *life-boats*" by filling them with water and had discovered that "copper or zinc cylinders, or air vessels, upon which the commanders and officers seemed to place great reliance," could not float flooded boats with crews aboard. Rough's message was brutally clear: too small tanks provided false security.[24] Not for years did Board of Trade regulations specify the ratio of tank volume to lifeboat size, and express lifeboat size in both linear and cubic feet.

The equipping of a lifeboat preoccupied the board, which carefully analyzed lifeboat-passage narratives to see what changes might prove beneficial. Captain Foster of *Trevessa,* having been torpedoed in the Great War, believed that tinned milk saved many castaways, and in the few moments before his steamship suddenly foundered he ordered his steward—who had sailed with him in wartime—to grab condensed milk rather than tinned meat.[25] This fact emerged at the Board of Trade inquiry into the loss of *Trevessa,* and by July 1, 1925, every British lifeboat carried a pound of condensed milk for every passenger the board certified the lifeboat to carry. But a lifeboat might be launched in tropic or polar seas, and its equipment had to be effective anywhere.[26] While the Board of Trade mandated that officers and seamen regularly inspect lifeboats and refresh water supplies, it never specified that equipment be altered according to the position or regular route of the steamship.

Every lifeboat carried the barest necessities, but to those necessities the Board of Trade devoted intense thought, and it expected certified lifeboatmen to do the same. Examiners expected candidates to enumerate and discuss lifeboat equipment, and Layton created a memory device to aid students: "PA SAW SOME BREAD MILK AND WATER BECOME PORK BALLS." This mnemonic was to help seamen remember the painter, ax, sea anchor ("SA"), warp, sails, oars, mast, and sailing gear ("E" for "etc."), bread, milk, and water, bucket and bailer ("BE"), compass ("COMe"), plugs, oilbag and oil, rudder and tiller, boathook ("K"), box of matches, ax, lamps, lights (red), and a set and a half of crutches—the last what American seamen called oarlocks. Layton specified the location in which Board of Trade regulations placed each piece of equipment and warned students that board examiners would ask about location.[27]

The painter, ax, sea anchor, and warp all lay forward, while the sailing gear and oars lay stowed lengthwise across the thwarts. Locating every piece of gear preoccupied some study-guide authors more than it did others; Blake's *Lifeboat,* for example, emphasized location as a way to remember every object. "An easy method of memorising the equipment of a lifeboat is to associate in the mind each article with the position it occupies in the boat," Blake explained, before detailing what items usually nested alongside or inside one another. To remember the location of one is to jog the memory about the existence of others. Testing applicants on the location of the gear assured Board of Trade regulators that lifeboats had a chance after sundown, especially if the men knew the equipment lay secured, he continued, "in a manner that allows the gear to be brought into use with the minimum of delay on a dark and dirty night."[28] Every item the Board of Trade required to equip a lifeboat might have to be found and used in total darkness, and the board intended that any British lifeboatman tumbling into any British lifeboat would find standard equipment in prescribed locations. Lanyards kept every piece of equipment in its prescribed place, even during the turmoil of launching and the chaos of storms.

Foster, sailing a *Trevessa* lifeboat across the Indian Ocean in 1923, in time discovered things about his boat that he had not learned in any examination, and his narrative includes not only an inventory of equipment required by the Board of Trade but musings on the equipment itself. Multiple water dippers, for example, provided for a choice in water-ration amounts, something vitally necessary in overcrowded lifeboats. And when the cap on the opening into one of the bread lockers broke and Foster found himself ordering his men to transfer the remaining bread into another watertight bread locker, he realized why Board of Trade regulations mandated that no more than seventy-five pounds of bread be stored in any one watertight locker, and that water supplies likewise be distributed among multiple breakers. Damage to one container, however serious, at least did not contaminate the entire bread or drinking water supply, although the damaged locker could no longer serve as a reserve air tank.[29] As a lifeboat crew ate and drank its way through its provisions and perhaps became less and less able to care for the boat, especially in storms, the Board of Trade thought the emptying provision lockers would become more and more useful as buoyancy tanks. Everything in a Board of Trade lifeboat seemed to have multiple uses. Foster made a point that the boat commanded by his first officer left the shipwreck with its lifeline intact, but it arrived with its lifeline replacing some parted rigging. Everyone abandoning ship, Foster and so many other passage-narrative authors insisted, got an equal chance, because everyone began with a Board of Trade boat equipped in Board of Trade fashion.

In its first minutes at sea the lifeboat might remain tethered to its ship by some length of the twenty-fathom-long painter required by the Board of Trade. Most of its would-be passengers would slide down the davit falls cast off a moment before by the two or three men who had ridden the boat down. Even if the crew lowers the boat filled with ship's passengers, the painter may help keep the lifeboat in position for a minute or two as the "efficient lifeboatmen" get out steering oar and pulling oars and begin to move the lifeboat away from the ship's side. Thereafter the painter may attach one lifeboat to another, and long after that, the painter is a source of line to make repairs to the lugsail rig.[30]

Board of Trade regulators always provided for the dark and dirty night first, and they expected lifeboatmen to neither sail nor row in storms. Instead of trying to keep a course, lifeboatmen should rig the sea anchor, a funnel-shaped canvas bag intended to sink beneath the surface and drag the bow of the lifeboat into the wind and waves. In extreme conditions, lifeboatmen should float the oil-filled bag on the sea-anchor line and let the tiny amounts of vegetable oil calm the seas before they struck the stem of the boat (something that seems near magic to landsmen). The sea anchor could keep the boat reasonably stationary in most weather, something useful if its crew knew help would arrive at the disaster location. Subtly and never explicitly, the Board of Trade emphasized that a lifeboat moved at sea in a way dramatically different from the way a steamship moved. In the first hours after shipwreck, when even the most experienced of mariners might suffer wretched seasickness in the lively small boat, riding to a sea anchor might be the best choice. After a few hours, perhaps at daybreak, the experienced lifeboatman could get his crew in order and proceed.

Examination books establish that Board of Trade examiners wanted not mere familiarity with equipment but real skill in boat handling. About the latter the board could only advise generally, but it most certainly expected efficient lifeboatmen to be able to raise and use the lugsail rig and to set up and use the compass provided in every lifeboat. As West learned in mid-Atlantic, using the lug rig well took a day or two of experiment, but in his boat, as in almost every boat, either an officer or a seaman knew something about it and the compass, having practiced a bit with both before passing the hands-on Board of Trade examination. "No amount of reading can take the place of actual experience in sailing, and it is advisable that the reader should take advantage of every opportunity of getting out in the boats," Blake cautioned, echoing so many examiners before him.[31] Only sailing a lifeboat taught much about sailing, and only working over a lifeboat compass taught much about navigation. The Board of Trade insisted on standardized boats and equipment, and it insisted on a standardized, practical examination that tested familiarity with the equipment in use: the standardized examination meant that shipping companies could direct masters to provide training and experience standard across the empire.[32]

By 1900 British and American steamship officers confronted stiffer tests than the Lifeboat Efficiency Examination. Intended to make officers understand the reasons underlying the design, construction, equipping, and sailing of lifeboats, the tests explained that metal hulls and other innovative failures accounted for some lifeboat characteristics. "In the United States Coast Guard Service, which probably has as much experience with beach or surf boats as any other similar organization, metal has been entirely discarded for wood," notes the 1941 edition of Austin M. Knight's *Modern Seamanship,* a book first published forty years earlier. Wood performed better than metal.

Updated by the Department of Seamanship and Navigation of the United States Naval Academy just before the United States entered World War II, the tenth edition of *Modern Seamanship* exemplifies the fierce preference of the United States Navy for durable, tough boats, but it reveals a growing ambivalence about modern material versus old, and especially about engines. *Modern Seamanship* insisted in 1941 on the same traditional wisdom it favored in 1901, even to quoting the Royal National Lifeboat Institution on the landing of boats through surf.[33] No matter how modern the warship, the manual cautioned young naval officers on the eve of war, some of its boats must be traditional, and some officers and sailors have to know how to work the traditional boats. But as the outbreak of World War II demonstrated, too few naval vessels sailed with any traditional small boats and—apparently—with far too few men familiar with them.

Merchant-marine officers absorbed a similar wisdom from the United States Bureau of Marine Inspection and Navigation, the American equivalent of the Board of Trade. In 1944 two master mariners, Edward A. Turpin and William A. MacEwen, published the *Merchant Marine Officers' Handbook,* a tome outlining the special responsibilities of United States merchant officers toward the traditional lifeboats so much used since September of 1939. Turpin and MacEwen advised readers that every lifeboat should be thoroughly overhauled and painted annually, that every three months every male crewman (stewardesses and other female crew did no lifeboat training) of every ship must be exercised at the oars of a lifeboat, that in emergencies "a deck officer or certificated lifeboat man must be placed in charge of each boat," and that the United States government allowed no more than ten minutes for launching a lifeboat in smooth water. Only in the tiniest of details did the bureau depart from Board of Trade requirements—the lantern had to burn for nine hours, not seven, on one filling, for example—and the bureau insisted as strongly as the board that deck officers understand the management of a lifeboat every bit as well as any lifeboatman. The only way to gain such understanding was to use lifeboats in sustained practice when ships were docked.[34]

By 1940 mariners everywhere knew—at some level—what West learned: in the most harrowing conditions, the Board of Trade boats performed ad-

Lifeboats and their equipment require continuous maintenance, even when the boat is built of steel, like this boat from *Carter Braxton*. (Courtesy of the Mariners' Museum, from the A. Aubrey Bodine Collection)

mirably. An impressive number made away, not from sinking ships, but from harbors and beaches falling into enemy hands.

Such were the boats that set out from Dakar, after the Vichy French interned so many Allied merchant vessels in the early years of World War II. Anxious to join their friends and countrymen in the war, small groups of men slipped out of Dakar at night in lifeboats, eluding the fort on Gorée Island and the patrolling gunboats. "They rowed as noiselessly as they could," recalled the Norwegian authors of *Flight from Dakar*, Eiliv Odde Hauge and Vera Hartmann. "The wind was astern, so they made good progress. When they thought they had got far enough out, they set the sail too, and then the boat ran with a foaming bow-wave."[35] Day after day, the *Livard* Number 2 lifeboat ran south along the coast, skirting miles of surf, dodging into inlets to escape French patrol boats, at times moving through calms under oars alone. Eventually the men found a British patrol boat and discovered themselves as safe as West found himself in Brazil. And like the British vice-consul, the British authorities at Bathurst took special care of the lifeboat, bringing it on shore and storing it in a shed with other lifeboats escaped from Dakar.

In World War II the world learned again what Foster learned in the Indian Ocean in 1923 and so many steamship men learned in World War I. The ordinary artifact carried on the boat deck of every United States, British, and European steamship might do extraordinary things. Once launched, it performed better than even officers and certificated lifeboatmen believed possible. But launching it successfully sometimes proved difficult.

DAVIT

Tense and preoccupied, Captain Von Henchendorf of U-151 grew increasingly exasperated as he watched panic spread aboard the halted steamship. Sunday morning, June 2, 1918, had begun well for the Imperial German Navy and had improved quickly, but Von Henchendorf knew well the risk he ran operating so close to the United States coast. He wanted SS *Winneconne* evacuated immediately, and panic only delayed the launching of its lifeboats.

Unlike many wartime encounters at sea, that involving U-151 and *Winneconne* is particularly well documented, not only because the master of *Winneconne* traveled to Germany after the war and interviewed the man who sank his ship, but also because U-151 had just captured a schooner when *Winneconne* steamed into view. Moreover, U-151 had aboard seamen from three schooners sunk earlier, and Von Henchendorf had assembled them on deck so that they might see both the might of his warship and his strict adherence to international law. Numerous witnesses and a postwar interview combined in time to make a sumptuous portrait of confusion.

The captured schooner *Wiley* masked U-151 as the steamship approached. Waldemar Knudsen, master of *Winneconne*, arrived on his bridge when his officers reported seeing a stopped schooner and perhaps a patrol boat alongside it. Between 7:30 and 8:10 A.M. *Winneconne*, carrying 1,819 tons of coal from Norfolk, Virginia, to Providence, Rhode Island, steamed slowly toward the stopped vessels, one more merchant vessel commanded by a master confident in United States Navy reports dismissing German submarine threats. At 8:10, U-151 fired a single warning shot and Knudsen turned toward shore, stopping two minutes later when a second warning shell landed perhaps two hundred yards ahead. Immediately U-151 drew nearer, and Von Henchendorf dispatched the *Wiley* motor launch to the collier, the boat ferrying an officer, two sailors, and four scuttling charges, the officer with orders to give the *Winneconne* men a half hour to abandon ship. The moment the officer ordered *Winneconne* abandoned, panic swept the steamer.

From the deck of U-151, seamen captured from the schooners *Hattie Dunn, Edna, Hauppauge,* and *Wiley* witnessed what followed. Stopped within hailing distance of *Winneconne,* Von Henchendorf watched the collier seamen prove unable to swing out the first lifeboat. After staring incredulously, Von Henchendorf "roared out a reprimand in plain and vigorous English for not maintaining proper discipline," according to one captive. Along with so many other master mariners, Von Henchendorf loathed lifeboat-launch panic, especially on a calm sea, in broad daylight, aboard a surrendered ship whose crew had enough time to gather personal belongings and leave long before the demolition charges exploded. Not surprisingly, the schoonermen watched the entire episode with the usual contempt sailing-vessel crewmen reserved for what they deemed incompetents masquerading as steamship seamen.[1]

In both world wars, a "small dark object" on the surface of the sea, often silhouetted against the twilit horizon, generally signaled imminent disaster. Merchant mariners learned to watch for the conning-tower shape, but they never learned to expect it. In 1941 shipmasters had the hard-won wisdom of an earlier war to guide them, and much of that wisdom involved fleeing from small dark objects, lifeboats or not. At least some master mariners had learned the lessons implicit in the *Winneconne* sinking, and in hundreds of similar ones. The small dark object might or might not signal the onset of attack, but it most certainly signaled the possibility, if not the likelihood, of panic.

Submarine attack forever altered the typical merchant officer's and seaman's conception of catastrophe. Between 1914 and 1918 the rules of ocean warfare changed dramatically, and not in a seamless chronological way. Three years before Von Henchendorf stopped *Winneconne,* another German submarine captain torpedoed the liner *Lusitania* without warning, killing some 1,200 people in an incident contrived by the British and abetted by American officials. British and American journalists who instantly labeled *Kapitan-leutnant* Walther Schwieger "a monster," "a war criminal," and worse neglected to examine the nuances of international law so differently understood by British, German, and other combatants, and by neutrals.

Months before Schwieger torpedoed *Lusitania,* indeed beginning in October 1914, the British Admiralty issued a stream of increasingly provocative orders to masters of British merchant ships. Not only did the orders make it an offense for a master to stop his ship when a surfaced U-boat fired a shot across its bows—several masters were prosecuted for so stopping—they directed masters of unarmed merchant ships to attempt to ram U-boats, and abrogated the "fire only when fired upon" policy by ordering masters of armed merchant ships to fire first whenever approached by a submarine. In order to increase the chances that U-boats would attack neutral ships, the Admiralty directed shipowners to paint out names and ports of registry and to fly neutral flags in British waters. In the archives of the Cunard Line the orders carry a manuscript annotation: "Pass the word around that the flag to use is the American." But perhaps most important, the Admiralty ordered British naval officers to treat captured U-boat crews as felons, and to fire on any U-boat displaying a white flag. In January 1915 a U-boat sank three ships off Liverpool with no loss of life, but it captured a complete set of the Admiralty orders.[2]

The German Admiralty knew that international law forbade masters of unarmed merchant vessels from attacking submarines and other warships just as it forbade a submarine torpedoing an unarmed merchant ship without warning. Masters who did attack, in international law, were indeed felons, and in 1916 the German navy court-martialed, then executed, a British merchant-ship master who had tried to ram a submarine with his unarmed merchant ship two months before Schwieger torpedoed *Lusitania*. In arguing that "while he was not a lawful member of a combatant force," the master "had committed a combatant's act against Germany," the German government emphasized the traditional distinction between legal and illegal combatants, something that in 1940 lingered in the minds of Nicholas Drew and other Dunkirk civilian lifeboat crewmen who knew they had entered a battle in civilian clothing aboard civilian lifeboats.[3] But once the German government had a copy of Admiralty orders, international law became largely moot, except in the United States' eyes. Top-level British decision makers manipulated Woodrow Wilson so effortlessly that the United States Department of State unwittingly helped conspire to create an incident that might bring the United States into the war.

Germany—and by January 1916, the United States Department of State as well—insisted that the British Admiralty had altered its policy concerning armed merchant ships from one of defense to one of offense, noting, moreover, that as early as November 1914 the British had sent to sea their first "Q ship," a warship disguised as a merchant vessel. By 1915 expecting German submariners to surface, approach merchant vessels, and order off their personnel seemed naïve at best.[4] In August of that year a U-boat that had surfaced and allowed the crew of the British freighter *Nicosian* to escape in boats was then sunk by gunfire from a ship flying the United States flag and carrying the Stars and Stripes painted on a board on its side. After rescuing the *Nicosian* crew, Royal Navy sailors aboard the Q ship *Barralong* methodically shot to death every German sailor still in the water. *Barralong* did the same thing to the captured crew of another U-boat a month later, and that time the first officer of the sunk steamship reported the incident to the Admiralty. Only because the eight *Nicosian* American crew members protested the atrocity did the incident come to light in the United States, and ultimately in Germany.[5] But in the case of the *Winneconne*, Von Henchendorf followed 1914 protocols—at least most of the time—and devoted much of that Sunday to painstakingly trying to sink ships without harming anyone.

Around six in the evening, after sinking *Wiley, Winneconne*, another schooner, then the SS *Texel*, U-151 encountered a liner, *Carolina*, en route from San Juan, Puerto Rico, to New York, carrying 218 passengers, 117 crew, and a cargo of sugar. Just as the master of the liner received his first radio warning from the navy, a woman passenger glimpsed a submarine astern and incorrectly identified it as American. Despite the extra lookouts, no one else aboard noticed U-151 until the submarine sent the first of three six-inch shells near the liner, which promptly stopped and began sending an SOS, but without giving a position, since its master feared that the submarine would sink any ship hastening to the scene. Again, Von Henchendorf had to cope with confusion, despite coming alongside the liner with his crew not attending to their guns but rather lined up, apparently to reduce passenger fears of indiscriminate shelling and machine-gunning, his ship flying the signal pennant "AB," silently ordering the master to abandon ship. Almost at once, the liner crew began putting first women and children and then the male passengers into the ten lifeboats, but before any of the boats could be launched, a group of engine-room and stokehold crew rushed one of the boats and had to be beaten back by seamen. Then the men lowering Number 5, loaded with twenty-two passengers, allowed the falls to run uncontrolled through the after davit, swinging the boat into a vertical position and spilling every passenger into the sea. To make matters worse, the crew then allowed the forward falls to run free, dropping the lifeboat upside down atop the twenty-two terrified, swimming passengers, including a nine-year-old girl. While *Carolina* crewmen righted the lifeboat and dragged into it all twenty-two passengers, Von Henchendorf waited patiently. He waited patiently while the nine-year-old girl discovered that the lifeboat had been launched before anyone had set its drain plug in place, which lack caused the righted boat to fill with water. And he waited patiently after ordering Lifeboat Number 1 to return to the liner and take off a man who had been left aboard. Seventy-five minutes after *Carolina* stopped, the lifeboats all safely away from its target, U-151 began firing into the abandoned liner. Light airs and smooth sea notwithstanding, on an even keel and despite extra lifeboat drills and a wartime mentality, *Carolina* took more than an hour to launch its lifeboats, and only Von Henchendorf's patience and thoroughness saved everyone's life, including that of the forgotten man.[6]

On that long day of June 2, 1918, Von Henchendorf learned a great deal about the inability of many steamship men to launch lifeboats in perfect conditions, and he learned a great deal more in the following two days, which for many Americans became the two most hectic, frightening days of the war. More than many other U-boat captains, Von Henchendorf took time to understand the ramifications of his attacks. While he is remembered best as the captain who released two ancient New Bedford whaling ships and a decrepit schooner owned and crewed by African Americans simply because their captains pleaded gut-wrenching poverty, his cruise makes clear the inability of steamship crewmen to carry out ordinary abandon-ship operations efficiently. It also underscores the importance of the davits between which each and every lifeboat hung suspended, supposedly ready for immediate use.

Not every U-boat captain had Von Henchendorf's thoroughness and humanity, and willingness to risk his own crew while superintending the inefficiency of another. More and more frequently after 1914, German submarine captains fired torpedoes without warning. Or even worse, they set mines, as U-151 itself did.

While a submarine about to submerge and attack might appear to sharp-eyed merchant seamen as a small dark object far off, a mine floating beneath the surface appeared not at all. Before it sank *Wiley* and the other schooners, then *Winneconne,* U-151 had been laying mines along the southern coast of Virginia and far up Delaware Bay, but not until June 3, when the United States Navy radioed orders for all merchant ships to make port at once, did its mines begin to add to the near panic caused by the first lifeboats being picked up by steamships or coming ashore, especially in Atlantic City.

The seven-thousand-ton oil tanker *Herbert L. Pratt,* racing toward the supposed safety of the Delaware Breakwater, broadcast its first SOS at 3:35 in the afternoon, its captain and crew certain their ship had been torpedoed without warning. About twenty minutes after the explosion, its crew at lifeboat stations, the bow of the tanker sank so far beneath the sea that the propeller rose above water. A minute or so later, the bow stuck fast in the sand of Hen and Chickens Shoal, and everyone settled down to await help.

Navy and Coast Guard vessels soon arrived to rescue the crew of the first United States vessel to be sunk by a mine in coastal waters. Men of the Fourth Naval District, ordered by their superiors to stifle the rising curiosity among newspapermen discovering the uselessness of a navy that had assured shipmasters of safe passage, arrived in Cape May, and at eleven in the evening took control of the local telephone switchboard and put the whole town under martial law. No further news concerning *Pratt* entered newspapers except through navy censors. But rumor spread fast.

U-151, and perhaps especially its mines, had sown confusion and terror everywhere along the Atlantic coast. On June 2 alone, Von Henchendorf had left 448 people aboard eighteen lifeboats, one of which, the beleaguered *Carolina* Number 5, rowed all the way to Atlantic City and landed amid a tumultuous welcome by parading, convention-happy Masons, thus sparking furious inquiries into government censorship. Another, *Carolina* Number 1, an engine-powered lifeboat, proved far less able, capsizing on the night of June 3 in a heavy squall and drowning thirteen of its occupants before it could be righted. And although the tanker *Pratt* sank in shallow water, many Americans assumed it had been torpedoed by a second submarine. Von Henchendorf monitored radio broadcasts identifying submarines from Cape Cod to Florida, and wondered himself if another U-boat had not arrived to wreak havoc. By that time United States warships and armed merchant vessels had begun firing at anything, especially small dark objects real and imagined, much of their frantic activity occurring so close inshore that coastal people from Massachusetts to Florida could see and hear it and report submarine attacks to an ever more frazzled federal government.

Van Henchendorf continued working south toward the Virginia Capes and quickly surprised a four-masted schooner. *Samuel G. Mengel,* en route from Africa to New York, carried no radio, and its master had no inkling of enemy submarines just off the Atlantic coast. In the evening haze, he mistook U-151 for a tugboat. Von Henchendorf's boarding officer arrived with scuttling charges, shook hands in the most formal manner possible, and when Captain H. T. Hansen argued that 125 miles from shore was a lonesome spot to launch a lifeboat, replied simply, "The weather is warm, the weather fine, and there are plenty of ships passing to pick you up."[7] All eleven schoonermen boarded a lifeboat, and next morning the Danish steamship *Paris* found them and brought them to New York.

Over and over, the captain of U-151 had to provide extra time for lifeboats to be launched, and sometimes too he had to provision lifeboats, or furnish their occupants with water, or provide medical aid to crewmen injured during launching, including those who had panicked and caused injury to themselves and to others. In several cases, U-151 radioed the position of lifeboats to the United States Navy, requesting that rescue vessels be sent, a measure of humanity that endangered the submarine.[8] While the cruise of U-151 offers an intriguing glimpse into the Great War moment when submarines warred on schooners and steamships alike, and when international law remained mute on unannounced attacks, it also provides a portal through which to study the launching of lifeboats, a portal opening on many other sinkings, in wartime and in peace.

Despite their supposed failings as deep-water seamen, coastal fishermen abandoned their schooners efficiently, without a trace of panic. On August 10, 1918, U-117 sank nine swordfish schooners on Georges Bank off Cape Cod. All day long, U-117 surprised, captured, and scuttled one schooner after another, its officers and men watching castaways in tiny open boats set sprit sails and lugsails and head off toward Nantucket Island and Cape Cod. Every single man arrived safely, including those who spent seventy-two hours making for Boston in a dory equipped with little food and water and no compass. No contemporaneous account of their small-boat seamanship makes the slightest mention of their skilled launching of dories.[9] Early twentieth-century fishermen still launched and retrieved dories daily, and everyone from journalists to historians of naval warfare accepted their expertise.

But students of Great War naval operations too easily discount the skill with which fishermen and coastal trade schoonermen abandoned ship when attacked by surprise, without the courtesies Von Henchendorf favored. Fishermen objected strongly to the navy sending armed decoy schooners onto the fishing grounds, fearing that submarine captains would begin attacking without warning. Five days after scattering the swordfish fleet, U-117 fulfilled their fears, opening fire without warning on the auxiliary schooner *Madrugada,* slamming round after round into the vessel. The crew abandoned ship in the midst of the attack and a few hours later found themselves rescued by a steamship.[10] Interpreting extant accounts of such sinkings proves difficult nowadays, partly owing to government censorship, but mostly because witnesses routinely expected superb seamanship from schoonermen. Yet experienced seamen offer some fragmentary insight into the harrowing experience.

"Nothing strikes panic into the heart more surely than the disarray of broken gear and a loose tangle of rope spread over the deck," writes Frank Mulville in *Schooner Integrity,* an inquiry into a shipwreck. "It is frightening when the order and logic of a boat are torn to pieces, suddenly and irrevo-cably and replaced by chaos."[11] But despite the order of things collapsing into chaos, the men of *Madrugada* got overboard in moments, minus extra clothing or provisions to be sure, but overboard and away from a vessel afire not in the way of *Hornet* in mid-Pacific, but in the twentieth-century way. One shell penetrated the fuel tank, blasting the oil through the hull and ig-niting it. Away rowed the crew in their old-fashioned boats, fleeing the twentieth-century horror of burning fuel oil.

One day later a mine sank the seven-thousand-ton British tanker *Mirlo*, a half mile off Wimble Shoal buoy, due east of Delaware, having first ignited its cargo of gasoline. What followed sickened even the Coast Guardsmen who rushed seaward in their engine-driven surfboat, having witnessed the preliminary explosion from their station. Three of the four lifeboats launched successfully, but the forward port lifeboat fouled its after falls and capsized in midair, throwing its occupants into the sea as instantly as had *Carolina* Number 5. Another lifeboat, in the charge of the boatswain, had almost failed to launch. "Even as we were lowering away hot flaming oil was already pouring from the leaking plates," recounted the boatswain. As he or-dered his men to pull inboard the men swimming around them, *Mirlo* ex-ploded in a candent moment, its flames roaring like those in a furnace, the rising gale driving a nightlike cloud of black smoke over the lifeboats, then sending long rivers of burning gasoline among the boats. Into the dark, flame-lit chaos probed the Coast Guard surfboat, its crew first finding vic-tims of the failed launch of the forward portside lifeboat—some men hideously burned but alive, others charred corpses—and in the midst of the holocaust, three lifeboats creeping away from death.[12]

But once afloat on the flaming, gale-tossed ocean, the lifeboats en-dured—filled to capacity or beyond, half-swamped, gunwales level with the sea, even the bosun's, which had stopped to take aboard burned men from the capsized lifeboat. If nothing else, the sinking of *Mirlo* demonstrated that the traditional ship's lifeboat functioned well in the most modern of calami-ties, the petroleum-fueled inferno. But also it demonstrated, once again, the absolute necessity of launching traditional lifeboats properly.

Conrad and other thoughtful observers trained in sail knew by the first years of the twentieth century that the very scale of steamships had produced worrisome problems ignored by naval architects. In his 1910 analysis of the last lugsail-rigged fishing boats of England, Stephen Reynolds juxtaposed several small fishing boats moving across a small bay with Royal Navy war-ships. "With their dark, bellying lugsails of an ancient cut, their beaminess, their high freeboard and their black and leaden paint, they looked like craft not so much from another fishing village as from a bygone age," a flotilla decidedly unlike "that vast floating machine-shop, the Navy."[13] Young men raised from boyhood aboard the lugsail-rigged fishing boats might spend years, even decades, in the Royal Navy aboard some dreadnought or destroyer, but they would always know how to sail a lugsail-rigged lifeboat.

Until the steamship era, launching ship's boats involved rigging lifts, something so traditional that few authors of seamanship manuals commented on it. Together, the masts and spars of any sailing vessel, even a lugsail-rigged lifeboat, constitute a lift or lifts already rigged, ready to raise or lower sail. On land, every crane resembles a sailing ship's mast and spars. Bits and pieces of terminology, especially the word *boom*, hint at distant common origins, but words do little to make onlookers imagine the masts and spars of a sailing vessel moving something other than sails. Yet the rigging of a square-rigged or schooner-rigged vessel can serve not only to load and unload cargo but to pluck a small boat from the deck, move it sideways, and lower it overboard.

"She is a small boat and not heavy, and you and I will cant her on to her bilge with handspikes, then I'll hook that watch-tackle round the foremost thwart and take the hauling part to the winch, and rouse her along to abreast of the gangway," a competent mate tells a woman aboard a sailing ship abandoned by everyone else. "'That gangway there unships, and we sit low upon the sea, and we'll tumble the boat through the gangway overboard, smack-fashion. If she proves too heavy we'll rig out a spar'—here he cast his eyes round—'with the watch-tackle made fast to her, and the winch will do the rest.'" Like so much turn-of-the-century sea fiction involving a strong man and a strong woman confronting danger, W. Clark Russell's *The Mate of the Good Ship York; or, The Ship's Adventure* emphasizes physical problems and physical opportunities, and presumes readers somewhat familiar with heavy manual effort, particularly that involving some risk of getting hurt. In the end, even readers unfamiliar with sailing-ship rigging encounter a detailed analysis of launching a small boat: "So he carried the block of the watch-tackle as far forward as its length would allow him and made a strop with a piece of gear round the thwart, to which he hooked the other block, bent a line on to the hauling part and carried it to the winch, giving Julia the job of hauling the slack in as he wound." In the end, they raise and swing the boat, lower it onto skids made of capstan bars, and after fastening a painter to it, run it bow-first through the gangway, where it lands perfectly, without shipping a drop of water. But of course the boat does not fall very far below the deck level of the waterlogged brig, and indeed the swell brings the sea momentarily level with the deck.[14] Unlike Stacpoole's castaway couple, Russell's two survivors aren't forced to kill hostile sealers, yet they accomplish feats involving great courage, skill, and strength. But launching the brig's boat—something Russell details accurately—is scarcely one of them. All the two characters do is lift the boat a few inches off its chocks, drag it forward using tackle, then swing it sideways by main force, then shove it forward and overboard. Only briefly does the boat hang suspended a few inches above the deck, and in the end the two launch it from a point only a foot or so above the sea.

In ordinary conditions, and in many emergencies, sailing-ship crews launched boats differently, swaying them up five feet or so and over bulwarks beneath a spar, then lowering them, something demanding four or five men experienced in the ways of rigging. By the end of the century, sailing ships carried boats specifically designed and equipped as lifeboats, often as quarter boats, one on either side of the stern, slung on davits outside the mizzenmast rigging. Such boats, Conrad recalled late in life, hung well above the sea but secured only by a line and toggle, ready for instant use. Only once did his ship *Torrens* launch one of the boats in haste, to recover a parrot that had flown overboard and landed in the sea. "I don't know how long all this took," Conrad wrote (he had been belowdecks at the start of the rescue), "but the parrot survived the experience; so we must have been quick enough to have saved a child, for instance, of which we always had several on board."[15]

For general crises *Torrens* had "two very roomy lifeboats" fitted out and provisioned to abandon ship on a moment's notice. Every morning, Conrad required the ship's carpenter to report "Davits and bolts free," four words that let him know that with one bolt knocked free, each boat need be raised only three inches to be swung out. The crew of *Torrens* grew accustomed to a routine test Conrad usually conducted at eight in the morning, as the watches changed, when he bellowed, "Both watches. Out lifeboats," four more words that in time made his crew able to swing out the lifeboats in seven minutes. This effort satisfied him that *Torrens* could protect itself against everything but sudden disaster, but in fog he ordered the boats swung out as a further precaution.[16] Not every master mariner worried as much as Conrad did, but the *Torrens*'s precautions were not unknown aboard other ships, especially in the crowded English Channel.

"Sudden disaster" denotes a remarkably small range of catastrophes likely to befall a nineteenth-century sailing ship. Gales and hurricanes precluded launching any sort of boat at all, and freak weather phenomena, white squalls especially, destroyed ships so quickly that no one moved toward boats, let alone began preparing them for use. Grounding frequently mandated that everyone stayed aboard ship until the surf subsided, or swam ashore grasping wreckage for flotation. Sudden, unexpected, and substantial failure of some part of the ship—perhaps losing yards aloft owing to a hidden defect in a mast, or taking on tons of water when a plank butt sprang free of its frame—prompted immediate repair efforts that usually lasted long enough for shipmasters to weigh the merits of releasing some crew members to prepare boats. A concatenation of smaller disasters, like that which struck *Serica* in 1868 and damaged all its boats, often led to abandoning ship, but even Cubbins had several hours to prepare himself for losing his vessel. Shipwreck history demonstrates that until World War I remarkably few sailing ships foundered instantly or even within seven minutes, although the thousands that went missing without trace may well form a wholly different cohort.

Fire and collision, often in combination, reshaped shipwreck after the middle of the nineteenth century, as a handful of observers noted well before 1900. Petroleum products increased the risks run in earlier days by ships carrying turpentine and other flammable liquids, especially whale oil. Kerosene caused many fires like the one that destroyed *Hornet* so suddenly that two of its three boats—the two that never reached safety—went into the sea damaged, apparently by falling spars freed by burning rigging. On a dry, hot night, especially in the tropics, a sailing ship moved as a sort of tinderbox, all its standing rigging covered with tar, its deck seams caulked with tar, and much of its dry woodwork heavily coated in either oil-based paint or linseed oil mixed with tar. Below in the cabin or cabins, passengers and officers relaxed under lamps burning whale oil or kerosene, while forward in the fo'c'sle the off-watch crew enjoyed their single light and their fug of tobacco smoke. On deck, in the caboose or cookhouse, the cook tended his coal-fired stove, perhaps baking bread for breakfast or doing passenger laundry. On either side of the forward standing rigging hung lanterns, a green one to starboard, a red one to port, each burning oil, and sometimes a white one hung on the stern as well. In the last decades of the nineteenth century, in the years before oil and gasoline moved across the sea in bulk tankers, holds contained tons of kerosene or old-style, highly flammable gasoline in five-gallon cans that manifests listed as "case oil." Stately, beautiful, and illuminated, nineteenth-century clipper ships and whalers, brigs and schooners sailed more and more under the threat of fire fueled by liquid spilling from a shattered lamp, then igniting dry, flammable material like tarred rigging, then exploding petroleum cargo.

Once rigging caught fire, launching almost any sort of boat—especially the ordinary large ship's boat usually carried atop cabins—proved almost impossible. Without lines running aloft through blocks, boats could at best be manhandled over the side, usually suffering damage, but even before running rigging burned away, the whole mast-and-spar crane frequently collapsed as standing rigging disintegrated. Explosive fire meant not only imminent disaster but the near impossibility of escape in boats, as everyone aboard the Boston packet ship *Ocean Monarch* learned a few hours after it sailed from Liverpool in 1848. Some claimed that one of the steerage passengers built a fire in a ventilator, but several passengers asserted that stewards stowing away liquor dropped a lantern that broke several bottles of spirits. Whatever the cause, three or four minutes after the first alarm the whole stern of the vessel burst into flames whipped up by a stiff breeze. *Ocean Monarch* carried some 380 people, and within twenty minutes, as the flames destroyed first the tarred rigging supporting the mizzenmast and then the mainmast, most of the passengers retreated to the forward end of the ship, even crowding out onto the jibboom, which collapsed into the sea as the foremast went over the side. Despite a well-nigh immediate order to get all the boats over the side, only two were launched before flames reached the others, and the remaining passengers, seeing the boats alight, stampeded over the side or flung themselves forward. In the end, 150 people lost their lives.[17] The burning of *Ocean Monarch* not only indicates the difficulties involved in launching boats using spars as cranes but also demonstrates how the destruction of boats precipitates panic among passengers.

Steam-powered propulsion only increased the danger of fire and produced disasters worse than that of *Ocean Monarch*. In the beginning, steamships carried masts, spars, and sails, both for use in emergencies and to conserve fuel. The half-steamer, half-sailing vessel worried officers, crew, and passengers, since the funnel directed smoke and sparks into the rigging and sometimes against sails. As steam propulsion proved itself, however, fears grew about the danger of spontaneous combustion in coal bunkers. Only slowly, as the century advanced, did another danger manifest itself: the proliferation of lamps aboard ever larger ships. For sixty years following the development of oceangoing steam engines, illumination came from lamps burning whale oil, then kerosene, then a more volatile fuel called lamp oil, a cousin to gasoline. A small freighter might have only five or six lamps lit at night, but a liner might well have a hundred times as many, not only one in each cabin but dozens lighting saloons, passageways, promenade decks, and other public spaces, and others lighting engine rooms, galleys, and storerooms. After the turn-of-the-century introduction of marine dynamos, traditional mariners like Conrad stood startled at the brightness of the illumination. Suddenly ships, especially passenger liners, moved through the ocean night as great concatenations of electric light, and the danger of fire receded.

As Conrad mused in his 1919 *The North Sea on the Edge of War* of sail-less steamers carrying "tame lightning" at sunset, "they spangled the night with the cheap, electric, shop-glitter, here, there, and everywhere, as of some High Street broken up and washed out to sea."[18] At least some of Conrad's magisterial understanding of light and shadow, which makes books like his *Heart of Darkness* so valuable today, originates in his living through the transition, not just from sail to steam, but from petroleum-fueled marine lighting to far safer electric marine lighting. He understood the old dark of the sea and the danger of petroleum fire, and he grasped the deeper significances of the new nighttime illumination. Electric lighting might dazzle the eyes of lookouts scanning ahead for anything from approaching ships to lifeboats to icebergs. And when it became ubiquitous, its unexpected absence might make emergency activities, especially swinging out and lowering lifeboats, more difficult. How would steamship seamen perform when the lights went out and the bulk petroleum exploded?

Robert Bennett Forbes answered the question in 1889, in a fierce compilation and analysis of shipwrecks entitled *Notes on Some Few of the Wrecks and Rescues during the Present Century*. Forbes held opinions contrary to those of many of his contemporaries, especially his fellow shipping magnates, and he had no fear of publishing his thoughts in newspapers and elsewhere. In a February 28, 1882, letter to the *Boston Journal*, he insisted that African Americans had the character and discipline to serve as officers aboard merchant ships, pointing out that "a full-blooded African" served as first officer aboard the ship *Washington*. He deplored the rising secularism that caused Mark Twain to deemphasize the religious faith of the *Hornet* men, and he championed the character and expertise of sailing-ship men as something that "should be kept before our young seamen as examples worthy of following." In his book, building on a list of causes of shipwreck begun by a friend of his on Nantucket, he tongue-lashed everyone from shipowners to merchant-ship officers to passengers for assuming that the nineteenth century provided anything approaching progress in seafaring. A century earlier, "ships were not abandoned when they met with grave accidents," he wrote of what might be his fundamental discovery among the patterns he discerned in hundreds upon hundreds of shipwrecks.

Steamships not only suffered hideous wrecks, they foundered from new causes like sudden conflagrations and boiler explosions, and perhaps even from sheer size. According to Forbes (who knew only steamships still carrying masts and sails), their design and crewing combined to make them unsafe *after* disaster. Published well before the perfection of marine electric-lighting dynamos, Forbes's book implies that steamship officers and crew could not manage the technologies they served, and that what had been repairable damage in 1840 had become by 1880 damage requiring the launching of lifeboats, if anyone could indeed launch lifeboats.

Forbes had before him the example of *Arctic*. When that steamship foundered in the North Atlantic in 1854, killing many prominent people, circumstances surrounding its sinking caused Americans and Britons to recoil in shock and revulsion, emotions that resurfaced two years later following the mysterious disappearance of *Pacific,* sister ship to *Arctic.* The loss of two liners destroyed the Collins Line and combined with the success of the Confederate navy to destroy United States mastery of merchant shipping for a half century.

Tradition and modernity collided in the North Atlantic on a foggy September morning in 1854 when the iron-hulled, steam-driven French bark *Vesta* slammed into the wood-hull steamship *Arctic.* Despite the fog, both vessels were moving at full speed, *Vesta* under all sail as well as full steam, and neither sounded a foghorn. While the collision barely shook the massive *Arctic,* it devastated *Vesta,* toppling its forward spars and rigging, destroying its bowsprit, and ripping off a full ten feet of its bows. At once panic swept the bark, and its crew of furloughed Grand Banks fishermen immediately launched the two lifeboats, capsizing one and drowning some of its occupants, and making off with the other toward the liner stopped in the fog. While the master of *Vesta* struggled to restore order, rescue the drowning, and cut away the mass of wreckage forward, *Arctic* moved slowly into view, then suddenly steamed away, causing many *Vesta* passengers to despair.

Yet *Vesta* had a good chance of survival. Not only had its builders divided the hull into watertight compartments, they had installed a collision bulkhead far forward, and that bulkhead held, keeping water from the rest of the ship. *Arctic,* however, had been built for speed, with a wooden hull open from bow to stern. Water rushed in from the gash in the starboard side at the rate of about a thousand gallons a second. The captain, James Luce, realized nothing could keep his liner afloat for long, and while his men tried emergency repairs from inboard and then attempted to slide sails overboard to fother the gash, he determined to race for Newfoundland, some fifty miles away. So suddenly did *Arctic* charge ahead that it left behind its first officer and several crewmen, who had launched a lifeboat to go to the aid of *Vesta*.

Almost immediately, everyone aboard *Arctic* realized they had needed that lifeboat, one of six high-technology ones built of corrugated, galvanized iron plating, and they knew too that they could have used the lifeboat launched in panic from *Vesta*, which *Arctic* had accidentally rammed and crushed a moment after starting for Newfoundland. As of April 30, 1852, the United States government required all passenger vessels to have aboard at least one metal lifeboat, a change in law resulting from the growing number of steamship fires. *Arctic* carried six, each built according to a patent awarded to Joseph Francis. By 1854 the Francis iron lifeboat had a place aboard most first-class American-flag steamers, and its simple design had become well known to passengers. Built from two large pieces of sheet iron formed into half-hull shapes and then riveted together, each lifeboat came equipped with air tanks at both ends that made it essentially unsinkable. *Arctic* originally carried six Francis lifeboats, four thirty-foot ones and two slightly shorter ones, not nearly enough for everyone and not exactly suited for the North Atlantic.

While his liner labored toward Newfoundland, Captain Luce ordered the five remaining lifeboats to be made ready for launching and as many women and children placed in them as possible before lowering. Panic then burst on deck from below as groups of firemen, forced from the stokehold by inrushing water, charged the boats. They and some male passengers mobbed one boat, which quickly fell astern, carrying perhaps twenty-eight people. Firemen rushed the next boat too, already loaded with twelve women and several male passengers, somehow snapping the forward falls, suspending the boat vertically, and flinging most of its occupants into the sea. The next boat, properly launched by the second officer, quickly filled in a horrible way with men jumping down to it, maiming its occupants. Confusion followed calamity as more and more men got drunk, as women and children found themselves thrust aside, as the chief engineer and his officers quietly launched the foremost lifeboat, having fully provisioned it and chosen to take no passengers. Even the final boat, alongside a raft being made of spars, departed only after being mobbed by crewmen, something Captain Luce hoped to prevent by stripping the boat of oars and thole pins. "It was every man for himself. No more attention was paid to the captain than to any other man on board," one fireman said a few days later to a New York journalist; "Life was as sweet to us as to others." But life ended for most of the 250 passengers left behind aboard *Arctic*. The London *Times* concluded that the crew members' desertion drowned the passengers, a denunciation of national character Americans found bitter indeed.[19] The handful of *Arctic* survivors praised the iron lifeboats but condemned the craven crew that had abandoned women and children. Overnight, American opinion turned against the male survivors, passengers as well as crew, who had saved themselves at the expense of women and children.

Recrimination began instantly and endured for decades. New York diarist George Templeton Strong traced its beginnings to the shock of hearing early morning newsboy cries and recorded its growing fervor day after day as survivor lists made clear the disproportionate number of crewmen saved. "To be sure, self-preservation is the common instinct of waiter and fireman, Wall Street broker and Fifth Avenue millionaire, and the strong-armed sailors and engineers can't be blamed perhaps for acting on their earlier information of the extent of the peril and using their superior strength to fill the boats and save themselves," he wrote on October 11, before remarking that Captain Luce must have been dealing almost single-handedly with "a crowd of terrified passengers; and worse, with the firemen and coal-passers, men with whom the officers are ordinarily not much in contact, and who swarmed up from their murky depths below intent on availing themselves of any chance of escape."[20] His diary entries reveal Strong's view of the firemen and coal passers as somehow less manly than ordinary seamen and less human too, relate what he learned of how the firemen prematurely abandoned their stations, and juxtapose such men against the women passengers, who relied on them for help and perished.

Strong understood the deeper social implications of cowardice among the first families of New York. "I shouldn't like to be Gilbert, or any of those men whose strength or coolness enured to their own benefit in the scramble for life and who secured themselves places in the boats," he wrote before admitting that he might have acted as they had—and adding that if he had and had reached New York, he would have had to run away or die of nervous fever. In subsequent entries, Strong analyzed the aftermath of *Arctic* foundering, noting the bitterness with which people spoke of the survivors and remarking that newspaper statements given by male survivors about their failure to rescue women resembled apologies by men who had not given up seats on a crowded omnibus. Upper-class men had behaved as badly as the "dastardly" firemen and coal passers, and in the end Strong condemned Captain Luce for losing control of his officers and crew.[21] The diarist confiding to himself presaged public opinion in newspapers and magazines. Luce had lost command over his crew, opined *Harper's Monthly*

Magazine a month after the disaster, "most of whom indeed abandoned their duty, seized the boats, and sought to save themselves, regardless of others."[22]

Yet *Harper's Monthly* had certainly contributed to the luxury-over-safety fad it suddenly condemned, for barely two years earlier it had published a lengthy feature article on a fast and luxurious *Arctic* passage. In "Ocean Life," in the June 1852 issue, John S. C. Abbott waxed enthusiastic about everything aboard the "majestic steamship," from the patent saloon lamps to the "very highest style of modern art" decorating a saloon of "almost Oriental magnificence" to the immense funnel, declaring that "never did there float upon the ocean a more magnificent palace than that which now bears us." Its 265–foot length was that of "four ordinary country churches," while the promenade deck was "as high above the water as the ridge-pole of an ordinary two-story house." The 135–man crew boasted a vegetable cook, a pastry cook, a baker, a butcher, and other specialists focused on passenger pleasure, as well as fifty-two men far below deck, tending the eight furnaces consuming eighty tons of coal a day. Despite foul weather throughout the entire passage, Abbott recorded his thrill at arriving in Liverpool, climbing down a long ladder to the waiting steam tender, and moving away from "one of the noblest of ships," the high-speed liner *Arctic.*[23]

The foundering of *Arctic* convinced many shipbuilders that watertight compartmentalization and collision bulkheads—such as were available on *Vesta*—offered a high degree of safety in collision, but few professional seamen inquired into the gross cowardice and selfishness of *Arctic* crewmen. Certainly *Griffiths and Bates Nautical Magazine,* in an 1854 article entitled "Ship-Masters and Mariners," remarked that the rapid growth of the American merchant marine—and the coming of steamships—had produced a shortage of trained seamen, but it did not link that shortage to the *Arctic* calamity.[24] It did, however, advocate changes to regulations governing numbers of lifeboats aboard large steamships. The editors suggested that governments should mandate lifeboats for everyone aboard, and urged that *Nautical Magazine* readers petition Congress to enact regulations far more strict than British ones.[25]

Just as no airline passenger expects a parachute, so nineteenth-century passengers accepted the lack of lifeboats for everyone, placing implicit faith in the Almighty and competent officers.[26] The common assumption was that officers and crewmen would save passengers, particularly women and children, following the example of the officers, crew, and soldiers of the British troopship *Birkenhead,* who in 1852 set off every woman and child first when their vessel struck a reef off Africa and sank within minutes. But after *Arctic* foundered, at least one faction of mariners questioned the idealistic notion that steamship crewmen would behave as nobly as sailing-ship men or British soldiers. Passengers would have to save themselves, or at least help to do so.

So argued Commodore Matthew Peary in March of 1855, in a brief but detailed *Hunt's Merchants' Magazine* article that used as an example *Baltic*, another Collins liner only slightly smaller than *Arctic*. Unlike commentators who suggested improvements in steamship construction after *Arctic* sank, Peary grappled with crew behavior during and immediately following collision or other disaster. The naval officer emphasized the obvious: the "nautical portion" of the liner crew consisted "of one captain, six mates, four quartermasters, twenty seamen, and two boys" in addition to the purser and surgeon; the rest of the crew served in the engine-room and steward departments. Since the liner frequently sailed with at least 385 people aboard, Peary suggested that it ought to carry ten lifeboats, at least six in davits, and all provisioned for ten days at the start of the passage. Some crewmen from every department should be detailed to the same lifeboat station at which they were stationed on earlier trips.

Male passengers should be detailed to certain responsibilities at particular lifeboat stations too, Peary argued, shifting the focus of the entire survival-after-shipwreck debate. Male passengers should caucus and choose a handful of "the most active" to help launch and crew the lifeboats, and that handful must train every day. Ten lifeboats, especially with six carried in davits, would ensure the safety of about three hundred passengers, leaving the crew to survive aboard rafts made from rigging. No longer would passengers expect to be rescued by crewmen.

Peary's "Notes in Regard to Safety on Steamers at Sea" appeared not only in *Hunt's Merchants' Magazine* but in *Debow's Review*, and it marks a turning point in the thinking of American businessmen, especially shipowners. Very subtly, the article stated what passengers like Abbott missed. Only a third, perhaps even as few as a fourth, of a liner crew knew anything much of seamanship: the bulk of the crew served the furnaces and engines, or else worked as stewards or bakers or porters. In any disaster, but perhaps especially in a collision, the deck officers and seamen would be almost wholly involved in damage control, while the engineering department would tend engines, start pumps, and manage such technological niceties as using bilge injectors to pump seawater directly into the boilers and convert it to steam. With the ship's surgeon attending to the injured, passengers might look only to the catering staff for assistance, something Peary knew boded ill. While he forthrightly admitted that his scheme meant that liners would still not carry enough boats for everyone and would carry some boats not in davits but in locations requiring seamen for launching, his article makes it clear that ever larger steamships would carry no higher proportion of seamen to other crew and that passengers must be able to help launch lifeboats and perhaps launch them unaided.

In the middle decades of the nineteenth century, roughly thirty years following the burning of *Ocean Monarch*, a current of moral uncertainty disturbed shipowners, governments, and the public—especially the seafaring public. The increasing size of steamships, the abandonment of auxiliary masts and spars useful in swaying out lifeboats stored amidships, and the sheer technological difficulties of providing enough easily launched boats for everyone aboard every ship swept away traditional assurance. Peary, Forbes, and others understood the *Arctic* calamity as the precursor to others, but no one knew how to fit steamships with lifeboats for everyone.

Governments and shipowners addressed the problem by assuming that most calamities happened in the presence of other ships. The lifeboats of the stricken vessel might make several trips to safety, and those of nearby ships might aid in rescue. In retrospect, *Vesta* could have carried every single occupant of *Arctic* to safety, but in the minutes immediately after collision Captain Luce did not know the condition of the ship that had struck his. The first officer and his men dispatched in the first lifeboat launched from *Arctic* never reached *Vesta*, and thus could not report back that the bark was far less damaged than it looked and offered safety for all aboard the liner. Indeed no one knows what happened to the men sent from *Arctic*, although two months later a schooner brought the lifeboat itself into St. John's, Newfoundland, all of its oars intact but its crew vanished.[27] But even if *Arctic* had stayed close to *Vesta*, it carried far too few seamen to simultaneously control hull damage, work the ship, and launch lifeboats, especially those stowed along the midship line. Aboard the davit-equipped *Arctic*, launching lifeboats proved far more cumbersome than it did aboard an ordinary sailing ship like *Torrens* years later. *Arctic*, after all, boasted a promenade deck almost three stories above the sea.

Inventors insisted that technology would soon overcome ignorance, clumsiness, and even panic, but a procession of subsequent disasters did little to reassure the public. When the steamship *Canadian* struck an underwater iceberg spur off Newfoundland in 1861, its watertight compartments proved useless, as did its "Clifford-patent" lifeboat-launching apparatus. Passengers surging around the apparatus blocked its use, and in a private letter made public in *Nautical Magazine* a few months later, one man explained how he followed his wife and daughter into a lifeboat against the captain's orders. "I can assure you, in moments like these, everyone looks to himself," he wrote, and offered the following proof: a moment later the captain boarded the lifeboat and ordered it lowered.[28] No one seems to have noticed the failings of the patented equipment. And inventors kept producing more gadgets.

After the Civil War so many American inventors boasted of their novel devices that a Treasury Department commission conducted a mass test of hundreds of them in New York Harbor, finding a great many useless, some downright dangerous, others practicable, and a handful of real use. Inventors demonstrated sail-rigged life rafts, lifeboats modeled on British shore-launched rescue boats, and even cork-filled mattresses intended to be used as personal flotation devices. Several offered improved devices for launching lifeboats, especially from slowly moving ships or ships in far from upright positions. About fifty provided either designs or working models of devices intended to disengage lifeboats from davit falls. Of the fourteen tested, the commission recommended only six "as being the simplest in design, strongest in construction, most trustworthy in operation, and most likely to withstand rough usage and the ordinary wear and tear of sea service."[29] But few inventions adequately addressed the complexities involved in launching lifeboats from very large ships, and some of the inventions produced serious problems after installation.

Even in a flat calm, launching lifeboats from decks towering up to six stories above the sea taxed the seamanship of many steamship crew members. If a ship listed to port, the portside lifeboats hung far out from their davits, suspended by lines running through blocks, forcing passengers to leap across a gap or be handed across by crewmen. At the same time, the port list jammed the starboard boats against the hull, causing them to scrape their way down the side of the ship during lowering and sometimes even to catch on a protrusion and overturn. When a ship settled by the bow or the stern, lifeboats hung angled in their davits, and while crewmen lowering the boats could level them through judicious use of the lines running from davits to bow and stern, the leveling required even more care than ordinary lowering, since the men handling each line had to lower it at precisely the same rate to keep from spilling the occupants into the sea. Clearly much of the safety problem lay with the traditional upside-down-J–shaped davits that swung lifeboats outboard and then lowered them on block-and-tackle-rigged falls.

At the turn of the century, a British inventor transformed lifeboat launching, winning worldwide acclaim even before his davits worked so well aboard *Titanic* in 1912. Axel Welin's "quadrant davit" used a worm-gear drive to lift a lifeboat from its chocks, swing it outboard for loading, then lower it quickly, eliminating much of the high-skill, brute-strength work older

After *Titanic* sank, many steamship companies, perhaps especially German ones, expanded lifeboat drills to include adding of extra provisions to lifeboats swung out under J-shaped davits, as J. Bernard Walker explained in *An Unsinkable Titanic*. (From Walker, *An Unsinkable Titanic*)

davits required. While a separate davit controlled each end of the lifeboat to ensure level lowering from ships down at head or stern, the Welin davit made possible the launching of a lifeboat in about three minutes.[30]

By 1910, when he had sold some two thousand sets of his davits, Welin knew they worked in emergencies, and he had the perfect example in the sinking of *Star of Japan*. When that steamer foundered off Africa, Welin davits carried half its lifeboats; round-bar J-shaped davits carried the others. Only the Welin davits worked. The lifeboats under them saved everyone aboard the freighter, giving Welin and shipowners a near perfect argument for the mid-weight, extremely reliable innovation that enabled lifeboats to be stacked, almost like dories aboard a fishing schooner.[31]

Stacking lifeboats under Welin davits not only increased the number of lifeboats a ship might carry but increased the promenade deck space available, something that mattered more to passengers—and steamship owners—than safety. After 1910 the British Board of Trade permitted a fourth of the lifeboats aboard a passenger steamship to be carried inboard of Welin davits, something that reduced edge-of-hull topside weight and incidentally saved considerable money, since a pair of Welin davits might serve at least two lifeboats.[32] *Titanic*, however, unaccountably carried only sixteen lifeboats, not the thirty-two its Welin davits could manage. White Star Line, the owners of *Titanic*, recognized and catered to steamship passengers' preference for palatial accommodations—and broad promenade decks—over adequate lifesaving equipment. Since in the twenty years before 1912 some thirty thousand Atlantic crossings had been made with the loss of only 150 lives, the decision to eschew sixteen lifeboats seems understandable, if subsequently wrong.[33]

Though operated by only a half dozen seamen, Welin davits worked almost perfectly aboard *Titanic*, as Third Officer Herbert Pitman subsequently testified before a congressional committee. "I thought what a jolly fine idea they were, because with the old-fashioned davits it would require about a dozen men to lift her, a dozen men at each end."[34] Pitman ordered his boat lowered level with the deck and quickly loaded it with women and children.

Not every launch went so smoothly, at least in the memory of other eyewitnesses. Fifth Officer Harold Lowe suffered much interference from Bruce Ismay, president of the White Star Line, finally telling him to get out of the way. Worried as he was about lowering the lifeboat levelly, and harassed by Ismay, Lowe refused to lower the lifeboat fully loaded with passengers, something that greatly interested the senators conducting the inquiry. Why could not a lifeboat certified to carry sixty-five people be lowered with more than fifty aboard, especially with new davits and falls? Lowe responded that he thought the lifeboat might break in half from the weight of a full complement of occupants. In his mind, and apparently in the minds of other *Titanic* officers, certifying a lifeboat to carry sixty-five people referred to the boat in the water, not hanging from davit falls. Although the Welin davits and falls could easily lower a much larger lifeboat than any *Titanic* carried, even one fully loaded, as Welin proved in publicity photographs, aboard *Titanic* no officer—except perhaps Pitman—believed the wooden lifeboats capable of carrying sixty-five people while suspended from falls.

Titanic officers assumed that some passengers (in Peary's terms, "active male passengers") would slide down the falls hand-over-hand to lifeboats afloat just under the great blocks. In almost no case did that happen. One yachtsman did so at the order of an officer concerned that a boat had aboard it almost no men at all. It seemed that *Titanic* carried very few active male passengers, at least in first class. Fifth Officer Lowe, off-watch when *Titanic* struck the iceberg, woke to voices outside his cabin, dressed, went on deck for a look around, and immediately returned for his revolver, explaining at the Senate inquiry, "You never know when you will need it."[35] He soon learned.

According to a Canadian survivor, Arthur Peuchen, the crew of *Titanic* performed badly after the collision. Not only did the ship seem short of seamen to lower lifeboats, Peuchen testified, but the appearance on deck of about a hundred stokers, each with a duffel bag, crowding the lifeboats astonished him. "One of the officers—I do not know which one, but a very powerful one—came along and drove those men right off that deck," he concluded.[36] Officer Lowe never explained why he immediately grabbed his revolver, or even suggested why the officers of a liner like *Titanic* might need sidearms, but he may have had an inkling that the "watch below," the so-called black gang, might rise from its engine room, stokehold, and coal bunkers and overwhelm passengers and seamen and officers launching lifeboats.

Firemen and other below-deck crewmen often acted as they did because they knew more about underwater damage than deck officers. They saw steel plates buckle, seawater blast up from bilges, watertight bulkheads glow dull red with heat from hold fires. By 1912 they had become such specialists that they knew almost nothing of the sea and seamanship, but only the care and repair of engines, boilers, fireboxes, and other machinery. "No matter how clever our designers, behind the sleek brilliance of the engine rooms were always the sinister appurtenances of ash-pits, fire bars, firebrick structures, ash-hoists, ash-shoots, wheelbarrows, and the everlasting problem of trimming coal so that the fires should be fed and the ship's trim remain unaltered," wrote one steamship engineering officer, William McFee, in 1934 of his early years at sea. Coal trimmers loading wheelbarrows in near-dark bunkers, coal heavers flinging it into fireboxes, wipers and greasers and oilers hastening about, all merged into the half-man, half-ape of Eugene O'Neill's powerful 1922 Broadway play *The Hairy Ape*. Theatergoers saw on stage what lived far below the decks of sleek liners, in parts of ships passengers almost never reached. "A ship like the *Mauritania* or *Deutschland*, with perhaps thirty boilers of four fires each, would carry two hundred men on her boiler-room staff because the speed at which coal had to be brought to the plates for the men on watch involved a large number of coal-passers," McFee recalled. The typical stoker, he mused somewhat witheringly, had "the peasant's mentality, being often a miner or agricultural labourer who had drifted out to sea."[37]

When oil replaced coal in the first decades of the twentieth century, great numbers of coal-heaving firemen lost their jobs, but the new breed of firemen was, if anything, even less steady. "From the day he joins the new-style fireman is aware of an invisible peril, the peril of petroleum," observed McFee, "and it works an unconscious but indubitable change in his personality."[38] While crewmen knew fuel oil itself to be barely flammable, they feared pockets of gas forming beneath boilers and in corners. They worried not about the sort of spontaneous combustion that burned the ship Conrad remembered in "Youth" but about explosion and liquid fire, especially after collision. Far above his stokehold, deck officers and "real" sailor men wondered what this new-style fireman would do when disaster forced him above, into the open air.

Welin davits and other inventions that simplified and quickened lifeboat launching produced new problems—lowering lifeboats so quickly that no one noticed missing drain plugs or broken planking—even as they solved old ones, causing some experts to support round-bar davits into the 1920s. "To absolutely condemn the principle of the ordinary round-bar radial davit is not justified by general experience," argued naval architect Ernest W. Blocksidge in an exhaustive treatise he published in 1920 at the behest of the Board of Trade. For one thing, seamen around the world had an easygoing familiarity with the traditional davits; for another, champions of patent davits almost invariably had a strong financial interest in damning all davits but their own. Blocksidge emphasized that the Board of Trade had not settled on any particular patent davit as reliable enough to be required aboard British steamships, and that a handful of inventors had marketed modest devices that substantially improved radial-davit operation while not altering the davits so much that seamen would find them perplexing.[39] In the end, Blocksidge maintained, trained seamen free of passenger intervention ought to be able to use traditional davits as efficiently as patent davits likely to produce novel problems in unexpected circumstances. For Blocksidge, a half century of mechanical invention obscured a seventy-year-old deterioration of expertise that began with steamships.

Inexperienced crewmen such as stokers might begin launching a lifeboat without preparing the boat—say, by not inserting the drain plug or failing to realize that something had gone wrong with davits or falls. And panic sometimes originated in well-grounded fears about postcollision hull stability in ships constantly trimming and balancing coal. Shipowners and merchant-ship officers sometimes ignored lifeboats and lifeboat equipment, especially in the years around the turn of the century when designers struggling to build inexpensive freighters produced what mariners soon called "coffin ships," vessels so unstable that they capsized in storms and after collisions, often drowning men caught belowdecks with inadequate escape routes. "The hands below could not get out of their forecastles; from the flooding engine-room and stokehold, engineers, firemen and trimmers—caught like rats in a trap—struggled up iron ladders in a vain endeavor to reach the deck," observed one master mariner of a 1900-era freighter design whose flaws led to denunciations in Parliament, radical redesign, and greatly improved inspection systems.[40]

Between 1890 and about 1910, crewmen aboard coffin ships witnessed some unsettling sights, as when one master ordered a lifeboat lowered to run mooring ropes on a windy day in a Spanish port. The second mate, finding that the davit tackles did not lift the boat, thought that dried paint had stuck the boat fast in its chocks, according to Frank C. Hendry in 1938. The master ordered the mate to run the lifeboat tackle to a steam-driven winch aft, and when the winch drum revolved and lifted the lifeboat to the ends of the J-shaped davits, everyone watching the rising lifeboat saw daylight through it and realized that its keel remained in the chocks. Aboard that ship, which the master and every other officer had newly joined, every lifeboat stood nailed to its chocks, useless in an emergency but safe from chafing damage that might anger the penny-pinching owner of a heavily insured freighter worth more sunk than afloat.[41]

Such incidents slowly overwhelmed the stalwart opposition of many British shipowning corporations to new sorts of davits and other safety measures, including inspections. But experiences aboard steel coffin ships crewed chiefly by below-deck coal stokers and trimmers may explain why a hundred or so stokers arrived on the first-class deck after *Titanic* struck an iceberg. For all those men knew, the liner on its maiden voyage might well be one more coffin ship, so ill designed and built it might capsize before lifeboats could be launched. Aboard a coffin ship, stokers and most other "watch below" crewmen became passengers once water or smoke forced them above deck, and like fare-paying passengers, they expected seamen—and modern mechanisms like the Welin patent davit—to save them.

Improved davits alone did little to make abandoning ship efficient or safe. After the founderings of *Titanic* in 1912 and *Empress of Ireland* in 1914, the British Board of Trade strengthened its requirements for equipment and training, but aboard many ships, especially tramp steamers working ports as unvisited as any Steinbeck found in the Sea of Cortez, inspectors rarely appeared. Textbook authors preparing cadets and seamen for Board of Trade examinations, and especially for Lifeboat Efficiency certification, emphasized what one author noted in 1912: "Boats are occasionally neglected on board ship, and often when required in a hurry, some of the gear is found to be missing."[42]

The inability to launch lifeboats efficiently in part explains the fierce and enduring loathing of Britons for unrestricted submarine warfare of the sort that sank *Lusitania*. Convinced since the Napoleonic Wars of its mastery of the sea, by 1915 Britain knew it risked losing the war to U-boats, and in its shame tried to brand as criminals the Germans who seemed likely to conquer. Yet running through the voluminous British literature condemning submarine warfare are traces of fears of their own shortcomings. British seamen seemed unable to launch lifeboats properly. British stokers worried continuously about escaping deathtraplike working spaces, especially aboard ships built before about 1910. "Get on with your work or by the Lord Harry, I'll put you down in the bunkers, where you'll be shut up like rats in a trap if she holes us," one first officer shouted to a group of seamen watching a German submarine coming up far astern of their old freighter vibrating at full speed. But the coal trimmers were already in the bunkers, flinging down coal to the stokehold, and there they stayed until the submarine landed a shell in the engine room. "The engineers, firemen, and trimmers were streaming from the alleyway when we reached it," an officer-cadet remembered, and the "sailor men" readied the lifeboats. "The shriek of the tackle blocks piped the boats to the water," and the men of *Monarch* joined thousands of others torpedoed or sunk in the Great War.[43]

Monarch launched its boats smoothly in part because its aged master and first officer had trained in sail, and now and then did things that amazed and puzzled the younger crew, especially the cadets, including George H. Grant, who later wrote a book about his experiences on board the freighter. On one occasion, the master ordered tarpaulins hoisted on derrick booms to catch enough wind to head the ship free of breakers; on another the first officer took the cadets sailing and rowing in the gig, making them pull miles in a dead calm. The two men used the freighter's own boats whenever they could, and they exercised the crew in launching and pulling the boats as well. In the beginning, the cadets dismissed the old men as cantankerous if efficient, but gradually they learned how deeply they distrusted modern mechanism, although they said little about it. On one occasion the first mate, watching a sailing ship far off, remarked longingly, "I suppose you'd rather be where you are, like the rest o' them . . . sodying-mudgying and chipping rust . . . learning the jobs of a laborer instead of those of a sailor."[44] Neither the master nor the first mate of *Monarch* believed that steam properly trained apprentice officers, and as the tired vessel plowed around the world they put their cadets through all sorts of exercises from times past. Mechanism, especially high-speed mechanism, might cause trouble that more mechanism and more expertise with mechanism might not cure, something Grant realized only after mechanism destroyed his ship and the seamen of *Monarch* smoothly launched lifeboats, put out oars, and pulled toward the U-boat surfaced a few hundred yards away.

The U-boat captain commended their expertise and directed them onto a course for Ireland, twenty miles off. Grant recalled how quickly it became apparent that his ostensibly outdated training was paying off. "Our mast was stepped and the sail was hoisted and set to the wind that had freshened with the night, and soon the lifeboat was clipping along with a weight upon her rudder and an exhilarating surge of white water along her bilge," he wrote.[45] Within a few hours they would raise the Irish coast, arriving under sail in a wooden boat, safe after U-boat attack.

A few months after the *Monarch* incident, another U-boat torpedoed another ship within sight of the Irish coast, and the entire world learned what the *Monarch*'s crusty mate taught: mechanism sometimes failed. Despite 175 watertight compartments and lifeboats for a third more people than it carried, the 32,000-ton liner *Lusitania* sank only eighteen minutes after being hit by a single torpedo, 1,198 of its 1,959 complement of crew and passengers dying in a collision of modern mechanism with modern mechanism.

With its lifeboats swung out and most of its watertight doors closed, its steam turbines driving it at the reduced speed of eighteen knots, *Lusitania* had almost reached harbor when the torpedo exploded between a stokehold and the engine room, setting off a secondary explosion that remains a state secret today in both Britain and the United States.

The torpedo struck at a particularly vulnerable moment. Many of the seamen had gone below to begin the complicated process of bringing up passenger luggage; they were killed instantly or trapped by the lift fallen down its shaft. But *Lusitania* had sailed with far fewer seamen than usual, perhaps with only forty-one aboard, not even one for each lifeboat. While documents state that its officers had conducted lifeboat drills in New York Harbor, the documents appear to have been falsified. But even with all forty-one seamen alive, the explosion of at least ten tons of high explosive—and maybe several hundred times that amount—left passengers with far too little help in launching lifeboats. In later years, *Lusitania* officers and crewmen revealed information quashed at the inquiries: that the fire alarm indicators instantly showed fire in all the forward cargo holds at once, that the cargo spaces listed as empty on manifests were jammed full of mysterious cases, that one lifeboat after another jammed in *Lusitania*'s J-shaped davits. Damage extended far aft, as deck officers learned when the engineers could not reverse the engines or begin to control the list to starboard. For almost eighteen minutes after being hit, *Lusitania* plowed onward, its momentum overcoming even the thousands of tons of seawater flooding its interior after the secondary explosion.

The loss of dozens of experienced seamen, the almost instantaneous fifteen-degree list to starboard, and above all the continuing movement made lifeboat launching difficult beyond imagination. *Lusitania* lowered only six of its forty-eight lifeboats, despite having them already swung out on the davits in the event of attack.

All along the port side, lifeboats killed people by the hundreds. With the ship pitched down about forty-five degrees and listing perhaps twenty-five degrees to starboard, the port lifeboats hung inboard on their falls, secured in each case by a heavy chain to stop them swinging as the liner charged across the Atlantic. As stokers, stewards, and almost anyone but seamen knocked free the steel pins securing the chains, the boats swayed inboard into packed passengers, then—as inexperienced crewmen and passengers tried to release the winch locks—tore free of their davits and slid forward down the inclined deck, crushing more passengers. Number 2 boat swung in, slipped free, crushed the collapsible lifeboat stowed beneath it, then slid down the deck, carrying a mass of passengers before it and mangling the people against the bridge superstructure. Number 4 boat followed suit, crushing a second collapsible boat and then sweeping another group of passengers to their deaths in the wreckage of Number 2. Despite the best efforts of Third Officer Albert Bestic, by then crying in rage and frustration, Lifeboats 6, 8, and 10 duplicated exactly the actions of 2 and 4. Each of the five-ton lifeboats, intended to carry at least sixty people, apparently crushed at least that many, perhaps almost twice that many, as each hurtled free.

At Lifeboat Number 12 another officer got the boat away, but the list prevented anyone at the davits from seeing the boat descend. As it passed the deck below, passengers mobbed it, someone shouted "Lower away," and one of the men at the davits lost control of the falls and dropped its stern, spilling the passengers. A moment later the forward falls parted, and Number 12 crashed into the sea atop its former passengers. The men launching Number 14 lost control, and it fell atop the wreckage of Number 12, Number 16 fell directly into the sea and splintered, and Number 20 bounced down the side of the ship and splintered to pieces on the rivets. Hundreds of rivets, each about two inches square and each jutting an inch from the steel plates, clawed at the boat. Despite one seamen ordering passengers out of a rivet-snagged boat with an ax while he freed it, launching the portside boats proved next to impossible. On both sides, crewmen repeatedly lowered one set of lines faster than another, upending boats and spilling their occupants into the sea. Despite the master's orders to cease lowering until the liner stopped, seamen attempting to launch Number 18 at the direction of a pistol-armed passenger succeeded only in loosing it down the deck, where it crushed thirty passengers at least.[46] "I always thought a shipwreck was a well-organized affair," one woman shouted to another as third-class passengers mobbed a boat. "So did I, but I've learnt a devil of a lot in the last five minutes," replied her companion.[47] Most of their education took place on the starboard side of the liner.

The two women—Lady Mackworth and Dorothy Connor—watched in horror as steerage passengers mobbed the lifeboat in front of them, shoving each other aside as they scrambled into the boat, but already another group had surged up from belowdecks. Almost immediately after the torpedo exploded, coal dust–begrimed stokers and cooks streamed onto Second Deck, striking one passenger as "a wild lot," terrified and confused. The two women turned away from a pistol-waving passenger and the lower-deck mob and suddenly realized that they could see no children anywhere, although they knew some 150 infants and toddlers were aboard. But they could do little to alert anyone to what they discovered. Already male passengers had begun helping to launch lifeboats, some joining a group fending off a band of frenzied stokers shoving women aside next to a boat filled with women who then were dumped into the sea as the forward davit falls ran out too fast.

Lifeboat Number 1 went down properly, but with only two seamen aboard, as did Number 3, which seems to have struck Number 1 in a collision that convinced the two seamen aboard to join the two in Number 3. But no one detached Number 3 from its falls, and although passengers quickly filled the boat, soon everyone had to abandon it as the liner slipped deeper into the sea and dragged it under.

Starboard lifeboats hung out some ten feet from the side of the ship, making the snubbing chains bar taut, and no one knows whether davits failed or seamen failed to release the chains of Number 7 (Number 5 had been blown to pieces in the explosion). The first officer of the liner managed to lower Number 15 filled with eighty passengers, but Numbers 17 and 19 spilled their occupants when davit crews lost control. Again and again, the J-shaped davits impaled lifeboats caught under them or else snagged passengers and crewmen as *Lusitania* slipped further under the surface.[48]

Several stokers, including one who had survived the foundering of both *Titanic* and *Empress of Ireland,* were literally blown through escape hatches by tertiary explosions. "A stoker was reeling about as if drunk, his face a black and scarlet smear, the crown of his head torn open like a spongy, bloody pudding," remembered Oliver P. Bernard twenty years later.[49] So grimy and scalded that their very appearance terrified onlookers, at least some stokers mobbed starboard-side lifeboats. Many of the firemen boarded starboard-side lifeboats as they simply floated off the listing deck, and two women who had slid down lines and into one just-launched boat found themselves surrounded by firemen. Discovering that the stokers had begun rowing before inserting the drain plugs, the women began bailing with their shoes as they beseeched the men to aid the victims swimming all around the boat. To their astonishment, the men refused.

Vociferous complaints by United States survivors proved difficult for the British government to counter, let alone squelch. A subsequent Board of Trade inquiry was conducted in a way that ruthlessly suppressed important evidence, including an affidavit by Third Officer Bestic that the davits had not been tested in New York. But the British government learned not only that the American survivors would not be silenced but that they had noticed a great deal in far too much detail. Moreover, at least some of the survivors had enough yachting experience to provide expert testimony. Only the port and starboard lifeboats opposite the promenade-deck entrances had new tackle, asserted one passenger. "On a great many others," he noted, "the cables were old and frayed and the tackle blocks in bad condition, as was proved by the blocks sticking and the cables becoming lodged and holding up the lifeboats." Decades before he published his autobiography, Oliver Bernard told journalists bluntly that the tackle was stiff from paint and so complicated that only seamen could use it—and he saw almost no seamen. A third survivor, Michael G. Byrne, reported that "we were really led to slaughter. An officer ran about the decks, telling passengers there was no danger; that the ship would be beached, although several passengers questioned the statement, knowing the torpedo had struck near the engine room."[50] After American newspapers printed such statements, the behavior of both crewmen and officers came under penetrating scrutiny, as did the orders to avoid launching the boats until the ship slowed to a halt.

A number of survivors published lengthier accounts, some almost immediately, like *The Lusitania's Last Voyage,* a pithy narrative by Boston bookseller Charles F. Lauriat Jr. "What I would have given for one real sailor man for'ard; we could have saved that boatload of people," he wrote of his efforts with one starboard lifeboat. Instead of a "real sailor man," Lauriat had only a steward using a pocket knife to cut jammed lines. As Lauriat struggled forward to help the steward, the liner abruptly pitched downward and the after davit slammed into his back, nearly knocking him overboard. A moment later he was overboard and swimming, urging the occupants of the davit-impaled lifeboat to jump into the sea before *Lusitania* dragged them all under. Lauriat maintained that *Lusitania* officers ordered passengers to get out of the boats that were being lowered, that the collapsible boat to which he swam lacked the pin to hold its sides upright and the wood and iron braces to brace its sides, and that in the end passengers had to launch lifeboats almost unaided. Lauriat lambasted the "lack of discipline among the crew and the lack of expert knowledge as to the handling of the boats, knowledge that can come only to the well-trained crew."[51] He also attacked the British assertion that passenger panic hindered lowering the boats. As he pointed out, about half of the crew survived, but barely a third of the passengers. And along with many other American survivors, Lauriat asked why two lifeboats left the scene immediately.

In his autobiography Bernard described the few stewards bravely trying to help passengers launch boats and how, a few minutes later, the "cavernous vent of the funnel appeared like a railway tunnel into which the sea poured its triumphant way" as his own swamped lifeboat—half its occupants stokers and coal trimmers—floated beside the hulk. He recorded that the *Lusitania* radio officer noticed a nearly empty lifeboat pulling toward land, the gold braid of an officer glinting in the sunlight, and he detailed how Lauriat's bellowed complaints at British authorities caused many of the injured to be landed at once, straight past military officers. But above all, Bernard related what Lauriat and other Americans noted but could not have explained at the time, the immediate pulling away of two lifeboats.[52]

Almost certainly, the Admiralty knew why the boats fled the scene, and perhaps Lord Mersey, who presided over the Board of Trade inquiry, suspected. In his book Lauriat concluded that the Court of Inquiry was a sham that masked the deliberate destruction of the liner as a political ploy.[53] Unwittingly, Lauriat had put his finger on a high-level British scheme to lure the United States into the war by killing a great many of its citizens, and his book may have done much to keep the United States from immediately entering the war. After the inquiry, Lord Mersey resigned from his judicial post and later privately told his children, "The *Lusitania* case was a damned dirty business."[54] The lifeboats that left the half-sunk liner carried mostly crewmen and at least one officer, almost certainly men who knew what New York longshoremen had whispered about for months before the sinking. *Lusitania* frequently carried massive amounts of high explosives and had

Managing davits takes on added urgency in wartime, as this 1943 image of the training ship SS *American Engineer* suggests, but by that time many merchant ships sailed with davits usually far more reliable than J-shaped ones. (Courtesy of the Mariners' Museum)

made the previous trip with its double-bottom filled with a hundred thousand gallons of diesel fuel. Below-deck crewmen might well have suspected that the foundering liner would not simply suck down nearby lifeboats but might explode again, catapulting flaming oil across the water.[55]

In 1927 Commander Joseph Kenworthy published his *Freedom of the Seas,* a book that includes his eyewitness account of what occurred in the Admiralty War Room in the hours before *Lusitania* sank. "The *Lusitania* was sent at considerably reduced speed into an area where a U-boat was known to be waiting and with her escorts withdrawn," he reported (Kenworthy originally wrote "was deliberately sent" in his manuscript, but after the Admiralty contacted his publisher the wording was changed).[56] American survivors asserted they had been led to slaughter, and they were right, but more than the British government slaughtered them. A lack of real sea-men manning the old-fashioned J-shaped davits *sixty-eight feet above the sea* slaughtered many. And despite hundreds of tragic examples in World War I, lifeboat launching changed only slightly after the war, and indeed throughout World War II.

7

WATERBORNE

"To my unspeakable relief, Lifeboat Number Seven was still intact and sitting safely on its chocks," Frank West remembered of the first minutes after the shelling of SS *Britannia* by a German surface raider in 1941.[1] All went reasonably smoothly at his lifeboat position: with only a bit of difficulty at one davit, the lifeboat went up from its chocks, then swung outboard. Under the command of a longtime merchant navy officer and a career Royal Navy ship's carpenter, the group of men thronged about the davits even had a moment to inspect their lifeboat, then stuff a handkerchief in one shrapnel hole and a piece of wood in another. No panic seized the men; only a desperate but controlled haste. They lowered the lifeboat quickly and evenly. Despite the attack and the heavy swell rolling the ship, the launching of Lifeboat Number 7 went almost perfectly—until the boat hit the water.

Once afloat next to the ship, in what World War II–era seamen came to know as the "waterborne position," West and the others experienced frantic terror. "Our station was so far aft that we were immediately crashing under the counter of *Britannia*."[2] Oars snapped as men used them to fend the lifeboat off from the steamship. Then a block dropped straight down from the davit high above, and only his helmet saved West from death. A few feet away from the ship at last, the lifeboat inched toward a rope ladder thronged with evacuees. The men aboard the lifeboat did manage to position the boat, hold it in place, and receive not only about seventy more men, who came down the rope ladder and a parallel knotted line, but a wholly unforeseen mass of suitcases, boxes, and other luggage, some tied to passengers already encumbered with overcoats and life belts. In the meantime, all around the lifeboat landed tables, planks, and other material that might float swimmers, then a mass of men diving from decks above, then a pistol and binoculars tossed by an officer. Over the protests of their owners, West and others began throwing suitcases overboard, and then suddenly the men who had eased the falls came down the ladder, joined the men "packed like sardines," and began shoving the boat away from the liner.

Moving the lifeboat away from the ship proved anything but easy, with the packed boat tossing in the waves breaking against the hull. Eventually the men got four oars at work and pulled the boat slowly away from *Britannia,* only to see the raider signaling its intention to open fire. Lying in a Brazilian hospital months later, making the notes that in time became *Lifeboat Number Seven,* West recorded his fear that shells missing the liner would land among the lifeboats, and remembered how animatedly the officer in charge of his lifeboat had urged the men to pull. "With a vocabulary I have rarely heard equaled, we were left in no doubt that we were a crowd of idle lubbers, possessed of unmitigated laziness, that we had no usefulness and a complete lack of skill, and when it came to pulling a boat, words could not be found, for we were not even fit to push wheel-barrows."[3] Twenty-seven-year-old William MacVicar had not trained in sail, but the third officer of *Britannia* knew a lot about managing a lifeboat in its first waterborne moments. A lifeboat so punctured by shrapnel that soon only its air tanks and continuous bailing kept it afloat, and moreover carrying thirty more men than its designers ever intended, inched away from danger in an episode repeated everywhere on wartime oceans.

Not every merchant ship lowered lifeboats quickly and levelly, as merchant-marine officer Derek C. Gilchrist noted in his 1943 *Blue Hell,* a blunt record of a long lifeboat passage begun about ten minutes after a submarine-launched torpedo struck the British freighter *City of Shanghai.* Arriving at his station, Gilchrist found his lifeboat lowered, its stern in the water and its bow hanging six feet above the sea, waves about to unhook the stern from its hooks, and men shouting that the forward davit tackle had jammed. He ordered the crewmen to take the dangerous step of cutting the falls, dropping the lifeboat into the sea. As the still-moving *City of Shanghai* dragged the now afloat boat along by its painter, Gilchrist grabbed some useful gear from his cabin and abandoned ship. Urging his men to pull as hard as possible, he tried to get the lifeboat away from the sinking hull. "To get clear of the sinking ship was my first thought," he wrote. "Once clear, we would be safe."[4]

What happened to *Britannia* and *City of Shanghai* befell thousands of merchant ships in both world wars. Lowering the lifeboat proved troublesome enough, but getting it clear from the dangling falls, steamship hull, and debris presented its own challenges. Suction fascinated turn-of-the-century mariners beyond reason, almost certainly because of more general-ized anxieties about ships so large and complex that no one knew what they might do. Fear manifested itself in a fierce desire to get lifeboats away from people jumping from decks high above and from swimmers who might swamp lifeboats in panic. As *Titanic* Fifth Officer Harold Lowe testified at the Senate inquiry, "What are you going to do with a boat of 65 where 1,600 people are drowning?" He presented his reluctance to risk his lifeboat as ordinary marine prudence, but the senators hammered away at him and other *Titanic* men who cited suction as the chief reason they kept their lifeboats far from the sinking liner and from swimmers.[5]

Passengers' testimony lambasted the behavior of crewmen like Quartermaster Robert Hitchens, who ordered his passengers to row away from the ship, refused to return in response to some sort of whistle, and disregarded the pleas of wives desperate to return the half-empty boat to rescue swimmers screaming for help. The other crewman in Hitchens's boat acknowledged that "we pulled away until we were clear of the suction of the *Titanic*," and at least one passenger in the boat testified that they had been told to "row as hard as possible away from the suction," an order Hitchens seconded.[6] In the end, the Senate testimony strongly suggests that all *Titanic* crewmen viewed suction not only as potentially fatal but as an excuse for escaping potential capsizing by throngs of desperate swimmers.

Fears of suction originated with the coming of iron-hulled steamships. While people abandoning *Hornet, Ocean Monarch,* and other burning sailing ships feared masts, spars, and rigging crashing atop boats—something that also worried people leaving ships damaged by storm—mariners knew that wooden sailing ships usually sank slowly. In many instances, crewmen and passengers launched ship's boats and lifeboats directly from decks as hulls sank low enough to bring decks awash. But all were aware that throngs of swimmers easily overwhelmed boats, something that explains the loss of life in the *Arctic* catastrophe. One survivor, W. P. Rathbone, whose partly filled lifeboat reached Newfoundland, announced that his party had deemed it "not prudent" to remain too near the derelict *Arctic* lest the boat be "engulphed" by waves as the ship sank. But Rathbone admitted that his group also feared swimmers mobbing the boat.[7] Remaining at what another surviving passenger called "a respectful distance" from the wreck angered many landsmen, including George Templeton Strong, who called "respectful" "a judiciously selected adjective certainly!"[8]

While *Lusitania* crew members seem to have feared further explosions of munitions, the passengers certainly feared suction, and many did as survivor Charles Lauriat did, swim as far as possible from the sinking ship. Lauriat later admitted that the lack of suction from *Lusitania* surprised him. But when cold water struck the hot boilers, terrific explosions of steam shot from the funnels, and as the ship slipped under the surface, a vast collection of wreckage lanced up. Lauriat shoved pieces of oars or boxes or anything else that floated toward swimmers needing flotation, then swam toward a collapsible lifeboat "floating peacefully around, right side up," and began the frustrating process of assembling its rusted parts. Once he made the lifeboat somewhat useful, he rowed it back into the wreckage and found hundreds of bodies, but he could not tell whether drowning or concussion had proved fatal.[9]

Luck saved Lauriat. The lack of suction may be explained by *Lusitania's* having its bow in the seabed and its stern settling far more slowly than anyone expected, but suction combined with subsequent explosion did blast wreckage to the surface in a great roiling of water.

When in 1940 a torpedo sank SS *Assyrian,* attacked as part of a British convoy by German U-boats, one of its officers learned firsthand about suction and secondary explosions, and about the dangers of lifeboats in the waterborne position. Having made certain his freighter could not be saved, Second Engineer William H. Venables reported to his lifeboat station. Held up by its buoyancy tanks, the swamped port lifeboat crept away from *Assyrian* while Venables and a few others launched wooden life rafts. Jumping into the sea, they swam to the rafts and got aboard, then waited for the lifeboat to rescue them. In the chaos of combined U-boat attack on the convoy, Venables watched another ship drift past and sink with a sort of whining noise. A few minutes later, from far beneath the surface, thousands of heavy timbers burst into view. The second ship had carried a full cargo of what Venables called "pit-props"—what Americans call mine-shaft timbers—and the balks of wood began smashing his raft to splinters. Tossed into the sea again, Venables and a handful of other men forced their way back aboard *Assyrian,* then set about rescuing other men from the trap made by thousands of pit-props. For hundreds of yards around the sinking ship, wood so blanketed the sea that no one surviving the up-thrusting timbers could move through it unless pulled by lines thrown out from the derelict. While Venables began work on a raft, other crewmen cut the falls of a damaged lifeboat, which landed upside down in the sea.

No one aboard any lifeboat would approach *Assyrian,* Venables soon learned, although many lifeboats passed near. As his ship began to shudder, Venables jumped onto the capsized lifeboat, but he and his fellow castaways could not force it away from the derelict: water rushing into the holed ship pinned the lifeboat against the hull. "Suddenly the ship's side seemed to jump at us. The boat-davit swooped through the air and cut clean through the boat between Bishop and me." The J-shaped davit had come crashing down on the lifeboat, like what had happened to Lauriat on the *Lusitania* twenty-five years earlier, and instantly Venables swam as Lauriat had swum, then turned back to see his ship founder. He found himself staring into the maw of the hole blasted into the side of the ship. "Quite impersonally, as if viewing it on a cinema screen, I pictured myself being sucked to the bottom of the sea behind the ship," he recorded. Then he heard a curious whine he decided was air escaping through ventilators as the ship went under, and found himself surrounded by bales of unidentified floating cargo. Ahead of him, from other sunken ships, floated massive crates too large for him to climb aboard. He looked around and considered his position.[10]

Over perhaps forty-five minutes, Venables witnessed an impressive demonstration of every reason why the master of *Assyrian* had ordered away the port lifeboat despite its being swamped. As in so many narratives of shipwreck and lifeboat launching, Venables's account, "The Torpedoing of the *Assyrian*," offers little in the way of analysis, but it does clarify the several meanings of suction. First there was the late nineteenth-century belief that great steamships created whirlpool-like holes in the sea which engulfed swimmers and lifeboats alike. But "suction" also designated the inrush of water through holes made by collision, torpedoes, and the explosions of mines. That sort of suction might drag swimmers and even lifeboats inside a holed hull and trap them as the ship sank. A third meaning refers to the rotating motion of waves alongside a drifting hulk that might keep a capsized lifeboat almost glued to the hull.

Inrushing water could suck lifeboats and swimmers against hulls, something many castaways discovered as their lifeboats touched the sea. But inrushing water alone did not always create suction. Steamships carrying massive superstructures might "turn turtle" as the tremendous weight of their upperworks caused hulls to capsize just before sinking. When three torpedoes sank the 23,000-ton liner *Orcades* in 1942, its master recounted a near perfect launching of lifeboats carrying 1,300 people, a procedure marred only by the dropping of one end of a lifeboat into the sea, an accident that killed thirty-eight people. "I swam as fast as I could from the ship, so as to avoid any downward suction as the ship sank," the captain related. "I expected to see her turn keel up with the weight of the funnel and upper works, but she sank on her side and disappeared beneath the waves."[11] Of course, *Orcades* and any other great ship sinking on its side presented an immediate danger of suction as its funnels touched the sea and water gushed into them, but the barrel rolling of a great ship created a gigantic disturbance of water that swallowed anyone or anything caught in it. Seamen appear to have carefully analyzed the terror of suction and generally assumed that everything—especially bodies—that shot to the surface after a steamship sank had been sucked down from the surface rather than carried down with the ship itself and released only as the ship disintegrated. What they saw, they saw: life jacket–wearing bodies surfacing after ships sank.

"None of us had any real experience in boat-sailing; those of us who had lifeboat certificates had gained them in the sheltered waters of some river or dock," mused one seaman after the war, of a harrowing post-torpedo lifeboat passage. Unlike most writers, Michael F. Page focused his account as much on lowering a lifeboat as on its subsequent passage to safety. He recalled the noises of the storm becoming indistinguishable from the noises of his ship breaking apart and the continual shouting of men lowering lifeboats. Then, as the men lowered a boat and began scrambling down lines into it, it leaped about so crazily that the men could not drop into it. Suddenly, without warning, someone cast off the painter, and the lifeboat whirled away into the gale.[12]

Not until well into World War II did governments inquire into the loss of life immediately following the torpedoing of merchant vessels, and not until the 1950s did researchers begin making sense of wartime and postwar narratives. As early as the autumn of 1941, fierce government antagonism caused the Chamber of Shipping, the chief British shipowners' association, to suspend distribution of its *Health Hints to Those in Lifeboats,* a document based on substantial medical advice. Despite protests by the Chamber of Shipping, the Ministry of War Transport did not issue its own guide until early 1943, when the British had already lost four-fifths of the merchant ships sunk in the war. The ministry's *Guide to the Preservation of Life at Sea after Shipwreck* contained the first how-to-abandon-ship directions derived from the experiences of men and women who had survived not only the lowering of lifeboats but the mental and physical horrors of maneuvering lifeboats in the waterborne position.[13]

In the 1950s researchers discovered patterns of traumatized mind-set among seamen and passengers in waterborne lifeboats and determined that mental stress sometimes resulted from the experience of lowering the boat and the waterborne moment, rather than from the lifeboat passage itself. Fear of deserting a great ship became fear of what the great ship might do as it sank. To the researchers, the shifting fears were as relevant as hypothermia and seasickness to an understanding of lifeboat survival. "When his ship goes down, a man's whole universe goes with it," asserted survival expert Alain Bombard in 1953. "Even if he reaches a lifeboat, he is not necessarily safe." Certainly the lifeboat may be crushed against the hull of the derelict or capsized by waves bouncing from it; but until the castaway shifts his loyalty from the sunken ship to the lifeboat, he risks being consumed by terror or despair. "He sits, slumped, contemplating his misery, and can hardly be said to be alive." In *The Voyage of the Heretique* Bombard focused on controlling terror and forestalling despair, killers he found more ruthless than physical dangers.[14]

Bombard missed—and Alan Villiers missed in *And Not to Yield: A Story of the Outward Bound*, which describes the first survival-school training for merchant seamen—the complex mentality of men and women still in touch with a deck. Rationally, many crewmen and passengers abandoning ship knew that the ship towering above their lifeboat might not sink after all, that it might become derelict and remain derelict for days. Less than rationally, lifeboat castaways hoped that the damaged ship would remain afloat, and feared that casting off falls and painter meant severing connections with possible safety.

Thirty hours after a torpedo sank the 18,000-ton liner *Laconia* almost exactly where *Lusitania* had sunk two years before, passenger and *Chicago Tribune* writer Floyd P. Gibbons filed a lengthy report from Queenstown in Ireland. *Laconia* sank at night, but despite calm seas and few passengers, its crew experienced harrowing difficulties with its davits, launching Gibbons's boat fitfully, at a forty-five-degree angle. Again and again, the boat pitched head or stern down, while seamen chopped at the falls with an ax. "Many feet and hands pushed the boat from the side of the ship and we sagged down again, this time smacking squarely on the pillowy top of a rising swell," Gibbons telegraphed. "It felt more solid than midair, at least. But we were far from being off." No one aboard Lifeboat Number 10 could disengage the falls. While the passengers hacked at the lines with an ax, Gibbons heard a shout from above and saw a passenger plummet down the sixty-foot side of the ship, landing three feet from the boat. After they dragged the man inboard, Gibbons looked up at *Laconia* and cringed. "Its ranking and receding terrace of lights stretched upwards," he told newspaper readers across the United States. "The ship was slowly turning over." For a moment the lights mesmerized the occupants of Number 10; then suddenly everyone began clearing the "tangle of oars, spars, and rigging on the seat" in order to get four "big sweeps" working on each side of the boat.[15]

Behind Gibbons a stoker began to panic. "'Get away from her; get away from her,' he kept repeating. 'When the water hits her hot boilers, she'll blow up, and there's just tons and tons of shrapnel in the hold!'" Gibbons swung around and saw the man convulsed with fear and freezing in his thin cotton undershirt. Panic spread to the ship's baker and to another fireman, "whose blasphemy was nothing short of profound [and who] added to the confusion by cursing everyone." As *Laconia* ever so slowly loomed over them, crewmen plunged into panic. "It was the give-way of nerve tension," Gibbons concluded. It turned out that no officer, not even a warrant officer, had joined Lifeboat Number 10. Cast against the side of a sinking ship, hypnotized by the lights from above, the lifeboat's crew sank deeper into terror.[16]

But while a lifeboat still attached to its falls might indeed perform like an amusement park ride gone mad, once free of lines and the waves that slammed it against the steel hull of the ship, the lifeboat would begin to behave as its designers intended: as a boat that might take care of itself, at least for a while, while its occupants reoriented themselves.

Getting lifeboats away, then, preoccupied responsible masters and officers. After the *Assyrian* was torpedoed, one lifeboat was quick enough to escape holing by the surge of pit-props that shot to the surface a few minutes later, but Venables's raft was destroyed. He was able to get back aboard the freighter only after floundering beneath a web of timbers, and had another man not reached out to him, he would have drowned under the wreckage. All aboard the derelict knew they had waited too long to abandon ship.[17] Hesitation about lowering lifeboats and hesitation about getting lifeboats away from sinking ships was fatal.

Many of the men aboard *Assyrian* had followed a World War II dictum that reverberates today: *The ship is your best lifeboat.* While undoubtedly this was true, especially for compartmentalized cargo ships such as tankers that sometimes broke in half and remained afloat, the dictum also encouraged seamen to make every effort to save every ship, something Allied governments knew as fundamental to victory. Unfortunately, however, authorities neglected to emphasize how thoroughly this dictum would have to be discarded once survivors boarded a lifeboat. The castaways would then have to shift to escape from the vicinity of a ship turned derelict.

Whatever panic surged across the decks of a foundering ship, panic aboard a waterborne lifeboat might be far more easily quelled by one or two men, especially men armed with a pistol, knife, or belaying pin, or with a particularly sharp tongue. Gibbons related how he found in the stern of his lifeboat an elder master mariner, a survivor of a sunk Nova Scotian fishing schooner. Captain Dear told Gibbons the lifeboat had lost its rudder during launching and proposed that he should steer with an oar, but only if Gibbons would relay orders. Dear had long ago lost his voice. With an over-whelming blast of profanity, Gibbons focused the attention of the castaways and announced that the lifeboat now had a captain. The old man issued orders to put out the oars to keep the boat bow to the rising seas and wind, holding it in position while everyone watched the sinking liner and initial panic and excitement subsided into disbelief.

Laconia took a long time to sink, and its lights continued to exert an almost magnetic attraction upon everyone in the lifeboat. The rows of electric light went from white to yellow to red before vanishing beneath the waves. "A mean, cheese-colored crescent of a moon revealed one horn above a rag bundle of clouds low in the distance," Gibbons wrote. "A rim of blackness settled around our little world." But still everyone watched the black hulk silhouetted against the overcast sky. Finally the liner stood almost vertical, then slid stern-first into the sea. Almost immediately Gibbons found himself tossed in a maelstrom, but not as severe as that closer to the spot where *Laconia* vanished. "Boat No. 3 stood closest to the ship and rocked about in a perilous sea of clashing spars and wreckage." Suddenly, amid the wreckage, a U-boat surfaced, almost directly alongside Lifeboat Number 10.[18]

And after it dived, men in his lifeboat having answered its commander's questions about the ship's destination and cargo, Gibbons helped locate the lamp and light it, performing the same task as people in other lifeboats tossing on the swell. But Lifeboat Number 10 kept its distance from the other boats: no one wanted a collision in the dark, and apparently everyone aboard the other boats shared the concern.[19] In an hour or so the boats had scattered over several square miles of sea, their tiny lamps replacing only a fraction of the electric light lost when the liner sank. However much the occupants of Lifeboat Number 10 wanted to remain near the liner so long as it floated, they were even more anxious to make away from the other lifeboats and anything else that might threaten their newfound safety.

What Gibbons had identified as the "give-way of nerve tension" others identified as the mental weariness or wariness immediately following panic or extreme excitement. Board of Trade instructors and examiners were well aware that even in peacetime, panic and excitement might cause as much difficulty as jammed falls or listing ships. In his 1938 *Ship's Lifeboats*, Layton warned that traditional davits reward forethought and common sense, and that if things go wrong, everyone must obey the order "Still," which means both keeping silent and stilling machinery. But neither he nor any other Board of Trade author extrapolated the order to the situation aboard a waterborne lifeboat. "As soon as the keel touches the water it is a good plan to order *Let Go*," he asserted later in his manual, but he admitted that it might be difficult to quickly unhook the falls and move the boat underneath a rope ladder so that seamen and passengers could climb down into it.[20] Layton knew the reality of the waterborne lifeboat, as did many readers of Alfred Noyes's 1917 *Open Boats,* a compendium of abandon-ship horrors intended to vilify unrestricted submarine warfare and—especially—inflame American sentiment against Germany.

When a surfaced submarine began shelling the unarmed steamship *Chic* off the Irish coast in 1915, its master ordered lifeboats launched into roiling waves churned up by a strong westerly wind. Noyes detailed what happened next. As the port lifeboat touched the sea, a wave lifted it so quickly that it went slack on its falls, then dropped heavily into the succeeding trough. All the after gear tore free, the falls nearly strangling one seaman, and the tumbling boat flung another into the sea, where he drowned. As men freed the boat from its forward tackle, it filled with water; and while bailing, its occupants discovered it to be so badly damaged as to be useless. Nowhere in their excitement did the men have occasion to hear the order "Still."[21] Once a lifeboat floated free of its falls, no one expected it to pause at the sort of command that stilled lines and blocks.

Lifeboat-launching manuals paid scant attention to the handling of a lifeboat floating alongside a ship, or even to a lifeboat hovering a few hundred yards away, issues that greatly concerned Noyes. Layton and other authors preparing seamen for lifeboat efficiency examinations assumed that lifeboats would immediately row or sail away from sinking ships, not merely to escape enemy action but also to escape damage by boiler explosion or wreckage spewed up from the depths after the ship had foundered. The authors and the examiners—who, as Michael Page knew, had worked in calm harbors and estuaries—failed to address the feelings of desolation Bombard and other mid-twentieth-century researchers began to report in World War II lifeboat incidents.

Perhaps no one could prepare seamen for the realities of wartime shipwreck. Too many mariners knew lifeboats only from calm-water drills, and too few seamen and officers wanted to face the likelihood of leaving their sinking ships by lifeboat. Early in 1942, for example, salvage master and diver Edward Ellsberg, then an officer in the United States Navy, found himself and some four hundred other soldiers, sailors, and civilians aboard a merchant ship headed from Virginia to East Africa. "As a vitally interested passenger who had some knowledge of the subject, I pointed out to General Scott, the senior officer aboard, the scandalous condition our lifeboats were still in, improperly stowed and questionably rigged for lowering, and with a new crew which seemed to know little about the boats and from their lack of attention to them, to care even less," Ellsberg related.[22] The conversation led to a drill that proved a disaster. Three of the lifeboats did not even lift from their chocks, two because the falls had been rove backwards and their patent davits had been improperly assembled. The remaining eleven lifeboats could be lowered, and while the mates did re-reeve the falls and refit the davits, no one did anything about provisioning the boats, or even seeing to it that they had oars. Finally Ellsberg, General Scott, and a squad of soldiers and interested civilians devoted a day to provisioning and equipping all fourteen boats, while the master of the ship simply ignored their efforts.

"I was amazed that the crew was so little concerned regarding their instant effectiveness in an emergency," Ellsberg recalled.[23] Had he known more about similar events aboard other ships, he might have recognized this as one example of a widespread refusal to accept the possibility that ships' lifeboats might be needed. That failure of imagination crippled seamen who later found themselves torpedoed, lowering lifeboats, and—especially—tossing miserably within them, all the while longing to be back aboard a ship sinking before their eyes.

In 1919 appeared the first written inquiry into the widespread refusal of seamen and officers to imagine themselves being at sea in lifeboats. The revised edition of *Practical Seamanship for Use in the Merchant Service* was published under the editorship of C. S. Mence, a decorated and opinionated master mariner and officer in the Royal Naval Reserve. Arguing that only a few steamship masters devoted much attention to maintaining lifeboats and training crewmen to use them, Mence announced that the era of sail-trained men had ended and that steamships—especially liners—now carried neat-appearing boats about which most officers and seamen were largely ignorant. According to Mence, *Titanic* had launched its lifeboats so efficiently because its davits and tackle were new; its boats performed so well in the water because the boats were new. Aboard many passenger liners, boats were stored outside of davits, often in ways that precluded their immediate use. Observing this, Mence implied that many masters ignored Board of Trade regulations in favor of giving passengers additional promenade space. He noted that several times he had taken passage in steamships in which not a single boat stood ready to rescue anyone falling overboard. And while he admitted that aboard first-class liners both davits and boats were usually in far more ready condition, he focused on the appalling problems caused by poorly maintained boats the moment they hit the sea.

Mence despised the canvas boat covers so many steamship owners thought protected wood boats from drying out in sunlight, and he attacked the widespread smothering of lifeboats in paint so thick it masked structural flaws. As every sailing-ship-era seaman knew, wooden boats shrink as they dry out, and a dried-out boat launched into the sea almost immediately fills with water. Mence knew his steamship-era readers intimately and warned them against the increasingly common practice of filling lifeboats with water and so crushing them into their chocks that ribs and strakes snapped. But he insisted that every morning the fire hoses used to wash decks should be directed at the uncovered lifeboats. The spray would keep the wooden strakes swelled tight, and the runoff would drain through the holes in the hull bottoms. With a little care, the boats would float watertight when launched from well-greased davits. Otherwise, they would founder thanks to multiple leaks belatedly discovered by crewmen struggling to free them from the sides of foundering ships.[24] Leaky lifeboats might destroy whatever remained of hope, confidence, and discipline.

Practical Seamanship is indeed practical, but it is a subtly written book too, gently validating the arguments of master mariners like Fred W. Ellis, who insisted that "deep water men got all too little of boat training."[25] Between its lines pulses not only the old debate about training in sail and a sophisticated awareness of lessons learned in World War I but also an understanding of the steamship-era fatalism that so astonished Ellsberg and other World War II writers. No matter how well launched, far too many lifeboats floated men—and women and children—stupefied that their ships were sinking before their eyes. Men accustomed to turning little dials suddenly experienced massive mechanical failure, only to fetch up at length aboard tiny boats rocking in the swell. As electric lights went out, as deck after deck submerged, as great steamships burned or blew up or capsized, traumatized lifeboat castaways lingered nearby, unwilling to believe that the tiny lifeboats they had so studiously ignored had become their only hope of survival.

"The important fact to remember about sailing is that the average double-end lifeboat, in good condition, is exceptionally seaworthy," counseled Phil Richards and John J. Banigan in 1942. "It is far better equipped to fight a storm than you are." Their *How to Abandon Ship,* the first of a number of wartime emergency books derived from both World War I and subsequent shipwreck experiences, and especially from the torpedoing of British merchant ships from 1939 on, is the first fundamentally modern guide to lifeboat management. Unlike previous authors, Richards and Banigan dismissed the traditional Board of Trade and United States lifeboat efficiency examinations and accepted the new seaman—the man Conrad and others so scorned as being unlearned in the ways of sail—as typical.

In a unique and intensely personal way, they understood how during World War II the waterborne moment presaged the success or failure of lifeboat passage making. Banigan, torpedoed very early in the war, spent nineteen days in a twenty-six-foot lifeboat, bringing his crew of ten 898 miles through "mountainous seas, tropical storms, doldrums, and blistering equatorial sun." From that experience, and from collecting the reports of scores of merchant-ship officers and seamen, he and Richards asserted that their "manual is the result of open-boat experience in the time of stress, danger, and sudden death. It contains no armchair theory. It is a digest of the lessons learned by the survivors of torpedoed ships." Published for merchant-marine men served only by prewar instruction, its stark prose reinforces its opening sentences: "This manual is concerned solely with human lives. Its purpose is to aid you to get off a sinking ship and to eventual safety in the best condition possible."[26] Before the war ended, hundreds of United States merchant ships would be sunk at locations around the world, and survivors would have to make their own way to safety.

Richards and Banigan explicitly warn readers against depending on peacetime regulations and training. And while they urge mariners to watch for bulletins issued by the United States Marine Inspection Service—the sort of information many seamen found tacked on shipping-office and union-hall walls—they emphasize that the era of Layton and other efficiency examination authors is gone. No longer could any crewmen assume that at least some other crewmen were expert in lowering and maneuvering a lifeboat. *How to Abandon Ship* tells seamen how to buy personal survival equipment not required aboard merchant vessels. It recommends that its readers lobby Congress and seamen's unions to ensure that every merchant ship carries skates—bowed, iron-shod timbers that let lifeboats slide down hulls heeling over as much as forty-six degrees, keep lifeboats from being wrecked on protruding fittings, and help prevent a waterborne lifeboat from being stove in against the side of the ship. But above all, the book warns seamen to acquire a new attitude. Fatalism must go.

Unlike every previous book published on lifeboats, *How to Abandon Ship* quotes scores of mariners on every subject associated with shipwreck and lifeboat use. After a few pages, the technique becomes ruthlessly effective. Richards and Banigan counsel some procedure and then quote a number of authorities, all men like those in their intended audience: "Arthur LaBarge, an oiler aboard the *Oneida,* reported on a North Atlantic trip in March: 'No lifeboat drills during the entire voyage, going or coming.'"[27] Perhaps Richards and Banigan intended to shame steamship owners and the Marine Inspection Service by naming names and making clear that the fatalism and negligence observed by Ellsberg and so many others suffused the entire merchant marine months after the Pearl Harbor attack. More immediately, they knew that identifying ships might warn seamen away from vessels so badly managed that a torpedo might well mean death for everyone aboard.

On one level, *How to Abandon Ship* offers a new psychology of disaster. Richards and Banigan quote B. A. Baker, third officer of the torpedoed *Prusa,* as saying that "the most important thing for any lifeboatman to do, is to school his own mind. . . . Make up your mind not to get excited and stick to it. Don't say you won't be afraid, for you will. When the torpedo explodes you will get a sinking sensation in the pit of your stomach, and your knees may become a bit weak. The best cure for this is action."[28] On another level, the manual offers practical, hard headed advice, from simple precautions like inspecting lifeboat equipment and provisions to step-by-step instructions on using J-shaped davits in the dark.

Slowly but surely, the information shifts from the personal to the communal. Richards and Banigan quote survivors on a wide variety of topics concerning the importance of personal preparedness and personal presence of mind. But starting with chapter 4, "Swinging Out," *How to Abandon Ship* directs its readers' attention toward group activities: first the procedures involved in launching a lifeboat, then those involved in maneuvering it once waterborne.

More than the dangers posed by wind and wave made getting away advisable. To be sure, a lifeboat might be slammed repeatedly against the steel hull of the ship or suffer irreparable harm as parts of the sinking ship fell. But in the end, the lifeboat had to move away so that its occupants might begin adjusting to their new vessel. *How to Abandon Ship*'s sixth chapter, "Waterborne," is pivotal. Having already shifted from personal to combined enterprise, Richards and Banigan probe an important psychological dimension of shipwreck survival: all attachment to the foundered ship and to shipboard life must be broken off immediately. The authors knew the waterborne position to be a temporary and unsuitable one in which most castaways suffered the mixed emotions Gibbons recorded after *Laconia* sank— what Bombard understood as the collapse of the universe. They also recognized that the waterborne moment, whatever its duration, had to merge into the lifeboat passage itself. But for many survivors rudely deposited into lifeboats alongside sinking ships, the waterborne moment seemed to endure forever.

As the accounts of Venables, Page, and so many other survivors attest, a ship's crew and passengers seldom abandon it under ideal conditions. Even *Orcades* suffered casualties despite near perfect circumstances. But only a very few writers have tried to describe the waterborne moment. The paucity of attempts within the genre of lifeboat narratives suggests that many survivors either deliberately ignored the moment or else remembered it so poorly that they could not describe it in detail.

Richards and Banigan depict the lifeboat as a secure refuge, one requiring—at least most of the time—nothing more than simple seamanship. "Away from the ship in a lifeboat that is riding nicely, you have little or no cause for panic," they reassure readers. So far as the authors knew—and Banigan had had firsthand experience of *Kriegsmarine* kindness after torpedoing—no Allied seamen aboard lifeboats had been deliberately attacked; in fact, the authors concede, enemy sailors will treat castaways well.[29] So jumbled in the traditional lifeboat, men sucked dry of will, common sense, hope, even the desire to live would be well advised to pause awhile before beginning passage aboard a craft even belligerent navies had agreed to honor. Once away from the confused seas pounding the sinking hulk and floating wreckage, the lifeboat would take care of its occupants until the castaways could ready themselves for going somewhere.

How to Abandon Ship presaged several government publications aimed at merchant seamen confronting wartime disaster. All follow the lead established by Richards and Banigan, emphasizing the fundamental necessity of imagining lifeboats launched and on passage, but all are remarkably simpleminded and upbeat. "It is a sign of courage and intelligence to face possibilities and then to figure out what you will do and how you will do it, if they should happen," begins *Safety for Seamen*, a 1943 publication produced jointly by the War Shipping Administration and the United Seaman's Service. The pamphlet emphasizes that men must recognize fear and consciously overcome it. Using terms like "morose" and "depressed," *Safety for Seamen* describes emotional behavior that indicates not just fear but technical incompetence. It contends that the only solution for depression and sleeplessness is continual drill, especially abandon-ship and lifeboat drill at least every four days: this repeated military-style training will in time produce men able to function without thinking.

The manual deals only briefly with many critical issues and often dismisses genuine dangers such as suction caused by water pouring into hulls. It advises swimmers and men entering lifeboats to get at least thirty feet away immediately, although it also cautions that hurry and panic are far worse dangers than suction. *Safety for Seamen* presumes seamen thoroughly

Safety for Seamen warned merchant seamen of the dangers implicit in lifeboat passage-making enterprises and explained, sometimes in the most elementary ways, how to avoid them. (From *Safety for Seamen*)

trained in the right way of launching lifeboats, but it says nothing about response to damaged gear, for example, for it tacitly assumes readers incapable of improvising in haste. In the end, it simply assures its readers that a lifeboat will take care of itself and everyone aboard once clear of wreckage. In the face of the detailed research Richards and Banigan published a year before, *Safety for Seamen* seems a cruel joke, its advice virtually useless aboard any lifeboat struggling away from a sinking ship toward the open sea, let alone toward a ship in distress.

Ship-to-ship and ship-back-to-ship rescues feature in a surprising number of novels published after World War I. Some, like H. M. Tomlinson's 1937 *Pipe All Hands,* deal only briefly with the critical operations, subordinating them to subtle themes of authority, responsibility, ownership, and class. Others, like Howard Pease's 1926 thriller *The Tattooed Man,* treat the lifeboat rescue at length, using it to introduce the strengths and weaknesses of several important characters. Novelists of all types saw lifeboat rescue as the harrowing coda to more modern facets of disaster at sea, such as radio calls for help. The radioed SOS might summon modern steamships, but in the end only old-fashioned seamanship in traditional lifeboats saved lives.

Tomlinson focused almost all of his lengthy novel on the events leading up to the near foundering of the tramp steamer *Hestia* in a gale, including the continuously changing responsibilities of its master and the lifeboat transfer of its lone woman passenger once the freighter seems likely to sink. The tramp steamer in Pease's *Tattooed Man* launches a lifeboat in a gale to rescue a man overboard. Pease meticulously details the procedure of uncovering a lifeboat, swinging it out, lowering it, and then the chaos involved in getting it away from the rolling hull of the steamer, *Araby.* Both novel and thriller use the lifeboat rescue to illuminate the clash of modern technology and attitudes and traditional skills and beliefs. Away from the modern, electric-lighted steamship, aboard the lifeboat tossing in the night, characters confront the collapse of twentieth-century order.

Autobiographers produced very similar accounts of ship-to-ship and man-overboard lifeboat work. In his 1944 *White Sails and Spindrift,* Frank H. Shaw described a lifeboat moving between his ship, *Dovenby,* and a sinking derelict pummeled by eighty-foot seas. The *Dovenby's* first mate, who years earlier had escaped on a raft from a burning ship and its useless, painted-in-place lifeboats, was "a fanatic about boats." The mate cared for the lifeboats in ways Mence recommended decades later, keeping them wet in dry tropical locales, in warm ports always putting them in the water and keeping them there until the ship sailed, making certain that the boats were never "painted-in by years of ship-sprucing, so that they refused to leave the chocks," protecting them with "stout wooden covers, not rotted canvas" in stormy latitudes. Despite massive seas and icy lines, the big lifeboat moved certainly if ponderously up and overboard, the men working waist-, even shoulder-deep in green water covering decks.

Ordered from his job of spilling oil to calm the waves, Cadet Shaw found himself aboard the lifeboat and bailing. Captain Fegan, like so many other sailing-ship masters, employed his apprentices in his gig and taught them what he knew of boat work. Shaw understood his task, and the moment the lifeboat hit the water he began struggling to release the forward block while the boat plunged sixty to eighty feet vertically and crewmen fended it off from the ship with oars. The first mate released the after tackle and Shaw, amazed, watched the forward tackle release itself. Some of the men began rowing as the tackles swung free, Shaw began bailing, and the first mate ordered the mast stepped and a reefed sail set. "That was a feat of no uncommon cleverness. But this ex-North Sea cabin-boy could have sailed a scow around the world, I think. We got up the diminished sail, and the boat plunged ahead, tearing the waves to foam and froth."[30] The mate steered with a great steering oar, since he feared such seas would unship the rudder. Then, as they approached the derelict *Minotaur,* the mate ordered the sail lowered and the oars out again.

So violent were the seas—despite gallons of fish oil dumped from *Dovenby*—that the lifeboat's floorboards tore out. But the mate conned the lifeboat around the stern of the wreck, into the mass of spars and rigging floating to leeward, then nosed the boat toward the hull. For the rest of his life, Shaw remembered vividly the danger of bringing a lifeboat alongside a wreck, an episode that shaped his career and his mind-set. A wave tossed the lifeboat onto the main deck of the *Minotaur,* then swept it back into the sea. Suddenly began "the wild, senseless scrabble" of men leaping from the wreck into the sea and being dragged aboard the lifeboat. One died, crushed by moving wreckage, but with fourteen aboard, the lifeboat began inching downwind, where the master of *Dovenby* had positioned his ship. As they gingerly approached their own ship, "the entire proceeding a blasphemous affair enough," heavy lines crashed down across them, producing more foul language. The men having handed up the rescued, *Dovenby* crept round to windward of the derelict again while the lifeboat crew swallowed rum. In worsening weather, the lifeboat made a second, equally successful trip, again part of it under sail, this time finding survivors almost wholly sheathed in ice and rescuing every one, and the ship's cat too. "The boat that had served us so well was hoisted aboard," and the ship turned toward the Falkland Islands to discharge its new passengers. "I know that the memory of how we saved the *Minotaur*'s hard-pressed men steeled me more than once to persist when hope seemed gone for good," Shaw recalled of this lifeboat rescue involving five waterborne episodes.[31]

By 1944 booklets were warning castaway seamen against doing anything immediately, advising them to wait a bit in order to recover their wits. Merchant seamen expected—and authorities warned them to expect—that the waterborne moment would last a little while, until dawn broke perhaps, certainly until the lifeboat designed to take care of itself contained men willing and more or less able to begin a long passage to safety. But such advice made ship-to-ship rescue almost impossible, because lifeboat rescues like the ones Shaw, Tomlinson, and Pease recounted involved no time for rest, no time to pause, no interval or interlude, however short, between launching and making a passage. The first mate of *Dovenby* had set sail almost the moment his lifeboat floated, and headed off into an Antarctic gale toward the foundering ship.

Ship-to-ship rescue involving ship's lifeboats twists and stretches the meaning of "waterborne moment" away from notions of escape and long-distance passage making toward issues of immediate rescue and return. But immediacy could sometimes seem interminable. Some lifeboats spent hours in proximity to ships in danger, or in the middle of convoys surrounded by multiple derelicts and masses of wreckage. Such lifeboat rescue mocks the contemporary love of life rafts. As so many yachting accounts demonstrate, ship's lifeboats endure not only to take people away from sinking ships but to rescue them from sinking vessels, particularly yachts.

Every ship's lifeboat may one day venture forth, often under atrocious conditions, to rescue survivors of shipwreck. Every ship's lifeboat may one day perform the humanitarian service for which the Royal National Lifeboat Institution and the United States Coast Guard are famed.

"Perhaps the captain had known all along that in the end, when all else failed, he would have to launch one of his own lifeboats," writes Frank Mulville in *Schooner Integrity,* a history of a 1970 rescue at sea accomplished by the Yugoslavian freighter *Bela Krajina* in the Atlantic during a winter gale. Having located the foundering *Integrity* by radar, the ship began ultramodern rescue efforts, its master approaching the pitching yacht from windward, making a lee with his ship, then drifting down toward it, his men having slung lines and a ladder down its side. But the people aboard the yacht, already exhausted and panicked, refused to grab the lines, or even to jump toward the ladder, and suddenly the schooner's masts snagged the freighter, then ripped from their place, crashing alongside the yacht. While *Bela Krajina* slowly circled again, the people aboard the yacht launched and inflated their life raft, which immediately tore to pieces in the gale. Then Captain Vlajki ordered his own twenty-man inflatable raft launched and floated downwind at the end of a line toward the wildly rolling yacht. But the gale tossed the raft upside down, and no one aboard the yacht seemed able or even inclined to right it. "To launch a lifeboat would call for skill, courage, and discipline of a high order. The Captain knew he was risking the lives of a whole boat's crew—as for the boat itself, he knew there was no chance of recovering it," Mulville relates.[32] In ordering his third officer to make the attempt, Captain Vlajki knew the chances, his options, and his responsibilities.

"Ship's boats are seldom used and often neglected. Their condition and efficiency are a barometer of the efficiency of the ship itself," as Mulville notes. The big engine-driven lifeboat of *Bela Krajina* was the responsibility of the third officer, and like the first mate of *Dovenby* seventy years earlier, he proved himself a fine ship's husband. Davits, falls, and lifeboat stood perfectly ready for use. But despite modern davits and Senhouse Slips intended to speed the release of the boat from its storage position, lowering the boat involved almost the same problems as Shaw witnessed. The after tackle jammed as the boat hit the water and hammered it against the steel hull of *Bela Krajin,* splintering the gunwale. Another wave lifted the boat by the after tackle, pitching the crew forward, and only the ax-wielding efforts of the third officer, propped in the stern like Pease's fictional hero, spared the boat from the fate of so many *Lusitania* lifeboats. Fending off with some oars and pulling with others, the men moved the lifeboat away from *Bela Krajina,* started its engine, then encountered the full force of the gale. Approaching the sinking yacht, they found it nearly impossible to convince its people to jump into the sea and swim a few feet toward the lifeboat plunging alongside. But the lifeboat crew rescued everyone, and with only one injury—to a *Bela Krajina* lifeboatman who slammed his back in a fall. For their efforts, the lifeboat crew were recommended for the Yugoslav Blue Ribbon, awarded for valor at sea; for his, Captain Vlajki lost an expensive life raft and an even more expensive lifeboat.[33] The worsening gale made it impossible to recover the lifeboat that saved so many lives.

Punctuating the vast literature of yachting are scores of rescues like that which Mulville analyzes. Aboard a fine yacht, something that experienced seamen might treat as a nuisance—say the loss of a topmast, as in the case of the foundered *Integrity*—becomes fatal. Contemporary accounts of yachting disasters often read like the novels of Stacpoole and other turn-of-the-century writers exploring the psychological and social transformations wrought among privileged people by shipwreck. Almost invariably, discipline fails; high-technology equipment fails next, or turns malevolent as it wreaks havoc on inexperienced crews. Aboard *Integrity* no one had the means of cutting free the high-tech wire rigging holding the wrecked spars against the yacht. But the modern lifeboat launched by the modern *Bela Krajina* lowered away on traditional manila line the third officer cut with an ax, required in every lifeboat for more than a century. Such juxtaposition of the modern and the traditional is now at the heart of the formal education of merchant marine officers.

To read a modern book like *The Cornell Manual for Lifeboatmen, Able Seamen, and Qualified Members of the Engine Department,* a 1984 publication intended to replace similar manuals discontinued by the United States Coast Guard in 1973, becomes an exploration of tradition welded to modernity. The reader slips from one decade to another, sometimes from one century to another. "Commands for boats under oars have not undergone change for many years," comment authors William B. Hayler, John M. Keever, and Paul M. Seiler, all master mariners teaching at the California Maritime Academy. "A seaman of a 100 years ago would be at home in a boat today." In a section entitled "The Lifeboat Waterborne," the authors reiterate the World War II–era truism that "after the lifeboat is in the water and the passengers and crew embarked, it is of utmost importance to get the boat clear of the side of the ship as soon as possible."[34] However, they emphasize, nowadays it is extremely unlikely that a ship will sink without shore authorities being notified. After all, the EPIRB (Emergency Position Indicating Radio Beacon) activates automatically as it floats free of the sinking ship, making known its position even if the radio officer has sent no distress signals. In short, help will soon arrive.

Always, however, the old possibility endures. "Undertaking a long ocean voyage to the nearest land should be considered only under the most dire circumstances," warn the authors. They end their section on the waterborne lifeboat with an eighteenth-century reference: the example of Captain Bligh.[35] Modern mariners must remember the great ship's-boat and lifeboat passages of the past, and recollect that even now the traditional lifeboat carries oars and sails and equipment and provisions for deep-sea passage making. Remember what World Wars I and II taught: when the diesel engine coughs to a stop, mariners must understand why oars and sails are in the lifeboat. So counsel the authors of a survival guide everywhere else focused on the highest of high-tech aid responding to electronic pleas for assistance.

Help does not always come to the waterborne lifeboat. Sometimes it is transformed into the passage-making lifeboat, its lugsail raised and its people hunkered down, bound downwind toward the nearest sea-lanes or land, across hundreds of miles of unfrequented sea.

PASSAGE

AFTER THREE WEEKS, JUNE 24, 1923—THE DAY THEY WOULD FINALLY spot land—struck the men in the twenty-six-foot-long lifeboat of the foundered steamship *Trevessa* as one more day of strong wind and confused seas, tough going for them in their weakened condition. Ahead lay the islands of Rodrigues and Mauritius, pinpoints of land easily missed, and well beyond them the island of Madagascar. Nearly fifty miles north of the latitude of these tiny islands, Captain Cecil Foster had to sail not only west but slightly south toward them, something well-nigh impossible given the wind and seas. Trying to make to southward proved frustrating, Foster recalled in *1700 Miles in Open Boats*. The Board of Trade lifeboat performed splendidly, but no one aboard it expected it to sail directly into the wind. Foster worried about overshooting Rodrigues and having to sail several hundred miles further to Madagascar, something only just possible given depleted water and rations. Yet from their accomplishments, he knew his men might well make Madagascar. An example inspired them.

Day after day, the men swapped waterfront and fo'c'sle stories of small-boat passage making, Foster wrote in his subsequent analysis of his June 24 log entries. The passage of the *Bounty* longboat, he recalled, "was not exactly a parallel case, but it was the nearest approach we could think of," and it gave the castaways hope.[1] They were modern men, but they sailed a boat from an era that seemed as remote as Bligh's.

Some freakishness sank *Trevessa* midway in the Indian Ocean as it was carrying zinc concentrates from Australia to England. A strange material resembling moist dust but with the consistency of half-set cement, zinc concentrate slime sets stiffer than mud and does not shift in storms. Despite its 5–10 percent water content, the dense slime crusts over immediately in a ship's hold, the crust becoming almost rock hard. The gale that slammed the newly inspected, carefully maintained hull of *Trevessa* was so severe that two of its four lifeboats came adrift from their chocks. While the same gale buckled the hull of *Port Auckland,* forcing that freighter, also carrying zinc concentrates, back to Port Pirie for repairs, it sprang enough plates forward in *Trevessa* to send water atop its cargo. The crusted slime proved impermeable and kept the water from draining into bilges and the bilge-pump intakes. A little more than two hours after the midnight report of water rushing into the hull and the unnerving discovery of dry bilges, the men of *Trevessa* lowered two lifeboats in the middle of a gale.

Only an experienced seaman could imagine the difficulties of lowering the lifeboats in a gale. But according to Foster, the men were superb and conducted themselves well in the subsequent waterborne moment. They labored at their oars to keep the lifeboats near the sinking ship, then next to each other until dawn, and at sunrise Foster decided to make sail to the northwest, toward land.

The only responses to the SOS sent an hour after Foster learned of inrushing water had come from far-off ships, and the *Trevessa* radio operator reported that in the past few days he had picked up no signals from ships anywhere nearby. Given the rapidity with which the gale drove the two lifeboats from the position sent by the radio operator, Foster determined that they had little probability of being found. A modern ship carrying a modern cargo and modern radio equipment had sunk in short order, throwing its officers and crew backward in time.

Foster was familiar with lifeboats. He had been torpedoed twice in the Great War, spent almost ten days in lifeboats, and landed on the coast of Spain a far wiser first officer. Unlike many merchant-ship masters, he had impressed upon his chief steward what to do if the abandon-ship order was ever given. Robert James, who had sailed with him in wartime, followed his wartime instructions when ordered to put extra provisions aboard the *Trevessa* boats. But despite bravery and haste, so fast did *Trevessa* settle that James could lug only some cases of tinned milk and biscuits from the flooding storeroom. Along with cartons of cigarettes and tobacco, the lifeboats went waterborne otherwise provisioned only with the water and rations mandated by the Board of Trade. Almost as suddenly as if it had been torpedoed, *Trevessa* sank, and Foster rediscovered the era of oars, sail, and wooden boats. Almost everyone else onboard was discovering the era for the first time.

James Stewart Smith, first officer and in command of one of the boats, recorded in his notepaper log not only details of a passage from 28°45′ south latitude, 85°42′ east longitude to Mauritius but also details earlier generations of seamen took for granted. "Noticed curious thing which had not struck me before," he wrote on June 8. "In a small boat, the wind, however strong, can be felt, but there is no noise except the creaking of the boat and breaking sea. Funny thing."[2] What Smith experienced and recorded derived in part from British merchant-service conservatism. *Trevessa* had been a German ship interned in the Dutch East Indies, then purchased after the war and rebuilt in Scotland, where its owners replaced its original steel lifeboats with four brand-new "British-built wooden boats." The crew heard the peculiar quiet of a lifeboat in high winds, but also the creaking peculiar to traditionally built, lapstrake wooden boats of the type familiar to Bligh, Mitchell of the *Hornet,* and so many other eighteenth- and nineteenth-century mariners.

So wonderfully did the boats perform that they amazed everyone, even the experienced Foster, who wondered if the Board of Trade had imagined that the boats really could make such a passage. "The mere fact that they withstood the test in itself speaks for the wonderful capabilities of these boats. The manner in which they stood the battering was a wonder to us all."[3] Directing the boats to land demanded a return not only to old-fashioned seamanship, however, but to a sort of proto-navigation. *1700 Miles in Open Boats* begins with the complex issue of abandoning both ship and many modern navigation techniques at the same time. Sextants, prized by ship's officers as the badge of navigational ability and purchased at considerable personal expense, frequently found their way into lifeboats when officers dashed to cabins for important possessions. But without chronometers, sextants provided only a reckoning of latitude, and the men had no accurate means of figuring longitude. Having chronometers aboard would have helped with water- and food-ration calculations, since the officers could monitor progress and adjust rations accordingly. Foster and the chief officer determined to find the latitude of their proposed destinations and then follow that line to safety, exactly as sixteenth-century mariners had done. Ordinary seamen viewed navigation as a complicated mathematical art, and Foster used their reverent mystification to maintain morale. "The mere sight of the instrument seemed to give them added confidence," he remarked of his precious sextant, while too aware of the crudeness of his calculations.[4]

Foster had to find latitude 19°55' south, then sail west. Even with a sextant—and a table giving declinations and times of rising and setting of the sun and various stars—the task proved difficult. For weeks the two lifeboats zigzagged slightly north and south of the line, Foster in one, the first officer in the other, neither man wholly trusting the compasses installed in the lifeboats according to Board of Trade regulations. Everyone watched the stars as the officers struggled to distinguish true from magnetic directions and to relearn the old haven-finding art embedded in seamanship.

Shifting mental outlook, indeed shifting worldview, seems to be a key ingredient in successful lifeboat navigation. As Conrad explains so succinctly in "Youth," for very young men a lifeboat passage involves instantaneous education simply because the lifeboat sails or pulls into novelty. Dislodged from a sailing ship or steamer, young men lacking experience in ship's boats undergo a profoundly enlightening experience, if they do not drown or die of thirst or despair first. Older men, especially lifelong steamship men, had to unlearn carefully honed skills and tacit assumptions, something that often took days.

On June 29, 1942, a Japanese submarine sank *Express*, an American freighter zigzagging south through the Mozambique Channel between Madagascar and Africa. Within minutes, Master William Kuhne found himself swimming, and a few minutes later learned that his ship made no suction at all when it vanished. But while the port lifeboat survived the waterborne moment brilliantly, the starboard boat washed aboard the capsizing freighter, then capsized itself. Not for an hour or so did the men of the port boat find Kuhne swimming and drag him aboard. Forty-one men filled the lifeboat designed for thirty-eight; fourteen others had died. After a thorough search of the wreckage, Kuhne ordered the mast stepped and the lugsail hoisted. With six inches of freeboard, the lifeboat headed for Portuguese East Africa, 130 miles away.

When the torpedoes destroyed *Express*, Kuhne had been at sea for forty years, and master for twenty-five. Of the more than thirty vessels in which he had served, half were square-rigged ships or topsail schooners, and his training in sail paid dividends. Certainly he had kept the lifeboat in excellent repair and fully provisioned, and certainly he knew how to sail it, as his crewmen soon discovered. Fortunately his officers had thought to bring with them a sextant, a chronometer, and a book of astronomical tables, everything needed to determine latitude and longitude both, as long as the instruments worked. But perhaps most important, Kuhne understood his position in sailing-ship terms, knowing, for example, that in summer a strong southerly current flows through Mozambique Channel, something trivial to a steamship mariner but important to a sailing-ship master. Unless he compensated for this current, his boat would be deflected far southward of the horn of Portuguese East Africa, toward a stretch of coast 260 miles off. Despite troubles ranging from a lack of cigarettes to temperatures that shifted from ninety-five by day to forty-five by night, Kuhne managed to keep his course even in a three-day gale.

Five days after *Express* sank, the men began to spot all the traditional signs of land: tree branches and a stump, broken eel grass, several fronds of kelp. That evening they saw the banks of cumulus clouds that form over tropic-zone mainland coasts, and shortly after midnight the beacon of a Portuguese lighthouse, at Barra Falsa. When no shore-based rescue vessels responded to the flares Kuhne fired into the night sky, he knew that he would have to bring the boat to land. By daylight, he and the men having sailed close enough to land to inhale the rich scent of jungle, Kuhne unshipped mast and rudder, put everyone in life jackets, ran out the big steering oar, and with the sixteen strongest men manning eight oars, aimed the lifeboat into the surf. It caromed along at twenty knots before grounding in the sand in the shallows. Everyone jumped out into thigh-deep water and pulled the empty, high-floating boat to safety.[5]

Once in command of the lifeboat, Kuhne wanted to reach the nearest coast of Africa as quickly as possible, and he understood summertime currents and prevailing winds. When he saw the lighthouse flashes, he suspected that they marked Barra Falsa, but even more crucial, he knew that unless the lighthouse stood far taller than most, it could be no more than twenty to thirty miles off to be visible from a lifeboat that put his eyes scarcely five feet above the sea surface. He knew that a parachute flare exploding a thousand feet above the sea on a clear night can be seen from about forty miles away by someone about ten feet above the waves, something that makes the flare visible over 5,275 square miles of sea.[6] He knew a subtle certainty derived from familiarity with sailing vessels and from his years of keeping watch aboard them.

Certainty depended on more than traditional knowledge, however. It depended on knowing the position of disaster, the starting point of the lifeboat passage. Ordinarily deck officers knew the approximate location of their ship at any moment, since they all had responsibility for updating the chart and log. Even off-watch, an officer knew roughly where his ship steamed, and wartime zigzagging and other enemy-confusing measures did little to diminish this confidence. But in wartime, especially on large passenger liners, prudent owners and masters took special care to make certain that all lifeboat occupants knew the position of sinking, in part because torpedoes often killed officers. "They used to have a flat cigarette tin nailed on the cleating at the side of every boat and every day at noon, or just after, a cadet would go round all the boats and put a card in it," recalled able-bodied seaman Bill Sparks, of wartime procedure aboard an elderly Shaw Savil passenger liner, *Themistocles*. "The card gave the position at noon, the course and distance to the nearest land, prevailing currents and any other useful information." The master of *Themistocles* assumed the lifeboats would be used and detailed one seaman to do nothing else but move from boat to boat, overhauling davits, falls, and boat equipment, and greasing the davit winches.[7] But creating and updating the cigarette-tin survival card involved moving a step beyond Board of Trade regulations. It made it clear to all crewmen that no *Themistocles* officers might live to get the boats lowered and waterborne, let alone navigate them to safety.

Aboard *Themistocles* operated a mind-set fully aware of catastrophes like that which befell the Holt liner *Calchas* in April 1941. U-107 torpedoed the ship, and the master ordered most of the lifeboats away but kept some men aboard, thinking that they might save the damaged vessel. A short while later the submarine fired a second torpedo, and within three minutes *Calchas* broke in half and sank, taking down all its deck officers and many of the crewmen. Hundreds of miles off the coast of Africa floated six lifeboats in the waterborne moment, their occupants first stunned by the destruction before them, then terrified as the submarine surfaced and its officers waved good-bye while it surged through the wreckage. At first the castaways determined to stay put and await rescue. Twenty hours later, the seamen in charge—not one of them an officer—decided to make for the Cape Verde Islands, which they believed lay about a week's sailing away to the south-southeast, rather than wait for a warship that might never come. Without sextants and charts, the men in charge relied only on their lifeboat compasses and their traditional understanding of the sea.

Three seamen—the chief steward, the ship's carpenter, and the bosun—commanded the six boats. Within a day the flotilla had partially dispersed, but three of the boats stayed together and eventually combined their occupants when two of them began leaking, probably from attack damage worsened by a three-day encounter with strong trade winds. Reports of the passages vary in detail, and some information is missing: no one recorded how overloaded some of the boats were. But all of the boats made land safely and with very little loss of life. Able Seaman Hughes brought his lifeboat 650 miles to the coast of Senegal, where a fisherman off St. Louis piloted the boat to land. Chief Steward Harvey descried high land on May 4, and after a failed attempt at landing through surf arrived a few hours afterward in the Cape Verde island of Sal. A day later, Seaman Vaight brought Lifeboat Number 8 into the island of Bonavista, and Carpenter Frost landed the un-numbered "spare" lifeboat on the opposite side of Sal. Having made passages of between fourteen and sixteen days, the men were all justifiably proud of what the Holt Company historian called a "miracle of extemporised navigation." Company president Lawrence Holt took special pains to piece together an accurate account of the *Calchas* castaways, partly because at least one passenger aboard the liner filed a complaint concerning the lack of officers in the lifeboats, but largely to make posterity aware of something the *Calchas* third radio officer had pointed out to him several months after the sinking. Shore authorities acknowledged receiving the radioed distress message, but the Royal Navy had no ships remotely near the catastrophe; Holt wanted that simple fact made clear to posterity.[8]

Knowing the latitude and longitude of a shipwreck gave experienced seamen—and experienced stewards and carpenters—a solid chance to live. Distance to land or, in peacetime, to the nearest sea-lane heavily traveled by ships was something worth knowing. But distance meant two things inextricably tangled together, distance in miles and distance in time, and any estimate involved some analysis of lifeboat conditions and provisions, something especially difficult in boats carrying far more than their regulation number of occupants or damaged by explosion or during lowering or in the waterborne moment. In the postwar years Holt and his company historian, naval warfare expert S. W. Roskill, used both company records and *Kriegsmarine* and Imperial Japanese Navy documents to piece together an astonishing amount of information about the loss of Holt ships, in the process ascertaining the distances traveled by lifeboats crewed by men who knew only the *approximate* locations of sinkings.

Holt and Roskill learned that of all the lifeboats lowered from sinking Holt ships, four from *Rhexenor* sailed the greatest distances, each making 1,200 miles in eighteen to twenty days. But two boats from *Medon* were longest at sea, sailing thirty-five and thirty-six days. That Holt Company lifeboats performed so well—the firm lost a far lower proportion of crew and passengers in sinkings than other lines—seems to have been a direct result of its corporate insistence on lifeboat maintenance and lifeboat training, and on taking extra steps like those taken aboard *Themistocles*.

Rhexenor launched its boats with charts and with updated data from the current *Nautical Almanac* aboard each, something that greatly contributed to its castaways making land in Guadeloupe and in Antigua after approximately nineteen days. "I could not believe my eyes, but there it was," the first officer wrote in his log as his lifeboat neared Antigua.[9] Another *Rhexenor* boat made land on Jost Van Dyke Island in the Tobago archipelago, and a third found a rescue ship just off St. Thomas, about 1,236 miles from where the freighter sank.

Such distances in nautical miles pale beside those made by the lifeboats of *Medon*. When picked up by the Portuguese freighter *Luso*, the first officer commanding Lifeboat Number 2 could give thanks for the wisdom he acquired when he survived the sinking of another Holt freighter in 1940, an experience that caused him to equip and provision the *Medon* boats especially carefully. For all his care, he learned that he had sailed, not to a position 383 miles southwest of the sinking, but instead to a spot 595 miles to the southeast. He had unknowingly encountered a current that set to the east approximately one knot per hour of the passage, producing an easterly drift of 830 miles. When picked up, Lifeboat Number 2 was 1,450 miles east of its destination of Cayenne, further away than when it had begun its passage. First Officer Edge could have used the sort of chart the master of *Rhexenor* placed in every boat or the sort of prevailing current information in every *Themistocles* cigarette tin.

But the *Medon* castaways did have some information, although their use of it suggests that by 1942 steamship officers had begun to lose the intimate familiarity with trade winds and currents which sailing-ship men had accepted as second nature. An Italian submarine torpedoed *Medon* in August nearly midway between Freetown in Africa and Trinidad in the Caribbean, about 9½° north of the equator. All four lifeboats successfully got away from the ship, but about an hour later two of them returned to the hulk so that officers might assess the damage by daylight and collect navigation instruments and additional food and water. When the submarine began shelling the hulk, the men back aboard lowered the *Medon* motorboat and escaped, while the oared lifeboats fled also. Once the five boats gathered on a sea empty of freighter and submarine both, the master ordered the motorboat abandoned, and the castaways made for British Guiana, about 1,200 miles west. Since the appropriate volume of *Sailing Directions* cited an east-moving current where they were, they intended to sail south 200 miles to get around it, in the process picking up the southeast trade winds and a west-setting ocean current. Every officer expected to find the trade winds at roughly 5° north latitude in the months of August and September, but each knew too that the lifeboats would have to cross the Doldrums. For some reason, perhaps simple disregard of orders, the second mate headed his boat directly to the west to safety, but the others headed theirs south, into a blazing hot, windless sea.

Equipped with only the lifeboat compasses and their watches (only Number 3 boat carried a sextant and chronometer, but the chronometer stopped), most of the officers could roughly check longitude by timing the changes in sunrise and sunset, and latitude by checking the height of the North Star. All of the boats encountered windless days, squalls that sometimes deflected them from their courses but nonetheless provided drinking water, and a variety of events ranging from a total eclipse of the moon to a large whale swimming alongside out of curiosity, something that further stimulated the visual acuteness of the castaways. But what impresses the reader in any examination of the passage of the four boats—which quickly dispersed—is the brutal work of rowing through the Doldrums and their vicinity while keeping some accurate understanding of position. Slow progress meant repeated reductions in rations—on the twenty-eighth day at sea, Number 3 boat determined to reduce rations again to provide for another fifty days at sea—and continuous attention to navigation. Anyone slightly familiar with eighteenth- and nineteenth-century sailing-ship passage making knows that *Medon* could not have foundered in a worse location, and anyone reading the lifeboat passage accounts must wonder immediately why its master abandoned the motorboat that might have towed all four lifeboats through two hundred miles of calm.

In fluky winds, navigation appears to have been inaccurate in all but the chronometer-equipped boat. When rescued by a passing freighter, the men of Number 4 boat, for example, had made good only 40 miles to the southwest in seven days but thought they had sailed 232 miles west by north. But when picked up by a passing ship, the men in the chronometer-equipped boat were only 7 miles off in their reckoning, having made good 313 miles to the southwest into the trade winds. But even at that, they were 800 miles from British Guiana.[10]

Throughout World War II, many lifeboats hewed to courses indicated by the compass every lifeboat carried, only to be deflected by winds, squalls, gales, and currents. The *Medon* boats made extremely long passages, rowing and sailing hundreds of miles, but they made little progress toward their destination. The men abandoned ship well—all the *Medon* personnel lived, including an army gunner who jumped overboard with a leg broken by shell-fire—and maintained morale, but they erred in navigation.

Perhaps they lacked a feel for the sea. Books like John Herries McCulloch's *A Million Miles in Sail* and others published just before World War II by sailing-ship masters suggested that steamship officers never acquired this sense. McCulloch knew the Doldrums well and understood them as the meeting of the northeast and southeast trade winds over a region of tricky currents. "The whole subject could be discussed, pro and con, at great length, for tomes of statistics bear upon it," he concluded after briefly describing the problems the Doldrums posed to masters of sailing ships attempting the shortest possible passages through them. But by 1942 no one had examined the tomes for decades. Steamships like *Medon* plowed directly through the Doldrums, their officers paying no more attention to fluky winds and slight currents than to the gentle breeze produced by a ship steaming at perhaps twelve knots. Moreover, McCulloch argued that far too few steamship officers recognized that massive atmospheric disturbances cause "all sorts of temporary and permanent water movements." A competent master in sail not only noticed such currents but used them to his advantage, however slight; getting the most use from the gentlest of variations distinguished great masters from ordinary ones. Such knowledge could now and then benefit even the master and officers of great steamships, McCulloch suggested: *Titanic*, after all, had struck an iceberg at a *temporary* edge of the Gulf Stream.[11]

When McCulloch wrote his long memoir of life aboard sailing ships, a feel for the sea was rapidly becoming a thing of the past. At the turn of the century, when he encountered "a dense cloud of dragonflies, butterflies, and small birds" on a calm day three hundred miles off the coast of South America and watched the exhausted insects and birds swarm aboard his ship to rest, he remembered that elderly sailing-ship men had told him that *in those waters* such a rarity presaged terrific danger. The exhausted animals indicated an intense, hurricane-like storm pulsing far behind the animals, even as a gentle breeze drove his ship along.[12] But in 1933 few steamship men cared what a butterfly at sea meant to old salts driving worn-out sailing ships south of the Doldrums.

One war earlier, such information would have meant the difference between life and death, especially in the winter North Atlantic. For soaked men in an open lifeboat, failing to make land as soon as possible meant enduring frostbite and a mental lethargy that descended with dispatch. William Morris Barnes suffered more in three days aboard a lifeboat in winter than any of the *Medon* castaways suffered in a month, and only old-time sailing-ship skill saved him and his fellows. A mile from his sinking freighter, concentrating on rigging mast and sail, Barnes watched a German submarine surface next to his lifeboat and heard its captain shout, "Steer about east-nor'east and you'll pick up a destroyer tomorrow or next day." As Barnes related in his 1930 autobiography, they most certainly did not pick up a destroyer or any other ship, but instead found only wretchedness in a lifeboat in which most men, especially the engine-room gang, had on hardly any clothing. Deeply divided over the advice of the U-boat commander— Barnes thought they were already slightly north of the main shipping lanes— the men gradually slipped into the past, Barnes singing them chanteys from sailing-ship days. But Barnes brought more than chanteys to the situation: he brought an uncanny ability to estimate distance *made good,* and knew that by the second night they had made only 50 of the 288 miles between them and Ireland.[13]

Despite being given a wide berth by two steamships that assumed his lifeboat rockets to be enemy traps, Barnes insisted that the castaways keep up their spirits. "What was doing for us was that the wind had set in steady like you will often see a nor'east wind. They say a nor'east wind has nine lives like a cat," Barnes recalled. The wind never failed them, for all that it blew astonishingly cold in February in 53° north latitude. Eventually the weather turned even better, at least in his view: the wind backed to the north, and he knew that "is the finest and smoothest wind that blows on the Atlantic in winter, although it is cold." By the beginning of the third night he reckoned he had made good some 200 miles toward Ireland, and despite the cold was optimistic that the Board of Trade lifeboat would do just fine. With a steady wind on its quarter, the lifeboat bowled right along. However much he welcomed the British destroyer that rescued him, Barnes had the satisfaction of learning that his lifeboat was exactly where he thought it was. He had the old-timer's feel for the sea.[14]

Dictated in his old age, Barnes's *When Ships Were Ships and Not Tin Pots* reads in a colloquial way that somehow seems more direct than anything McCulloch or others wrote. By 1930 Barnes and his contemporaries had seen steamships at war and had acquired a grim view of wartime catastrophe. For Barnes, being rescued by a steamship from a dismasted sailing ship in World War I meant exchanging one peril for another. A day later the six-thousand-ton steamer encountered a hurricane, began flexing scissorlike in a way that terrified him, snapped a steam pipe that meant a loss of steering for four hours, then lost three lifeboats from its starboard side. As soon as Barnes arrived in port, he shipped out in a sailing ship, simply because he saw sailing-ship problems, even fires like the one that destroyed *Hornet* in mid-Pacific, as *manageable* problems that close observation, common sense, and courage usually solved.[15]

Making good in a lifeboat meant managing navigation by reducing most of its complexity to seamanship and a proto-navigation that no term, not even *dead reckoning,* precisely designates. Steering by the North Star worked fine on clear nights but in many regions proved impossible. Night after night lifeboats scudded along under cloudy skies or battled gales in which the night sky merged with the night sea in fearsome confusion. Many castaways sailed for hours in blinding snow, trusting only in their compasses, and when oil for the binnacle light ran out, discovered that in the dark they had to steer according to the set of the wind and swells. In the end, most castaways—at least those who survived to tell their tales—trusted compasses if they had them.

But not every lifeboat lowered with a compass aboard, despite Board of Trade regulations. In December 1942 the third mate of a freighter sunk by an Italian submarine told British reporters that the Italians had surfaced and sportingly handed over three bottles of cognac, thirty gallons of water, and a compass.[16] How the lifeboat came to be without a compass remains one more minor mystery of wartime, but given the paltry instructions found in pre–World War II textbooks aimed at seamen preparing for lifeboatman examinations, the Board of Trade regulations seem faintly ridiculous. Why include a compass and not teach seamen how to use it?[17]

The unlikelihood of inexperienced men being able to cope with variation and deviation accounts for part of the answer. Not only does magnetic north lie some four hundred miles south of geographic north, but it is constantly and predictably changing its location. The amount of separation between magnetic and true north is expressed in degrees and varies with the locality east or west of the true meridian (at times the two coincide, so that the amount of variation is nil). On coastal charts, governments denote variation by printing compass roses with three rings, an inner, traditional one marked in points, and two outer ones marked in degrees, the outermost one oriented toward true north, the one within oriented toward magnetic north. In order to find the variation in the year the chart is read, the mariner multiplies the annual increase or decrease noted on the chart by the number of years elapsed since it was printed. As Carl D. Lane remarked in his 1942 *Boatman's Manual,* "In practice, normal errors in steering and in estimating drift and windage will greatly outweigh the annual error in variation" in ordinary coastwise passages.[18] Unless a coastal mariner uses a very old chart, he can safely disregard the three-ringed compass rose. But ocean navigation is something else: Lane warned that long sea courses demanded an awareness of annual variation. A slight variation at the outset of a long lifeboat passage like that undertaken by the *Trevessa* castaways might, for example, mean missing Mauritius.

Lane's warning precedes a lengthy analysis of deviation—compass error caused by the magnetism of the vessel itself. Anything ferrous might cause deviation, including the knives and marlinespikes he insists must be removed from helmsmen's pockets. Given the massive overcrowding so common in Great War and World War II lifeboats, ferrous objects undoubtedly surrounded a great many compasses. In *Compasses and Compassing,* R. D. Ogg explains that "hand tools, nails, screws, fish hooks, belt buckles, even some kinds of eye glass frames and many other objects will act like small magnets and disturb the compass if they are close to it."[19] And while deviation can be corrected aboard any vessel, steel-hulled steamship or wooden lifeboat alike, correction begins with permanently mounting the compass.

In a ship's lifeboat, the compass is stowed securely away from the harm likely to befall it in the chaos of lowering, and only at the close of the waterborne moment do castaways unpack it and lash it where it can easily be seen by the helmsman. Always it may suffer from greater or less deviation, in part from having been stowed in a wooden lifeboat carried aboard a steel-hulled steamship. By itself the steamship produces "induced magnetism," which fluctuates wildly in duration and degree but often grows stronger the longer a ship sails on a given course. It might also suffer the "temporary magnetism" that Edward A. Turpin and William A. MacEwen warned in their 1944 *Merchant Marine Officers' Handbook* accompanies shocks made by artillery fire.[20] But without mounting the compass at least temporarily, then anchoring the lifeboat in a bay with perfectly charted seamarks and "swinging" the lifeboat (taking at least eight bearings as the bow is pointed round in a circle), no one knows the particular deviation of any lifeboat compass.[21]

For many naval and steamship officers at the start of World War II, variation and deviation were part of the deepening obscurity created by reliance on the shipboard gyrocompass, the electric-powered instrument that replaced magnetic compasses aboard many large vessels, especially liners and warships. For young men unknowingly grown dependent on gyrocompasses, unpacking a lifeboat compass meant grappling with issues they likely presumed old-fashioned until they faced the enormity of navigation "by magnetic." Some of their difficulties originated in mistakes as simple as putting steel boat lanterns too close to compasses at night.[22] Others developed over days, when they understood some fragments of traditional compass knowledge far better than others.

Managing navigation in a lifeboat making a long passage meant above all evaluating the worth of the compass not only in the waterborne moment but repeatedly thereafter. But how to evaluate the worth of the compass escaped most writers, especially in wartime. Writing during World War II, Richards and Banigan advised readers of *How to Abandon Ship* to sail east or west if they knew of no land close to the north or south. Exactly what they hoped to communicate to lifeboat castaways on this point proves difficult to discern, since they asserted that the lifeboat compass gives general indications only. "*Watch the sail,*" they insisted. "If you are making good speed, do not worry about keeping a course." However opaque, such advice makes slightly more sense when read against its givers' directions for securing the unpacked compass: they advised using the steel lifeline staples to mount it. How the steel staples might be used the authors did not specify, nor did they remark upon the deviation likely to result. But they did warn castaways to "note the lines of variation" plainly shown on charts, and in one sentence defined variation as the difference between true and magnetic north before telling readers, "Don't worry your head about what variation is.

Concern yourself merely with what you have to do about it."[23] No reader of *How to Abandon Ship* can escape the message implicit in the highly simplistic four-page chapter entitled "Navigating." Richards and Banigan clearly assumed that the inexperienced men aboard a typical lifeboat would be rescued soon after the waterborne moment or else make a long passage trusting to more than precise compass readings intelligently interpreted.

The authors insisted that every seaman purchase the six great pilot charts of the oceans sold by the United States Hydrographic Office and disregard the government warning that the charts must not be used for navigation. Citing his own example of nineteen days in a lifeboat, all the while plotting his position on a pilot chart, Banigan showed the document's usefulness for passage making if not for navigation. A pilot chart shows prevailing winds and currents, and Banigan reproduced his, detailing his passage near St. Paul Rocks through a part of the Atlantic few seamen had visited since sailing-ship days. Given the chart, and its instructions regarding wind roses, even inexperienced men can plot and keep a course, although the authors suggested that such men ignore the marks showing currents. In a moderate breeze with the lugsail kept full, a lifeboat makes three or four knots, they noted. Knowing that and using the scale marked on the chart, castaways ought to be able to guess reasonably accurately at their progress.[24]

Pilot charts and lesser aids appeared in the lifeboats of well-run steamship firms well before governments began to require them. In September 1942 yachtsman Chanler A. Chapman regaled *Life* magazine readers with tales of his eight-day lifeboat passage in the central Atlantic. His lifeboat lacked sextant and chronometer, but it had a compass and a child's outline map of the world. Despite some very inaccurate hunches about the location in which a submarine had torpedoed their unnamed ship—estimates varied from 450 to 650 miles from land—the second mate called upon his twenty-three years of sea experience to remember that the Brazil Current joins the Equatorial Current, and that both run north and west along the eastern coast of South America. However much the mate feared missing Trinidad, Tobago, and the Windward Islands and sailing deep into the Caribbean, he chose to sail to southwest and southwest by south. Once decided upon, this course became sacrosanct. After making certain of Chapman's reliability, the mate assigned him the night watches. Chapman paid close attention to the compass, wind, waves, and sometimes a star for four hours at a time. By day he and the mate studied the outline map of the world. A merchant-service officer and a yachtsman navigating a lifeboat filled with merchant seamen never asked or expected them to help read compass or map.[25]

In the chaos of attack and lowering lifeboats, obtaining accurate position information frequently proved impossible and made all subsequent compass work suspect. The shock of attack and sinking badly confused some officers: Derek C. Gilchrist freely admitted in *Blue Hell* that the torpedoing of *City of Shanghai* while he was on watch caused him a sort of temporary amnesia that lasted for several days. Even as he ordered the mast and sail set aboard his lifeboat and swung the vessel toward Africa, he could not remember the position he gave to the radio officer when he first spotted the submarine, nor could he recall the latitude of Freetown in West Africa. He told his men that Africa lay about three hundred miles east, and in the first day he fought highly variable winds that pushed his lifeboat southeast, something he later realized only by paying scrupulous attention to the compass. Gradually his head cleared. He learned that his lifeboat sailed well enough to keep a course and that it made good mileage each day to the northeast. Encouraged by the behavior of the boat, he boldly asked if anyone aboard it knew the latitude of Freetown. No one did.[26]

After a few days Gilchrist made a sketch map of Africa and asked each man aboard to improve it. He feared that their initial wanderings had set them so far south that they were sailing into the Gulf of Guinea, and he knew that they would have to make for the coast of French Equatorial Africa, the Gold Coast, or Nigeria. Eventually a schoolboy passenger aboard, having been challenged to remember his geography lessons of a few months before, scrambled aft and improved the sketch map enough for Gilchrist to guess that 1,200 more miles of sea lay ahead. But he continued to worry about the vague location of their starting point and the first days of aiming perhaps too far north in a south-setting current. Especially at night, when—despite continuous efforts at cleaning its burner, trimming its wick, and carefully refilling its reservoir—the binnacle light flickered and went out, he was concerned that winds or currents had subtly shifted the boat off course. Despite a fervent desire for rain squalls to quench their thirst, the *City of Shanghai* castaways longed for clear nights too, for only with the North Star in sight could they make certain that the compass continued to perform with a four-point error. Frequently when off-watch at night, with the ship's carpenter or someone else holding the tiller, Gilchrist found himself awake and scrutinizing the mast in the forward end of the boat, willing it to stay in line with a chosen star ahead and wondering how their course fitted into the reality of world geography.[27]

Merchant seamen adored the United States Hydrographic Office pilot charts when they first appeared in 1883. Determined to make the Hydrographic Office more useful to everyday mariners, its new director, John Russell Bartlett, began with the standard Mercator-projection North Atlantic chart featuring a compass card, magnetic variation curves, the 100-fathom line abutting coasts, and small black arrows that indicated the tendency of ocean currents. Bartlett had this black-printed data overlaid with innovative blue arrows with crossbars showing the direction, frequency, and force of winds, and blue lines showing the limits of trade winds and the location of the Doldrums, along with a variety of other information such as the southern limits of icebergs and the locations of seasonal fog. In red the office printed another range of information, especially steamship and sailing-ship routes, the location of lighthouses, and the position of the belt of Newfoundland fog.

In 1884 Bartlett convinced Congress to open branch offices in six United States seaports and began collecting information from arriving shipmasters, all of which the Hydrographic Office reviewed and disseminated monthly on new charts distributed free. While the five-year Bartlett administration knew other successes as it shifted away from pure research—it published a useful pamphlet about using oil to calm breaking seas—its creation of pilot charts proved a triumph. On the back and around the borders, Bartlett provided all sorts of additional information ranging from weather summaries to directions of winds around low-pressure areas. As the Hydrographic Office deciphered patterns from the information provided by shipmasters, it became possible for mariners, and especially for sailing-ship masters, to make use of subtle variations in winds and currents. Soon the Hydrographic Office issued charts for other oceans too, tweaking the pride of the British Admiralty and providing shipmasters, seamen, and interested passengers with a free education. In time the cost rose to ten cents per sheet, or sixty cents for the full course.[28]

Within five years, the pilot charts had attracted widespread public attention, in large part because each chart was so jam-packed with information displayed in visually intelligible ways. "To one who is not familiar with the subject it would seem almost impossible to publish on one chart such a variety of information of such a diverse character, and yet have a chart that can be of practical use in plotting a vessel's track," Everett Hayden told a large audience assembled at Philadelphia's Franklin Institute in a lecture later reprinted in its *Journal*. "It would be very difficult to do so without the distinction of colors."[29] Hayden showed a lantern slide of a chart specially printed in black to make his point that colored inks transformed the visual interpretation of data such as the fluctuation of the equatorial calms centered just north of the geographical equator which caught the *Medon* boats decades later.

The use of color almost coincidentally enabled the Hydrographic Office not only to plot the drift of derelicts north from the Caribbean but eventually to note that the myriad tracks implied changes in the velocity and course of the Gulf Stream. In only five years, the charting of derelict tracks prompted the dispatch of a Coast Survey research vessel to the edge of the Gulf Stream. There investigators discovered a monthly variation in speed linked to phases of the moon, an effect especially strong off Florida. What had begun as an effort to help the merchant marine took on a larger life as the data prompted scientific research that alerted Hydrographic Office personnel to the need for more attention to less-frequented ocean routes.

By dividing oceans into squares bounded by five degrees of latitude and longitude and then collating data submitted by shipmasters, Hayden and his colleagues learned about a region of equatorial rains they marked in blue, but also about regions no one seemed to cross. But as more and more sailing-ship masters followed the pilot-chart suggestions for fast passages, the little known regions became less and less well known, through no intention or fault of the Hydrographic Office. Paradoxically, the superbly pithy pilot charts in time caused some regions of the oceans to become almost unvisited, except by derelicts.

Tracking drifting derelicts taught Hydrographic Office experts what a handful of adventurous sailing-ship masters already knew: as steam replaced sail, whole regions of oceans grew less traveled by the year. Derelicts offered the only information about winds and currents in certain regions, Hayden explained in his lecture, citing examples like that of *Ada Iredale*. Some 1,900 miles east of the Marquesas Islands in the South Pacific, the sailing ship caught fire from the spontaneous combustion of its cargo of coal in a manner like that Conrad described in "Youth." As did the *Hornet* castaways, the *Ada Iredale*'s crew abandoned the ship in three boats, the flotilla arriving in the Marquesas twenty-five days later with the loss of one man and one boat. Nearly eight months after its abandonment, the still-burning derelict reached Tahiti, 2,423 miles from where its crew left it. But until the French warship *Seignelay* found the ship adrift just off Tahiti, no vessel had sighted it on its long, slow passage. By 1888 the Hydrographic Office experts not only knew the frequency of ship abandonment and the track of many derelicts but had begun to predict the trajectories of derelicts through unvisited seas according to the season and weather.

Knowing, for example, that "vessels wrecked between the eastern edge of the Gulf Stream and the Bermudas seem to take no direction in common until they are driven westward and enter the Gulf Stream north of the thirtieth parallel; they circle about for months, influenced mainly by the winds," the Hydrographic Office *expected* derelicts to show up in particular locations.[30] Derelicts adrift in frequented regions produced many reports and finely detailed pilot-chart tracks like that of the schooner *Mary E. Douglass,* which drifted haphazardly off the Bahamas for ten months, being reported again and again; or the schooner *Twenty-one Friends,* which drifted from a point about 160 miles off the capes of Chesapeake Bay to a point about 130 miles north-northeast of Cape Finisterre. *Twenty-one Friends* traveled approximately 3,525 miles, roughly 2,130 of these in the Gulf Stream and the remainder easterly toward Spain. Shipmasters reported it twenty-two times.

Other derelicts, like *Ada Iredale,* materialized like nightmares from unvisited regions. The sailing ship *Oriflamme* caught fire in the Pacific at about 18°12′ south latitude, 92°42′ west longitude in June 1881. On October 24, the steamship *Iron Gate,* en route from Adelaide, Australia, to Portland, Oregon, encountered *Oriflamme* at 13°27′ south, 125°19′ west and sent its lifeboat across to it. The following February the hulk drifted ashore on Raroja Island in the Low Archipelago, having covered 2,840 miles in about eight months, although during that time it was reported only once.[31] The Hydrographic Office well knew the consequences of any vessel, even a steamship, plowing into a dismasted, awash, iron-hulled derelict, something especially likely to happen at night. After *Waratah* went missing, the possibility of collision with a derelict began to attract attention, and after *Titanic* foundered, experts insisted that however freakish its collision with an iceberg, colliding with a submerged derelict would have sunk the liner in even less time than striking the iceberg.

In 1912 J. Bernard Walker, the editor of *Scientific American,* warned that derelicts endangered liners speeding far south of the iceberg zone. Lookouts high aloft could not be expected to see derelicts at night, he asserted. Colliding with such an object would destroy almost any vessel, especially great liners moving at speed, even those designed to survive ordinary collisions between ships. As Walker explained in his treatise *An Unsinkable Titanic: Every Ship Its Own Lifeboat,* at twenty-one knots, *Titanic* moved forward about thirty-five feet a second. Given fifty-three feet as the average length of its watertight compartments, its moving inertia of more than one million foot-tons would carry it hundreds of feet across the derelict, ripping open one compartment after another. Any steamship, including ones compartmentalized against damage from mines, would sink fast after such a collision.[32]

Although the Hydrographic Office was well aware of the dangers posed by derelicts, it had very little data about derelict tracks in many seas. Its vague warnings made some unfrequented regions even less traveled: prudent steamship masters took their vessels away from any possibility of danger, especially if they commanded vessels steaming at full speed at night.

The bulky volumes Tomlinson read in the 1920s as *Altair* plowed through the Red Sea and that Steinbeck studied as *Western Flyer* prowled the Sea of Cortez two decades later were reminders that steamships avoided almost all but the great harbors and rarely strayed from what mariners called "steamer lanes." As Tomlinson repeatedly pointed out to his nonmariner readers, almost no one but shipmasters and officers ever *read* the bulky tomes issued by the British Admiralty and the United States Hydrographic Office, and even they rarely read them cover-to-cover, instead opening the volumes to the pages dealing with particular harbors. Only the most experienced and thoughtful master mariners and the rarest of landsmen distilled from these books a general, pilot-chart view of stretches of sea. Tomlinson argued that the sailing directions provided updated information only on the limited destinations of steamship masters, but for less-frequented areas repeated in issue after issue the nineteenth-century findings—often extremely vague—of masters-in-sail.

By 1940 pilot charts had ceased to be navigation tools and had become merely ephemera on which officers sketched a course day after day, then discarded at the end of a passage. Not until lifeboats suddenly cruised the oceans did any authority realize the deeper value of the pilot charts, and by then many castaways had died. In the middle of 1943 the first pilot charts begin appearing in United States lifeboats, and in mid-1944 the United States Coast Guard began to require that every lifeboat carry aboard pilot charts fitted into a metal container.[33]

In the spring of 1943 the National Maritime Union displayed the lifeboat perfect at its New York headquarters on West Seventeenth Street, incidentally attracting the arch attention of the *New Yorker.* The author of the account observed that among the most important pieces of equipment on display was a metal tube filled with pilot charts "intended to help any uneducated castaways to grasp the fact that heading for the nearest land isn't always the quickest way to get ashore." For example, castaways off the southwestern African coast should use prevailing currents to head toward South America, 2,500 miles west.[34]

By 1955 modern lifeboat charts would put maritime academies out of business, Jan de Hartog half-jokingly asserted in 1955 of the oilskin-wrapped parcel labeled "Charts for Ships' Boats" by then found aboard every lifeboat under Board of Trade jurisdiction. The waterproof parcel held simple navigating instruments and charts of all oceans, along with sailing directions from any point to the nearest land. In *A Sailor's Life,* de Hartog, a Dutch tugboat master and veteran of numerous wartime attacks on Arctic convoys, argued that the charts offered an effective short course in world navigation, one solidly based in the age of sail and developed from much earlier pilot charts. Pilot charts and subsequent lifeboat charts offered a summary view of what to do in a lifeboat, how to do it, and how to keep track of it.[35]

World War II lifeboat passage making threw into sharp relief the disparities in prewar training among various steamship lines. The lines that valued training usually provided extra equipment aboard lifeboats, if only an information card shoved in a cigarette tin or an outline map of the world. Canadian Pacific and Cunard officers—the latter presumably having learned a thing or two from the *Lusitania* debacle—devoted much attention to lifeboat training and went beyond Board of Trade regulations to organize lifeboat rowing and sailing races among crewmen of the deck, steward, and engineering departments. The lines owned by Alfred Holt all issued orders mandating that crewmen become familiar with sailing and rowing lifeboats, and the peacetime orders paid dividends in war, when castaways of Holt ships survived at a far higher rate than average.

As Tony Lane shows in his incisive *Merchant Seamen's War*, the 1913 Board of Trade suggestion that British merchant seamen needed formal training in sailing lifeboats fell on deaf government ears even as some British shipping firms listened and acted. What the British government failed to accomplish, these firms did, and Lane's meticulous statistical analysis illuminates how training saved lives not only during lifeboat lowering in post-attack confusion but on lifeboats at sea for days and weeks. According to Lane, rescuers found two-thirds of British merchant lifeboats within twenty-four hours, and almost one-half within two days. Only three in every hundred lifeboats made passages longer than three weeks.[36] Training often made the difference between survival and death.

Beneath the surface of so many wartime shipwreck and lifeboat narratives—and sometimes of peacetime ones too—lies a partial explanation of Great War and World War II tales, today largely discredited, of German and Japanese barbarism toward lifeboat castaways.[37] No pilot chart exists to plot the undercurrents of wartime propaganda and rumor, but just as Hydrographic Office personnel learned to track drifting derelicts, any careful reader of lifeboat narratives and maritime and naval autobiography and history discovers how quickly enemy action came to mask the inadequacy of lifeboat crewmen.

The U-boat captain who shouted directions to William Barnes in World War I was more or less typical. While surfaced to verify the identity of the sinking ship, U-boat captains generally found a moment to explain to the castaways what course to follow to the nearest land. Such behavior lingered well into World War II, until Allied aircraft launched from new airfields and small aircraft carriers blanketed the skies everywhere in the North Atlantic, and there is no reason to believe that *Kriegsmarine* officers provided fraudulent positions. After U-159 torpedoed the freighter *East Wales* in December 1942, it surfaced and approached the waterborne lifeboats, its machine gunners aiming at the British seamen as a standard precaution, since U-boat men feared hand grenades lobbed from lifeboats. The U-boat captain asked some questions about the vanished ship, its officers, and its cargo, and received some highly imaginative answers. Seeing wounded men in the boats, he ordered bandages passed aboard and then shouted, "Steer southwest for 420 miles, until you come to Brazil. Merry Christmas!" Before submerging, the captain told the castaways the real name of the torpedoed ship, its route, and its cargo, confirming merchant-seamen rumors of German spies on Allied wharves. In the end the *East Wales* castaways were lucky, for their second officer had grabbed a sextant, nautical almanac, and table of positions before abandoning ship, so they not only knew the truth of the German advice but knew their position as they sailed toward Brazil.[38]

When a U-boat torpedoed *River Afton* in July 1942 four hundred miles north of the Arctic Circle, it surfaced and provided food to the castaways and directions to the nearest land. Soon after the men were rescued by a British destroyer.[39] An able seamen from the torpedoed *Ripley* told Admiralty interrogators that the captain of the attacking U-boat advised the men in the lifeboat to check their drinking water breakers. "We did this and in one of them it came out black and stinking. The German captain said we should maintain our boats better, that he'd seen this sort of thing before. Then he shouted something down the conning tower and up came two cans of water."[40] And in other cases, as when a U-boat sank the freighter *Fabian* in November 1940, off the east coast of Africa, U-boat commanders told castaways to stay put and keep sharp lookout, for their lifeboats floated in the middle of frequented sea-lanes.[41] Nevertheless, not until long after World War II ended did veterans, survivors, and historians begin to publish reports contradicting wartime news bulletins of submarines machine-gunning lifeboats.

No one can know how many U-boat men murdered castaways aboard lifeboats, for evidence and U-boat crews alike might well have perished soon after the incidents. But the sole known incident involving a German submarine resulted in convictions and executions at the Nuremberg trials, and naval historians now classify it as what *Kriegsmarine* officers claimed, a horrendous but unique event.[42] Historians know of four authenticated instances of Japanese submarines machine-gunning castaways, but Allied warships subsequently sank the four submarines and the victors found no one to prosecute later.

Yet not until the late 1980s did merchant seamen testify to what Allied governments had known in World War I and World War II. Newly published reports—often disseminated by oral-history projects like that conducted by the Liverpool Public Libraries—revise older accounts of the war at sea every bit as thoroughly as post-1960 historians revised the events of the Dunkirk evacuation. British Admiralty debriefings between October 1939 and March 1944 revealed fifty-six encounters between U-boat crews and lifeboat castaways and not a single instance of attack. But throughout the war, Allied government officials, journalists, novelists, and filmmakers denounced the barbarism of submarine warfare in terms originating in Great War propaganda aimed at boosting morale, masking the gross failure of the Royal Navy to protect merchant ships, and bringing the United States into the war on the Allied side.[43] In an eerie way, World War II denunciations appear almost seamless with those of the Great War epitomized in Alfred Noyes's *Open Boats* and those Herbert Russell described in his *Sea Shepherds,* a book about the horrors of unrestricted submarine warfare. While sometimes lifeboat castaways might have confused machine-gun fire aimed at ship radio aerials with gunfire aimed at themselves, no castaway reports exist to explain the subsequent denunciations.[44]

The U-boat captain who told the *Ripley* survivors to check their water supplies acted out of humanity and experience: as he remarked, he had seen that sort of thing before. But other submarine commanders had not, and when they first encountered it they were aghast. In December 1942 the third mate of a ship torpedoed by an Italian submarine told the London *Times* that, after giving the castaways a compass, the Italian first officer burst into tears. Several weeks later, the *Times* printed an account of a German submarine crew that gave the survivors of a torpedoed ship food and drink aboard their submarine and set and splinted one survivor's broken leg before returning the men to a lifeboat, which the Germans had righted from its capsized position. As historian Tony Lane notes in his *Merchant Seamen's War,* this sort of report not only did nothing to stop the organized denunciations in Britain but found particular disfavor among United States correspondents. Walter Winchell, for example, rebuked a survivor from the torpedoed troopship *Laconia* who told him of the humane actions of the U-boat men who sank the liner in April 1942, then spent nearly five days shepherding overloaded lifeboats toward Africa under a Red Cross flag— and being attacked by American aircraft whose bombs killed only castaways.

Such actions, Winchell insisted, could not have occurred.[45] But Winchell avoided hard issues of lifeboat passage making.

In *Blue Hell*, Derek Gilchrist mused that most everyday activity aboard his lifeboat connected him and his men with an earlier era.[46] For Gilchrist, adventure meant escaping death by keeping his lifeboat on a compass course. It would have been nice to have had a pilot chart as he wondered about the reliability of his compass and his memories of latitudes and the rough outline of the west coast of Africa, but he knew he had a well-equipped Board of Trade boat and he remembered what the *Trevessa* castaways remembered in the 1920s: Bligh did it, as did other sailing-ship men of old. And they did it without pilot charts.

What he did not remember, or at least what he did not recall remembering when he wrote *Blue Hell*, were Great War disasters like the 1918 lifeboat passage away from the American freighter *Dumaru* which presaged so many World War II catastrophes, pilot charts notwithstanding.

In 1918 the American freighter *Dumaru,* bound for Manila from Guam with munitions, barreled gasoline, and barrels filled only with gasoline fumes, steamed into a fierce tropical squall late in the afternoon of October 16. While the squall deluged the ship with rain so heavy it overwhelmed scuppers and filled the well deck, it brought little wind and scarcely raised a sea, in part because it struck so close inshore. The lightning flashes were so massive and so close to *Dumaru*, First Assistant Engineer Fred Harmon later recalled, that they distorted the electric dynamo. After the engine-room telegraph jangled "Stop engines," Harmon received a telephone call from the bridge ordering him to reduce steam in the boilers, then get to his lifeboat station. He thought lightning had struck the ship and torched the barreled gasoline in the forward hold, but perhaps some static electrical charge fired the empty barrels stowed above the full ones or ignited the fumes lingering in the hold. The chief engineer, long suspicious of anarchist activity aboard, thought sabotage possible, indeed likely; the second mate insisted that everyone heard two near-simultaneous explosions, one of a lightning strike, the other of gasoline; the radio operator told everyone that a single strike caused an instantaneous electrical explosion. Moments later, a lightning strike on the wireless antenna destroyed the radio equipment and almost killed the young operator, and men—especially the farmers and others recently recruited into the wartime merchant marine from civilian occupations—began to panic.

Without orders, the third mate and six other men lowered one twenty-man lifeboat and began rowing away from the ship. Their desertion and the rampaging fire forward, moving toward the explosives stowed aft, caused

Captain Ole Borrensen to give up lowering the second leeward lifeboat and instead try to lower one on the windward side. Men got in each other's way, and the davit falls twisted and jammed. The storm had by then raged just long enough to raise big waves against the steamship hull. Harmon, the bosun, and two other experienced men rode down the lifeboat as others lowered it, fended it off with oars, and maneuvered it toward the stern. Man after man jumped from the deck of *Dumaru*, and Harmon and the others pulled them aboard.

Harmon had seen a white-uniformed man leap into the sea and be pulled into the lifeboat. This United States Navy lieutenant took command immediately from the first mate, and as soon as Captain Borrensen shouted his intention to put over a raft, ordered the painter cut and the boat away under oars. The order, which meant abandoning men still swimming in the flame-covered water, so enraged one elderly, tropical-schooner seaman in the lifeboat that he stood up and demanded Lieutenant E. V. Holmes turn back. Another seamen knocked him down. When he got to his feet again, the old man grabbed the mast and furled sail secured amidships and flung it overboard, shouting that he hoped everyone would die. No one stopped for the equipment. Moments later *Dumaru* erupted, sending a massive wave at the lifeboat and hurling flaming pieces of wood all around it before releasing a flood of flaming gasoline. Struggling to avoid the burning wreckage, the men rowed away from the gasoline and tried to find the other lifeboat and any rafts that still floated. Quickly they discovered the uselessness of burning flares in a night illuminated by burning debris, and Lieutenant Holmes ordered the boat turned toward the coast of Guam, a few miles away. At first rowing, then raising a sail on two oars lashed together, the lifeboat wallowed downwind toward the nearby coast.

The lifeboat wallowed as only Lundin patented lifeboats could wallow. Essentially two cylindrical air tanks bent to form the shape of a boat and equipped with a self-bailing floor and other patented inventions (including self-releasing gear that released Harmon's boat from the davit falls before he and his fellow seamen intended), the Lundin boat lacked a keel and did best when floating castaways about to be picked up by nearby rescue ships. The unhappy *Dumaru* castaways soon discovered that its steering oar could not control its course. Jammed with thirty-two men, the Lundin boat proved incapable of speed and limited in maneuverability. After four hours' pulling and sailing, Holmes ordered it hove to for the rest of the night, fearing its inability to avoid a reef jutting from the beaches. Riding to its sea anchor in seas greatly diminished by the end of the squall and the heavy rain then falling, the Lundin boat seemed near the end of its passage.

Dawn brought land in sight, men grumbling with the work to be done at the oars, a tin of rations opened for breakfast, no deep fear at all, just discomfort and the aftershock of shipwreck. But dawn also brought Harmon to his first distrust of Holmes, who turned out to be a far younger man than his voice indicated during the night, and to a far greater distrust of the Lundin boat, which sailed downwind at such an angle that its wake streamed from midships. When the boat moved into coastal crosscurrents, it began to spin around. In thirty minutes of sailing by daylight, within sight of land, Harmon began to seriously worry about the capabilities of both the officer in charge and the patented lifeboat.

A few minutes later worry only deepened. The wind gently shifted, then gathered force, and the men could not row the flat-bottomed boat against it, as Harmon recounted later. By nightfall the wind had dropped a bit, and the Lundin boat, stopped from skittering haphazardly by its sea anchor, lay filled with quiet men listening to the conversation among the officers aft. The men heard the officers discussing their mistake in forgetting the trade-wind shift, Harmon reported, and deciding that it was impossible to correct.

Overnight the trade winds had begun their seasonal shift of direction. Harmon and most of the other men forward knew that the trade winds blew from the same quarter smoothly for months at a time and that the Lundin boat would never make Guam against them. Eventually Harmon, himself an expert in coal-fired boilers, engines, and propeller shafts, realized that Holmes and the *Dumaru* first mate and chief engineer had begun discussing passages to Saipan, the Anson Islands, or the Caroline archipelago, all places far downwind.[47]

The nightmare passage—*drift* more accurately describes the process—of the Lundin lifeboat to the island of Samar in the Philippines devolved into a grisly horror of death by dehydration and suicide, and then cannibalism. What happened to the thirty-two men aboard reveals much about group dynamics, the collapse of modern technology, and the failure of men reliant upon modern technology, all subjects for the next chapter. Holmes's first-night decision to stop rowing points up his ignorance of climate and weather patterns and the seamanship sailing-ship men took for granted; it also demonstrates his inability to understand the sailing qualities of the Lundin lifeboat.

Almost certainly, a Board of Trade lifeboat would have made land, or at least the line of surf, by daybreak, rowing or sailing from the waterborne moment to the end of passage. Nothing more clearly confirms the wisdom of the Board of Trade traditional lifeboat design than the *Dumaru* catastrophe. A steamship confronts ocean force with force, Jean Merrien writes in *Lonely Voyagers,* a study of single-handed, deep-sea passage making, while a small boat yields as the swells roll beneath it. But wind changes the equation. "If the wind surfaces are great and the sea resistance small, the object is carried along, but without undue stress," he asserts of floating debris and poorly designed small boats.[48] Almost all of the *Dumaru* Lundin lifeboat, even packed with thirty-two men, floated above the water. Its patented buoyancy eliminated any likelihood of it making a passage or doing anything but skittering jerkily downwind. Unlike the Board of Trade boats of *Trevessa* and a thousand other foundered vessels, the Lundin boat could not care for its occupants. It defied ordinary rules of traditional small-boat seamanship.[49]

Only a year before *Dumaru* exploded and sank, F. W. Sterling, a retired lieutenant commander in the United States Navy, published his compact *Small Boat Navigation,* a terse guide to small-boat passage making, including those castaways might make in lifeboats. Sterling pointed out that the periods of ebb and flow do not match low and high tide, and that in some places—especially off lone islands and among archipelagoes—topographic restriction makes ebb tides endure long after low tide. While explaining how to use a compass accurately, navigate by dead reckoning, and chart a dead-reckoning course, all things the navy taught in World War I and Holmes should have known, Sterling insisted on traditional observations too, describing ten types of clouds as meteorologists know them, then using words like "scud" to explain their significance to uneducated but knowing seamen. *Small Boat Navigation* anticipates the tragedy of the *Dumaru* lifeboat: it presumes traditional small vessels making passages according to tradition.[50]

Sterling expected his little book to instruct Great War naval and merchant-marine officers and men in managing small craft within sight of land and during short passages at sea. *Small Boat Navigation* is not a manual for lifeboat passage making, although the *Dumaru* castaways would have profited from its analysis of coastal currents. It is a sort of retrospective book, for all that it mentions the latest in navigational instruments and techniques, and it insists on the value of traditional thinking and wisdom. *Small Boat Navigation* appeared in one printing in 1917 and it appeared again in 1942, at a time when another generation of steamship-trained officers and men made the passage-making errors epitomized by the *Dumaru* castaways, and when thousands of landsmen found themselves in naval or merchant-marine service. Not until 1943 did something vastly better supplant it.

Almost forgotten in an electronic age, *The Raft Book: Lore of the Sea and Sky* endures as a curious amalgamation of ancient and traditional navigational expertise and wartime necessity. Compiled by Harold Gatty, an Australian scientist expert in Polynesian navigation techniques, the sumptuously illustrated little book appeared in two editions, one slipcased for sale to individuals, the other sealed in rubber pouches for installation aboard lifeboats and rafts. Supplemented by a foldout table for navigation computations, folded paper scales for measuring distances, and a unique worldwide chart, *The Raft Book* demonstrates the usefulness of Polynesian star-based passage-making technique. Even more important, it insists on the significance of sailing-ship-era close observation for World War II castaways.

Gatty explains that Polynesians "viewed the stars as moving bands of light and knew all the stars of each band which passed over the islands they were interested in," and that "their method of navigation by these heavenly beacons was to sail toward the star which they knew was over their destination at that particular time."[51] Using an innovative chart printed on one side with a star map and on the other with the oceans of the world dotted in red with the location of specific stars, *The Raft Book* not only analyzes the theories underlying Polynesian navigation but tells how to make a crude instrument out of string and a stick of wood called "the harp" with which lifeboat castaways can measure the height of stars above the horizon. Equipped only with a good watch set to Greenwich Mean Time and kept in a rubber pouch, a castaway unfamiliar with navigation could plot and keep an accurate course to land.

The Raft Book comprises far more than Polynesian celestial navigation techniques modified for wartime castaways. The book details how to use birds, insects, sea snakes, and other animals as indicators of position, and especially as harbingers of nearby land. It contains charts showing where specific bird species can be encountered (and with what frequency), along with identification pictures of birds in flight. Gatty found what he called the "lore" of birds especially useful for castaways and devoted entire paragraphs to the meaning of species behavior. For example, the land-based frigate bird, a predator that catches boobies returning to land with bellies filled with fish, cannot catch fish on its own or spend a night at sea. In late afternoon and early evening, frigate birds fly toward land at an altitude that allows them to see land invisible from lifeboats. In numerous tables labeled "Land Indications from Sea Birds," *The Raft Book* explains the meaning of bird sightings by season and even month. From December through February, for example, castaways seeing six or more stormy petrels at the same time may conclude that their lifeboat is within seventy-five miles of land, while in any month of the year castaways seeing twelve or more masked (blue-faced) boobies should assume land within twenty-five miles. Especially when combined with the known habits of other species—sea snakes, Gatty observes, rarely venture beyond the 100-fathom line along coasts, and their appearance should hearten castaways—a familiarity with the lore of birds will pay dividends.

While Gatty outlined how seventeenth-century European explorers used the appearance of birds, sea snakes, and bonito to warn them of land nearby—he quoted the seventeenth-century Dutch explorer of the Pacific Abel Janszoon Tasman at length—he emphasized the importance of scrutinizing seaweed. Even more important, he asserted, is the scent of land. The smell of burning peat can be detected far off the Falkland Islands, the scent of orange groves wafts from the Cape Verde group, and "the sweetish rancid odor of the drying coconut may be detected far out at sea in tropical regions."[52] Gatty recollected that he smelled new-mown hay eighty miles off New Zealand, and he warned castaways to be especially alert for the sweet smell of earth after heavy rain, for it is often the only indication of nearby land passed by a lifeboat at night.

In one other way, *The Raft Book* welds the ancient Polynesian navigation method to the traditional methods and interests of sailing-ship masters. Gatty asserts that scrutinizing the sky and the sea proves useful in finding land, since tiny clouds hover over the lee side of atolls and reflect the bright turquoise of lagoons. Sailing-ship seamen knew all this, Gatty maintains, and if they did not, they had only to read mid-nineteenth-century sailing directions. *The Raft Book* focuses on the Pacific, but its author notes that the Vikings knew how to watch for the "ice-blink" clouds in the subarctic, and he quotes nineteenth-century explorers of the Antarctic who found them there.

Castaways mindful of the sky and wind can easily make the jump to watching waves and swells, and Gatty details how the Micronesians made sophisticated charts from shells and coconut-palm ribs that depicted the swell pattern across reaches of the Pacific. The stick charts depicted the ways in which landforms—including very small islands and reefs—distorted long-distance swells, something everyone in every lifeboat ought to know. Waves are not, in Gatty's words, "meaningless mass." *The Raft Book* teaches castaways how to genuinely and acutely perceive the colors of seawater that indicate everything from depth to salinity. It urges lifeboat castaways to put their hands in the water and feel temperatures and temperature differences, to hang over gunwales in fog and listen for sounds carried beneath the fog layer, to recognize exhausted butterflies as portents.

The Raft Book also touches upon an experimental eighteenth-century Euro-American form of navigation discarded after the development of accurate longitude fixing by chronometer. Benjamin Franklin and other eighteenth-century thinkers, Gatty notes, nearly perfected a passage-making method they called "Thermometrical Navigation," one based on sampling sea temperatures and referring to charts giving locations of warm and cold currents. While realizing that no lifeboats carried thermometers, Gatty nonetheless insists that Polynesians sensed temperature differences with their fingertips, and his book includes pages of ocean-temperature charts for the months of February and August.[53]

Moving concisely and certainly over a great range of material, Gatty's book has a certainty of tone similar to that of *How to Abandon Ship* and other books derived directly from experience. Nothing theoretical interrupts its arguments, and while the pages of extracts from pre-nineteenth-century voyages of discovery may exist in large part to occupy the minds of lifeboat castaways making passages as long as those of the *Trevessa* boats, in the end they reinforce the keel-deep message of the book: before this time of modern navigational instruments, mariners looked around acutely and succeeded in making land. *The Raft Book* derives its message from the traditional realm of the monthly pilot chart, the world of sea and sky, animals and plants, sounds and smells that the lifeboat castaways either know as they lower their lifeboats from the davits or else must learn, especially if they lack compass and chart. It is a realm intimately close, quite unlike that described in another book published in 1943, *What to Do Aboard a Transport*, written by "A group of science writers and artists."

Despite its long, detailed, and well-illustrated chapters, the tone of *What to Do Aboard a Transport* is observational and distant and detached in the usual way of academic prose. James Cuffey, a United States naval reserve lieutenant, suggests to readers of the chapter on navigation that lifeboat passage making is a straightforward affair: "You select the nearest suitable land and head for it. A good compass, and a knowledge of directions as indicated by the sun and stars, particularly the north star, are essential to this sort of navigation. Next in importance is a set of small-scale charts showing the entire oceans." How anyone, let alone a naval officer, could issue such simplistic advice in 1943 remains unsettling.[54]

At least one traditional seaman wondered about the preparedness of American forces arriving in Britain. The year 1943 found Harry Bagshaw commanding the ancient Thames River sailing barge *Scone* at the Clyde Anchorages Emergency Port in Scotland. Bagshaw mused on the juxtaposition of *Scone* and the modern warships to which it lightered ammunition and other supplies. As the war progressed, *Scone* also ferried American fighting men directly from *Queen Mary* and *Queen Elizabeth* to Glasgow. The shift from immense troopships to the eighty-eight-foot, elderly *Scone* came as a shock to the American troops. Many of the men came from far inland, and their reading had taught them nothing of crossing oceans. "To help them to understand that new experience they were issued with a little book that was called *What to Do Aboard a Transport*," he snorted.[55] Maybe the book belonged aboard the liners-turned-transports that towered above *Scone's* masthead, but it had little relevance aboard the tiny barge plowing toward Glasgow, and none aboard a lifeboat.

Not until 1951 did anyone distill wartime lifeboat passage-making lessons into something young steamship officers might find useful, and even then the tenth edition of *Navigation and Nautical Astronomy* misses much of Gatty's wisdom. When Edwin A. Beito updated the book, first prepared in 1926 for instruction of United States Naval Academy midshipmen, he knew that dead reckoning formed the core of lifeboat navigation. But despite providing all sorts of information about a variety of scratch-built navigation devices, including a timekeeping pendulum and a cross-staff wrenched from the medieval past, Beito warned his readers against relying on crude instruments. Lifeboat navigation demanded more than substitute instruments: it demanded traditional passage-making skills.

But by 1950, his eight-hundred-page book ignored all sorts of information he admitted lifeboat navigators had to know. No longer did he assume even the most general knowledge about currents and stars, for example. He warned his readers that weather lore might be valuable in ways they never imagined aboard ship: "The particular information of value in emergencies," he wrote, "is a knowledge of prevailing winds at different seasons in the operating area, and the ability to detect early signs of approaching storms and predict their paths relative to the course of the lifeboat." Still, he offered few specific examples. Above all, he urged readers to consult pilot charts before deciding on a destination—despite the fact that federal authorities had stopped issuing waterproof pilot charts as ordinary lifeboat equipment.[56]

In an age devoted to electronic-signal-summoned rescue and life rafts incapable of any progress but drifting, Howard Gatty's *Raft Book* is forgotten, even by ocean-survival experts. "*The sure successful conclusion to the voyage should always be given priority over the random chance of rescue,*" asserts Dougal Robertson in his 1975 *Sea Survival: A Manual*, in a key italicized sentence that explains the decision of the *Trevessa* officers to make for islands far distant from where their ship foundered. *Sea Survival* is a book accompanied by unique charts showing ocean currents, prevailing winds, rainfall estimates, even magnetic variation, all of which make sense when the reader discovers that Robertson assumes raft-borne castaways principally need to know where they are drifting, rather than where they intend to go. Robertson realizes that all too many late twentieth-century castaways will find themselves aboard inflatable rafts which cannot be steered much of anywhere. He warns that destroying the protective canopy to make a sail is permissible only when land is nearby.[57] *Sea Survival* focuses not on lifeboat passage making but on surviving in a raft long enough to drift ashore or be rescued: thus even though it provides charts, it is a far less sophisticated manual than Gatty's.

Despite its title, *The Raft Book* is a book about lifeboat passage making based on the most traditional Western and Oceania passage-making wisdom. While castaways survived World War II disasters by staying alive on both inflatable and wooden rafts—performing feats of endurance recounted in books like Robert Trumbull's 1942 *The Raft* and Kenneth Cooke's 1960 *Man on a Raft*—in the end, only traditional lifeboats prove capable of keeping a course toward somewhere castaways intend to reach.[58] Outside of books aimed at mariners sitting for the Lifeboat Efficiency Examination, *The Raft Book,* published in 1943, remains the latest and finest statement of what lifeboat castaways need to know to make passages like those made by the *Trevessa* men.

By the middle of World War II, government handbooks advised officers to grab their sextants before abandoning ship, something that suggests a recognition of how badly merchant-marine officers and certificated lifeboatmen often performed in lifeboats making long passages. During the war, catastrophes like the drift of the *Dumaru* castaways perhaps occurred as frequently as successes like the extemporized navigation of the *Calchas* castaways, but in many cases no one lived to record the catastrophes. Lacking officers, even pilot charts, but secure in Board of Trade lifeboats and traditional knowledge, the *Calchas* castaways did as well as Bligh's men and the *Trevessa* castaways. A good lifeboat and a solid knowledge of traditional passage-making skills went a long way toward making interpersonal tensions manageable.

9

BAGGAGE

AT THREE IN THE MORNING OF MAY 24, 1942, AN ABLE-BODIED
seaman in Lifeboat Number 2 of the torpedoed British freighter *Peisander*
heard what he thought might be a cow mooing. He woke F. A. Brown, first
officer of the steamship and commanding the boat, who soon discovered the
sound to be that of surf breaking on Nantucket Island beaches. A few hours
later, trailing a pail astern on a line to slow their progress through the surf,
the castaways aboard the double-ended lifeboat arrived on the sand un-
harmed after six days, fourteen hours at sea, having navigated some 340 miles.

As did every surviving Holt Line lifeboat commander, Brown sent a report to company headquarters. He found the Board of Trade lifeboat excellent, and the additional provisions—he had aboard "raisins, prunes, and
peanuts to vary the diet"—extremely valuable. Morale remained high
throughout the passage. Three days after *Peisander* sank, he and his men
unanimously refused a rescue offer from the freighter *Baron Sempill* when
they learned it stood toward South America, a decision that surprised everyone aboard the steamship. All aboard *Peisander* survived the attack and subsequent sinking, and all three of its boats made land with no loss of life.
Brown credited such success to the maintenance of "a careful discipline."[1]

World War II prompted a refitting of lifeboats with everything from
pilot charts and raisins to signaling mirrors and rubber-packaged wool blankets, but by the time U-653 sank *Peisander* experts had begun focusing on
discipline as fundamental to castaway survival. Indeed it was the most important commodity aboard any lifeboat, asserted the *American Merchant Seaman's Manual: For Seamen by Seamen* in 1942.[2] Still, discipline proved harder
and harder to define. Gatty resurrected the lore of wind and sky and sea in
The Raft Book, but reviving old notions of discipline, especially *self-discipline,*
vexed World War II experts and castaways alike.

While the authors of *How to Abandon Ship* advised merchant-ship crewmen to grab "abandon-ship" bags and assured them that even torpedoed ships floated long enough for men to collect such equipment with little risk, no one knew how to toss discipline aboard lifeboats lowering under davits. Very early in World War II experts admitted the likelihood of panic, and *How to Abandon Ship* regarded it as a certainty, especially at night.[3] Training might alleviate panic, especially among inexperienced seamen, but did little or nothing to strengthen discipline. Discipline might originate in the heart of every lifeboat castaway or it might begin aft, in the attitude of the officer or lifeboatman in charge of the boat, or again, it might casually appear as part of what Americans—although not Britons—called "morale." On the other hand, it might be engendered and maintained by the coarsest possible brutality.

Four days into a July 1942 Indian Ocean passage, violence erupted among the cigarette smokers aboard a lifeboat jammed with forty-two survivors of the torpedoed American freighter *Express*. An oiler awoke to find his last cigarette butt, rescued from the bilge and placed carefully on a thwart to dry, gone. The former coal miner accused his neighbor of theft, and within seconds the two men had each found allies, stood up, and pushed other men aside to make room for a fight. Captain Carl Kuhne half rose in the stern and shouted at them to sit down. Nothing he shouted worked. Then with one fluid motion, he slid the tiller from the head of the rudder, took two steps forward, and using both hands swung the heavy timber directly at the oiler's head. The man fell into the bilge as though dead, and in absolute silence the castaways watched the captain step backward, slide the tiller into the rudder, and ease the boat back on course. Kuhne later revealed that he had tried to maintain "the necessary mask of calm authority," but secretly he feared he had killed the man. When the oiler came to with nothing worse than a massive lump to show for his experience, the keeper of discipline felt immense relief.[4] In several seconds Kuhne had shifted from the master of a modern steamship to master of a sailing vessel, but not the sailing ships in which he had trained and first had command. He had shifted into the mode of an eighteenth-century master governing his crew with discipline alien to most World War II merchant seamen.

The possibility of tension of the type that erupted in the *Express* lifeboat had been growing since the last years of the nineteenth century; sailing-ship men had warned that something—perhaps character, perhaps discipline, perhaps seamanship itself—had gone missing from steamship men. By the 1930s many traditionalists were ignoring the future of seafaring and instead contenting themselves with glorifying the past and condemning the present. Everyone employed aboard a liner, from master to coal trimmer to elevator operator, supposes himself a seaman, lamented Alexander H. Bone in 1933. "It was not always so. In my apprenticeship days everyone had to be a sailor, and it was on his worth and ability as such that he gained the respect of his shipmates." *Bowsprit Ashore* demonstrates that *seaman* had become essentially a legal term, and that *sailor,* already vaguely anachronistic in its connotation of sails, might even so be the only term appropriately applied to men who might perform well in marine emergencies. Aboard a liner, Bone argued, most crewmen had become specialists at essentially indoor jobs, and no thoughtful old-timers expected much of them in shipwreck or in lifeboats.[5]

World War I narratives suggest that lifeboat disaster often resulted not only from a lack of traditional seamanship but also from the inability of crewmen to act in concert or even to tolerate each other in the lifeboat. World War II experience confirms what Great War narratives imply, and also strongly suggests that lifeboat disaster frequently involved the failure of crewmen rather than passengers.[6] The example of the *Titanic* sinking suggests that in 1912 officers expected weakness and worse from their own men, and perhaps from passengers too, as in the horror of the *Arctic* foundering. A surviving *Titanic* passenger, Mrs. J. Stuart White later testified that a woman held the tiller of her lifeboat during the long night, that only women rowed, and that the several crewmen aboard not only knew nothing of rowing but argued among themselves constantly. "I settled two or three fights among them and quieted them down," she remarked disapprovingly of crewmen she found incompetent, divisive, and dangerous.[7]

As Steven Biel demonstrates in *Down with the Old Canoe: A Cultural History of the Titanic,* American newspapers manufactured rousing stories of gunplay aboard the foundering ship, most of them involving first-class passengers shooting men from steerage. But the *Baltimore Sun* reported an illuminating event in the radio cabin. The assistant radio operator discovered "a grimy stoker of gigantic proportions" about to stab to death the chief radio officer, preoccupied with sending an SOS message, so that he might steal his life belt. According to the *Sun* story, the revolver-wielding assistant saved his superior by shooting the stoker dead. Whatever their reliability, such lurid newspaper tales circulated widely, became accepted as factual by early historians of the disaster, and perhaps gave rise to the tales of revolvers aboard *Titanic* lifeboats.[8] On the other hand, surviving officers Lowe and Lightoller both apparently carried their revolvers with them into their respective boats.

British newspaper accounts of the disaster differed from American ones in an important respect identified by Richard Howells in *The Myth of the Ti-*

Top: In his 1837 compendium of ocean disasters, *Interesting and Authentic Narratives of the Most Remarkable Shipwrecks*, R. Thomas recounted the heroic efforts of men to save women and children first. *Bottom:* Thomas recounted as well the most selfish of acts during shipwreck.

tanic. British journalists immediately united to defend the behavior of *Titanic* officers and crewmen. Howells's analysis reveals not only the unwillingness of the British press to investigate the use of revolvers during the sinking and the behavior of crewmen in lifeboats but also its greater unwillingness to implicate the White Star Line, *Titanic* officers and crewmen, or the Board of Trade in the loss of life.[9] The orgy of self-exculpation, which stemmed from the massive blow to British self-esteem, finally prompted Conrad to explode in the pages of the *English Review* that those who drowned in the disaster died as ignominiously as anyone poisoned by a can of spoiled food and that the master of *Titanic* ought not be honored for having gone down with his ship. "So did the cat," snapped Conrad.[10] But the elite readers of the *British Review* were few, and the handful of essays on the topic by Conrad and other thoughtful and courageous Britons did nothing to impede the successful mass circulation effort to valorize the sinking.

"Be British." The words end the statement on the monument to Captain Edward John Smith at Lichfield in England, and they form the core of the British perspective on the disaster today. Immediately after the catastrophe, and despite subsequent testimony to the contrary from a stoker, the British press insisted that Smith spoke the words in his last order to his remaining crew. "'Be British,' was what we would have expected and wanted him to say," wrote J. E. Hodder Williams in *British Weekly*. "He belonged to the race of the old British sea-dogs. He believed with all his heart and soul in the British Empire. He had added that to his creed. I am glad he recited it at the end."[11] Unsurprisingly, subsequent allegations of misbehavior were perceived as an insult to Britain.

The probing of Biel and Howells into the mysteries occluding the *Titanic* disaster explain much about the contemporary fascination with the shipwreck and even more about the early power of modern mass media. Howells addresses explicitly not only the ways James Cameron's 1997 film *Titanic* deals with the darkest legends about the shipwreck—especially the gating of exits leading up from steerage—but also the power of cinema to perpetuate such untruths. "In view of the millions who have seen this film," Howells writes, such legend "is increasingly likely to have entered the recent social memory as fact."[12] The film *Titanic* is an American film demonstrating the enduring nature of a cultural divide. *A Cultural History of the Titanic* and *The Myth of the Titanic* resemble Richard Collier's *The Sands of Dunkirk* in emphasizing how mass media distorted fact into what became legend. The defeat of Dunkirk became the victorious evacuation and, in time, the myth of the "little boats" that crossed the Channel and saved the day. The *Titanic* lifeboats vanished before the Senate investigators began to piece together a picture of ill-equipped lifeboats crewed in at least some instances by incompetent men. The night after the boats arrived in New York, souvenir hunters stole all their equipment, according to Walter Lord, and the following day White Star Line employees sandpapered the name *Titanic* from each.[13]

Revolvers drift through lifeboat literature as fitfully as any derelict ever tracked on a pilot chart. *How to Abandon Ship* warned readers to make certain no seaman had a pistol, explaining that firing it at a surfaced submarine would only draw return fire from the enemy. And in any case, federal law prohibited merchant seamen from possessing firearms.[14] As early as 1903, Congress forbade merchant seamen to wear sheath knives, in the first of a series of acts that in time prohibited guns.[15] The acts speak loudly of the declining importance of sail—how a seaman aloft on a stormy night is to open a clasp knife while hanging onto rigging with one hand must perplex anyone who routinely goes aloft with a sheath knife—and whisper of shipboard violence between crewmen. More subtly still, they suggest incipient violence against officers.

The 1872 federal shipping act abolished flogging in merchant ships, and the 1898 act further protected merchant crewmen from corporal punishment, but violence like that recounted in Howard Pease's 1926 steamship novel *The Tattooed Man* remained routine. From the last decades of sail until well into the twentieth century, physical punishment of the sort Captain Kuhne used in the *Express* lifeboat remained ordinary in extraordinary circumstances. Shipping companies, officers and men, and passengers alike accepted it right through World War II, in accordance with an 1899 federal decision: "Prompt obedience by the crew of a ship to the commands of the officer on deck is essential to the safety of the vessel, and may be enforced by the officer, even by blows, when necessary; and a court will not hold him liable in damages therefore, where he uses no weapons, and there is no evidence of malice or excessive punishment."[16] Judges understood that at sea an emergency often demands instantaneous response, and that men refusing duty, especially men beginning to panic, must be managed with whatever means an officer has available.

Officers were ordered to keep their pistols, remembered Royal Army officer Neil McCallum of his passage aboard a World War II liner-turned-troopship bound from Scotland to the east coast of Africa via Cape Town. First the officers were told to store their side arms in the ship's armory, but a few hours later orders changed: the pistols would be useful if the ship had to be abandoned. *Journey with a Pistol: A Diary of War* differs from many accounts of wartime in that it derives from a detailed journal in which Mc-Callum recorded even conversation. The troopship sailed with more than four thousand troops aboard plus its crew, but had lifeboats for less than half that number. McCallum's commander expected panic if the ship had to be abandoned and issued a simple order in advance: "Keep your pistols and shoot anyone who dives for a boat. Officers or men." McCallum implied that such concern resulted less from lack of faith in the British troops than from lack of faith in the civilian crew, but overcrowding destroyed discipline from the start. Solitude proved impossible, he wrote of the men packed belowdecks, queuing for meals, for lavatories, for fresh air, even to be seasick, all worried by the constant threat of U-boat attack and irritated by the substantial overcrowding.[17]

When torpedoes or bombs strike troopships, the loss of life is terrific, warned the authors of *How to Abandon Ship*. "*Arcadian*, 279; *Aragon*, 610; *Louvain*, 224; *Santa Anna*, 638," begins the list of torpedoed Great War troopships and numbers of dead. Although convoy escorts make troopships unlikely to be attacked, let alone sunk, the authors continued, in the event that soldiers must abandon ship, merchant seamen should expect to see devolution of army discipline among enlisted men and officers alike.

By the middle of World War II a number of experts considered troopships to be disasters waiting to happen, exemplifying conditions otherwise encountered only in lifeboats. In the final chapter of *What to Do Aboard a Transport,* R. W. Gerard briefly explained not only the mental and emotional effects of homesickness but the depression, anxiety, and aggression that frequently come over passengers aboard troopships. Simple crowding and other physical constraints aggravate such conditions, and hitherto tiny annoyances can flare into serious emotional disturbance. A professor of physiology at the University of Chicago, Gerard presented pathological mental and emotional states largely as the result of severe crowding, inactivity following extremely active outdoor training, and continuous anxiety concerning attack that could not be directly repulsed. Because almost no one aboard a troopship has gone aboard willingly, everyone has already lost all the psychological support of peacetime life, and homesickness and seasickness alike originate in wrenching change about which nothing much can be done. Before concluding his remarks, Gerard briefly turns his attention to lifeboats. "The hazards of a small open boat at sea deserve some attention, unlikely though you are to experience them," he wrote, before dealing with the physiological effects of cold, heat and sun, hunger and thirst.[18] The shock of losing the troopship is a worse shock than losing peacetime status, and the crowding, inactivity, and anxiety in a lifeboat work even more strongly than the same forces do aboard a troopship, for in the lifeboat all are exacerbated by the small size of the boat.

In World War II, merchant-ship officers and others relearned some of the difficult lessons forgotten since the end of the age of sail, and many of the lessons involved human nature as it manifested itself aboard lifeboats.

Nowhere is the reorientation more clearly displayed than in the changing lifeboat catechism presented in Felix M. Cornell and Allan C. Hoffman's *American Merchant Seaman's Manual,* a book first published in 1938, hastily revised in 1942, and reissued in a newly revised edition in 1964. The 1942 version presents not only detailed information about the design, construction, launching, and sailing of ships' lifeboats but a detailed question-and-answer summary intended to make candidates for lifeboat certificates think about information too easily memorized without being understood. The drill differs only slightly from those of Layton and other British Board of Trade textbook authors and emphasizes that aboard a well-run merchant ship every lifeboat contains all the stores necessary for a safe passage. But after the war Cornell and Hoffman added a new question about grabbing additional equipment.[19] Its answer demonstrates how much merchant-marine officers and men had learned about passage making, for while it specifies the grabbing of sextants and charts if possible, the third thing a *seaman* should grab—a revolver—has nothing to do with the extemporized navigation of lifeboat passage making. Aboard troopships and other merchant vessels carrying large numbers of troops, military officers kept their side arms, and the revolvers went aboard lifeboats as they went aboard the *Titanic* boats.

Derek C. Gilchrist retrieved his automatic pistol from his cabin moments after a U-boat torpedoed *City of Shanghai,* and snatched up his haversack too, which contained spare ammunition for the pistol and other personal survival necessities, although he forgot his sextant. Within a few days, Gilchrist threatened to use the pistol when some of the crewmen aboard his lifeboat began agitating for an extra biscuit ration. His simple threat to kill the next man who complained startled everyone, himself included, into silence. He also used the weapon in a vain effort to kill dolphins for food. After being rescued by a Dutch freighter, he hid the weapon for a few hours until he was certain the Netherlanders were not Germans masquerading as allies.[20] His pistol had become giver of both authority and self-protection.

The pistol appears in numerous lifeboat-passage narratives; the half-concealed, half-brandished weapon, almost invariably aimed at men with knives.[21] Seamen, the 1903 law notwithstanding, habitually carried knives, which, after all, were integral to the traditional seamanship required in launching and sailing lifeboats.[22] But no one knew if those knives made lifeboat mutiny more likely, any more than anyone knew if discipline depended on firearms aft.

Here lifeboat design intrudes into complex issues of human nature, morale, and discipline. By age-old custom, the commander of a small pulling or sailing boat sits aft, as did Bligh and Josiah Mitchell, not only holding the tiller and closely supervising men immediately about him, but facing the men forward, keeping an eye on rowers and passengers alike. Less often remarked is that the tiller slides easily from the rudder head and becomes a useful sort of club. The hand that holds the tiller holds a daunting weapon. But the design of the lifeboat rig diminishes the power of the men aft, for the lugsail rigs can be manipulated only from the bows of the boat. Given a marked change in wind direction or speed, the big lugsail must be tacked or reefed *from forward,* something West and the other experienced officers of *Britannia* Number 7 lifeboat forgot.

In the waterborne moment West and his fellow officers determined to make not for the African coast some 600 miles east but for Brazil, 1,200 miles downwind, and having made their decision, they set about organizing the boat for the passage. The British officers put the Indian crewmen forward, the Indian passengers amidships, and the Britons in the stern, something they soon realized was a mistake. The Indian crewmen knew nothing of the lug rig, and West and the others were forever scrambling over people to reset or repair it. Moreover, the Indian passengers persisted in standing on the thwarts while hanging onto the shrouds and scanning the horizon for rescue ships.

The Indian steerage passengers, landsmen en route from Britain to India, irritated West and the other officers because they did not take orders well. The Indian crewmen were not seamen, and the officers expected no more from them—and no less—than they did from engine-room personnel of any ethnicity. On the other hand, engine-room men or not, the black-gang castaways of *Britannia* know enough about small-boat dynamics to refrain from standing on seats and hanging onto backstays, and to follow orders from officers. Trouble in Number 7 began among the passengers, and in time the Britons dealing with the fracas aroused the ire of the Indian crewmen forward, who resented the treatment of their countrymen even if they were landsmen, and who fell under the sway of holy men who resented the authority of the officers. Dealing with human trouble occupied West and the other officers every bit as much as did issues of navigation and seamanship, and sorting out the human dynamics in *Lifeboat Number Seven* and other lifeboat narratives eventually preoccupies any reader.

Gilchrist remembered separating his crewmen by specialty, putting the black-gang men in amidships because he knew they lacked most skills useful in a lifeboat. In separating seamen from engine-room men, Gilchrist divided Indian men into two groups, but unlike West, he went a step further toward ensuring British control of the boat. "Four of the nine Europeans slept aft, and the other five under the bow-cover forward, with the twenty-one Indians between," he recorded in a passage underscoring the importance of controlling the lugsail rig.[23]

Like Conrad and other British merchant navy officers, Gilchrist delighted in knowing men around the empire, and he prided himself on distinguishing among them as they distinguished among themselves. He learned easily from them, especially about their religions. He was aware from the start of the lifeboat passage that the Indians divided into Hindus, Muslims, Roman Catholics, and a topass, and he had difficulties apportioning the lifeboat rations appropriately. The Hindus and Muslims refused the tinned meat absolutely, but grew angry as they watched the Britons and other Indians eat it. Eventually, religious squabbling broke out: at one point, the battle concerned the right of anyone to pray aloud for relief while others were trying to sleep. On three occasions Gilchrist withheld condensed-milk rations from those who prayed aloud all night, and he implies that the men so punished were all Muslims.[24]

Unlike West, who began his passage thinking that differences between passengers and crew might prove the worst social problem, Gilchrist confronted only crewmen, but of different ethnicities and religions. As days stretched into weeks, he found himself struggling to understand how their several religions affected them and how conflicts of faith entangled some of the men in difficulties he labeled as fatalism. On the other hand, he knew nine of the Indians as experienced seamen he had served with at length, and he trusted their ability to subordinate religion to seamanship. Five of the Europeans, chiefly Britons, proved rather more difficult to analyze and manage, since they knew nothing of the sea and merchant-marine discipline, something the Indian seamen recognized all too clearly and dealt with accordingly.

British merchant navy officers and seamen encountered a major problem in managing lifeboats crewed by non-British seamen. They faced men of the empire and beyond—Indians, Chinese, and others—whose confidence in the British merchant navy (and Royal Navy for that matter) plummeted when steamships sank, and dropped even further when the men in the stern seemed uncertain at lifeboat handling. Under these circumstances, discipline proved hard to establish and maintain.[25]

Non-European crewmen sometimes behaved so well during shipwreck that their behavior unnerved Europeans, officers and fellow crewmen alike. When a U-boat torpedoed *Viceroy of India* northwest of Algiers on a dark night, the Peninsular & Oriental men performed so exactly according to their training that they terrified at least one passenger interviewed subsequently. No fuss or panic marred the abandon-ship process, wrote George F. Kerr in his official history of the company at war, and the crewmen from the coasts of the Indian subcontinent behaved correctly during the lifeboat passage. However, one passenger remembered that after stumbling up three decks in blackness, hands before him, he touched the face of a man who neither moved nor spoke. Then he touched the face of another man standing - stock-still. Against all orders, the passenger turned on a flashlight and saw a "party of Lascars who had fallen in at their boat stations and were now standing perfectly motionless in the dark and at attention, waiting for their orders."[26] Following Board of Trade instructions forbidding unnecessary talking,the men refused to speak to the passenger, perhaps because the latter had been fool enough to shine his flashlight, but perhaps because, after all, he was only a passenger. Under the orders of their petty officer, the men intended to prove themselves more unflappable than any Englishman, and they succeeded.

Nowadays so many of the words defining Asian crews drift beyond the reach of even specialized dictionaries. *Serang* and *seacully, seacunny* and *tindall* correspond roughly to traditional British terms like *bosun* or *quartermaster*. However Anglo-Indian English differentiated among such words over decades of imperial rule, the actual role of the serang or seacunny paralleled that of European warrant officers. A British freighter ordinarily carried a European bosun even if it had a lascar crew managed by a serang. *Lascar* itself, used without definition in nineteenth- and early twentieth-century British Merchant Shipping Acts to mean crewmen from the East Indies or India, meant different things to different companies, officers, and writers.

To Conrad, it appears to have meant the Malay crewmen of *Patna,* the steamship of his novel *Lord Jim,* whose protagonist, after the ship collides with a derelict at sea, learns the depths of European-officer depravity. Knowing they have only seven boats and eight hundred Arab pilgrims bound for the Red Sea, the five white men abandon ship in an act of panic and cowardice that orders the entire novel. Conrad frequently focused attention on the relations of Europeans and people from everywhere else, and he was especially attuned to the nuance of authority at sea. Ancient and rusted, *Patna* carries only European officers for its all-lascar crew, in part because the ship is "owned by a Chinaman, chartered by an Arab, and commanded by a sort of renegade New South Wales German." In other words, it is a "country ship," one owned on the fringes of the empire. Its quartermasters are lascars, and they answer to a serang who is actually the bosun, and a competent one. When a French warship subsequently discovers the steamship drifting on a flat sea, the serang has already hoisted the British ensign union-down as a distress signal and the cooks are feeding the passengers.[27] The serang has acted correctly, but he and his cohorts cannot navigate the ship, even in an emergency, for they are dependent on the skill and authority—what Conrad calls "wisdom"—of the European officers.

Shipwreck frequently shattered lascar belief in European wisdom and competency, and on British and American ships, ended pay as well. Aboard British lifeboats, officers and European crewmen confronted men with whom they communicated only through serangs or tindalls, and who accepted the mechanical order and strict discipline of a steamship as given, as part of the proof of European superiority. When the ship sank, officers had to prove their worth in front of men suddenly critical. "There is something peculiar in a small boat upon the wide sea," opines the narrator of *Lord Jim*. "Trust a boat on the high seas to bring out the irrational that lurks at the bottom of every thought, sentiment, emotion."[28]

No predictable patterns of culture clash emerge from wide-ranging reading of open-boat and lifeboat narratives written by Europeans, nor from recent analyses of twentieth-century merchant marine labor history. Even nineteenth-century authors like Frederick Benton Williams, whose *On Many Seas: The Life and Exploits of a Yankee Sailor* describes the seeming lack of feeling expressed by Hawaiians whose brothers die at sea, do not quite touch the cross-cultural mysteries of at-sea behavior in adversity.[29] Sometimes Asian crewmen are fatalistic, sometimes blessed with an optimism opaque to Europeans; sometimes Asians are silent, sometimes noisy; sometimes lascars are intrepid seamen, bold with the lugsail rig, their discipline and enterprise recalling the original military meaning of lascar. Sometimes they remain for days in a stupor that makes them more useless than landsmen passengers and puzzles not only European seamen but Asian passengers too. Often Indian and Chinese crewmen die early of frostbite and exposure, far earlier than their serangs, but sometimes only engine-room men die early. At least occasionally, engine-room lascars fare better in lifeboats commanded by engineer officers they know and trust, and understand to be as unfamiliar with sailing open boats as themselves.[30] In all too many cases, Asian crewmen deprived of their serangs behave in ways that suggest mixed awe and scorn toward the European officers holding tillers and struggling desperately to manage lifeboats from another era.

Racism suffuses many lifeboat accounts, but the racism is not easy to categorize. In *1700 Miles in Open Boats,* Foster reports the specialist contribution of an old seaman experienced in sailing ships and calls him "a real white man." But what does "white man" mean to the master of *Trevessa?* Perhaps the important word is "real." Certainly it is a Briton who convinces some of the "colored men" that they can drink seawater, and it is another Briton who drinks alcohol from the broken compass; both white men grow ill and require care thereafter. The *Trevessa* crew included men from Britain and Sweden, Singapore and Burma, India and West Africa, the Cape Verde Islands, and from wherever originated the men Foster designated "Arab," but trouble seems to have begun among the British and then spread to non-Britons.[31]

But crowding explains more of the human tension that characterizes so many castaway experiences than the simple spread of misinformation and misbehavior. Company is harder to tolerate in a very small vessel than solitude, observed Jean Merrien in *Lonely Voyagers.* "Propinquity remains a problem, and great efforts of self-control are required to avoid serious disagreements—quarrels, even." In a lifeboat, idleness exacerbates crowding. Most of the time, only a few of the castaways have much to do, and sometimes only the person holding the tiller is occupied. Years into World War II, a few steamship lines, seamen's unions, and eventually government agencies decided that boredom might be slightly alleviated by cardplaying, and packs of playing cards found their way into survival bags and lifeboat equipment boxes. But cards alone did little to assuage thwarted creativity. As Merrien noted, "Man is not made to live in small masculine (or—dare I say?—feminine) communities without domestic or artistic outlet."[32] Unless they looked at their surroundings with the attention Gatty recommended in *The Raft Book,* most castaways had no outlet whatsoever.

Thirst and hunger alone fail to explain the intense absorption with which so many lifeboat castaways monitored the distribution of water and food. As Foster points out in his narrative, distributing water and rations became a sort of entertainment and the domestic highlights of every day, especially after the tobacco supplies gave out, thus eliminating the entertainment of lighting up pipes and cigarettes. Only a few of the men in his boat had the skills to do much, and only the old seaman had the know-how to keep himself busy with improvised tools and materials.[33] His busyness made for him a tiny zone isolated from the crowding. Beyond that, Foster realized, it also made for valuable entertainment, an ongoing demonstration of the traditional skills so necessary in managing a long-distance passage in a boat from the age of sail.

Crowding precludes any confidential discussion; equally so any masking of error. On the other hand, it facilitates the transmission of misinformation by whisper and gesture and furthers any opposition by strong-minded men forward to the strong-minded men aft. Nevertheless, crowding in itself does not automatically create opposition to authority. Benson's *Log of the El Dorado* demonstrates that good luck, fast thinking, and old-fashioned seamanship can carry a greenhorn crew to safety with little dissention. Benson and others imply that opposition rarely arises when officers are alone in their authority and performing in ways that reassure crewmen. Only when discussion or failure draws opposition does crowding worsen matters and discipline—or morale—decay.

Back in the early 1800s, William Vaughan had begun analyzing shipwreck and open-boat narratives for the good of mariners everywhere, suggesting that when properly studied, misfortune offers useful lessons. So many narratives had appeared toward the end of the eighteenth century that Bligh's passage fitted into a much larger framework, albeit one little known to mariners frequently away from London and other bookselling centers, and often too poor to purchase books in any case. An open-boat narrative by David Woodard, which Vaughan edited for publication, prompted him to place small-boat survival in context. Woodard and four seamen, dispatched from one becalmed sailing ship to another for provisions, became separated from their vessel, then lost for days before landing on the island of Celebes. Vaughan examines this incident to learn how discipline and morale contribute to survival. While offering all sorts of advice on physical matters, Vaughan focuses chiefly on open-boat philosophy. "Nothing can be more destructive to life, or to perseverance, than permitting the depression of the mind or spirits," he asserts.[34]

After launching boats and getting adjusted to being at sea in small craft—Vaughan does not use the phrase "waterborne moment" but he clearly recognized the post-lowering interlude as a discrete period of time—the real dangers to morale arise from fair weather, lack of provisions and equipment, and poor direction. Fair weather, although it reduces the need for exertion, ends seasickness, and enables sleep, also produces a sort of emptiness. Smooth sailing saves work, but it quickly erodes physical condition because the cramped men have little else to do. Exposure and a dearth of provisions physically weaken small-boat castaways, who while thirsty and hungry must also suffer from cold or heat, wind and wet. Vaughan argues for putting spare sails and blankets in ship's boats, and for insisting that men sit in the bottoms of boats, even if awash in saltwater, because the sides break the wind. But loss of direction remains the chief evil, even for experienced seamen, and it is the duty of officers to maintain direction, on which depend discipline and morale and the tamping down of what Conrad called the irrational. "Direction" connotes more than the course the lifeboat sails: it means an emotional vector.

According to Vaughan, Bligh and other successful small-boat passage makers "always discouraged despondency" by directing crewmen away from gloom. As does Stacpoole in *The Beach of Dreams,* he emphasizes that however well seamen do in ordinary adversity, shipwreck empties them of something officers or other strong-minded individuals must restore. "Despondency, insobriety, and insubordination—qualities that canker hope and induce vexation—have often proved the seeds or secret springs of mutiny and disaster," Vaughan concludes.[35] Officers must prevent such wickedness by realizing that discipline in a small boat differs fundamentally from discipline aboard ship.

In the boat, discipline derives from the primary need to survive, and officers must form new bonds with their men. Vaughan suggests that the ordinary hierarchical structure of command must give way to a team approach in which the officers command by example. Punishment means not individual disgrace or loss of privileges but dismasting, capsize, death for all. From the waterborne moment onward, therefore, the officers must encourage the men's religious strength by praying with them for common deliverance from common danger. Prayer produces tranquillity and encourages both hope and exertion while forestalling despondency and despair.[36]

As a sin, despair fits into the sophisticated religious construct of meaning and hope that ordered the notebook entries of the *Hornet* castaways. As a problem in leadership, however, despair is something much vaguer. No chapter on leadership opens steamship-era books. The *Merchant Marine Officers' Handbook* advises young officers that among their new duties will be taking charge of lifeboat maintenance, but about castaway leadership it is silent.[37]

In steamship-era handbooks, intimate knowledge of equipment is the prerequisite to proper leadership of men, all authors assuming—in the old sailing-ship-era way of training cadets—that leadership develops "naturally" as a sort of coefficient of seamanship. How long leadership ability might take to develop in lifeboats no one addressed. By putting young officers in charge of lifesaving equipment, especially lifeboats, shipowners and -masters enforced intimacy with artifacts that whispered of nasty situations and a lack of superior officers. About despair, faith, and above all direction, steamship-era manuals are mute.

Adding passengers to the castaway stew complicated any lifeboat passage, especially when the passengers were women and children. The *Britannia* castaways were all grown men; Gilchrist's *City of Shanghai* lifeboat carried all males, its sixteen-year-old schoolboy finding himself abruptly promoted to manhood. Yet long before World War I, lifeboats crowded with women and children pushed away from sinking ships, and in wartime such incidents multiplied dramatically. Nevertheless, rarely do women and children figure in narratives written by men.

A small number of women castaways kept journals, and among the best is that of Emily Wooldridge, who chronicled the burning of *The Maid of Athens* in February 1870 and its subsequent grounding on Staten Island, south of Tierra del Fuego. Wooldridge; her husband, master of the ship; and eight other men eventually reached the Falkland Islands in a modified small boat, and the journal describes seamanship every bit as fine as any exhibited by Bligh or Mitchell. But about herself, Wooldridge says little, describing herself at the end of the muster list only as "on board ship a No-body, so it cannot matter whether I am good or bad tempered." Despite actively assisting in the small-boat passage—she doled out water and rum rations, managed the store of sacking used as blankets, and kept a candle burning steadily night after night in atrocious seas in order that her husband might see the compass—she describes herself as being shut up under a canvas cover almost as baggage.[38]

Shipwreck disorder worsens in the presence of women and children simply because shipboard order essentially ignores them to begin with. Only a handful of experienced seamen remarked the essential problem posed by women passengers aboard sailing ships, and while James Fenimore Cooper and Conrad near the subject in novels like *The Red Rover* and *The Rescue*, perhaps Alan Villiers strikes more closely, in large part because he wrote after the collapse of the Victorian and Edwardian sensibilities that so restricted his predecessors. "It was a man's world, a bachelor's world, run at sea (but not in port) very like a monastery," he writes in *The Way of a Ship*, of sailing vessels at the turn of the twentieth century.[39]

Villiers offers two crucial insights into everyday sailing-ship culture. First, most vessels carried no "surplus people, to get in the way," except occasionally the wife of the master; and second, merchant seamen favored spontaneity and innovation over the drill-based ethos of large-crew warships and, later, steamships. Aboard a typical sailing ship, all but the newest hands knew their duties by experience, and even the neophytes had begun learning by doing. But as great steamships supplanted sailing ships, drill came to be identified with a new shipboard order, and the word *drill* became peculiarly associated with the word *lifeboat*. Lifeboat drill, in Villiers's opinion, "was retained in the passenger vessels, to impress the passengers."[40] A bit of drill made everything seemingly naval and reassured passengers, perhaps especially women, that in an emergency things would proceed according to drill. Nothing, of course, was further from the truth.

For generations, seamen considered women aboard ship bad luck, as Villiers explained in a 1932 *New York Sunday Herald Tribune* article on marine superstition in the age of ocean liners. "Hoodoos Still Haunt the Sea" recounts the true story of a girl stowaway aboard *Herzogin Cecilie,* a grain-carrying square-rigger racing others from Australia to Germany. Discovery of the young woman filled everyone aboard, master included, with foreboding: no one expected to make a good passage, and everyone recalled that the last time they had a woman aboard they had gone aground in Denmark. Interviewing Jeanne Day after the passage, during which the master made her work her fare as a cabin boy, Villiers had the opportunity to learn something of her thinking. The stowaway believed that "high-strung and somewhat nervous females" irritated seamen by destroying their routine. "The sailor must have no nerves; the woman always has them," Villiers concluded as a summary of her thinking, not his.[41] More than simple sexism characterizes this passage. Too many seamen found passenger nervousness—male and female—unsettling.

By 1932 it had become a commonplace that the acceleration of American living since the 1880s had produced two generations of "nerved-up" men and women whom the psychologist William James termed "bottled-lightning" types.[42] The cultural velocity engendered by automobiles, cinema, air travel, radio, and other novelties only heightened the nervousness physicians, magazine columnists, and journalists soon called "the American jitters."[43] Many passengers boarded ship so highly strung that even two- or three-week-long passages did little to calm them, and seamen gradually learned that such passengers caused all kinds of trouble and in an emergency behaved badly indeed. From all accounts, women became more "troublesome" after the close of the sailing-ship era, in part because loosening sexual mores encouraged onboard teasing of seamen like that O'Neill portrays in *The Hairy Ape.* Too many women passengers knew only protected environments—the safe if busy world of suburbs, department stores, office towers, and highways—and they failed to recognize the dangers implicit in seafaring. Traveling by liner seemed to them merely an extension of traveling by express train, and indeed some women writers equated the two.[44]

Great War propaganda complicated the already problematic position of women. British writers condemned U-boat warfare in large part by describing the horrors to which it subjected women castaways. By 1915, certainly after the sinking of *Lusitania,* the English-speaking world was witnessing the rebirth of Victorian sentiments in the service of British naval victory. Writers on the "rape of Belgium" merely shifted their attention seaward and began to describe women thrust backward in time—women frail, weak, and defenseless, and rather incompetent too, for all that in lifeboats they kept stiff upper lips. The same framework reappeared in World War II, in morale-boosting books like Elspeth Huxley's *Atlantic Ordeal: The Story of Mary Cornish,* although in that case an undercurrent of resilience and good sense animates the text. Not only does the propaganda framework reinforce the ancient-mariner superstition of woman as "hoodoo," but it camouflages the demonstrated abilities of women aboard lifeboats. The propaganda framework essentially demeans the accomplishments of women castaways. Many women learned in lifeboats what Jeanne Day discovered aboard *Herzogin Cecilie:* seamen accepted women as equals when they abandoned the expected "nervous frailties" and accepted the same discipline as the men about them.

Shipwreck itself unmasked hidden facets of class and gender inequality. The importance of clothing to all but upper-class women, for example, figures so subtly in so many shipwreck narratives as to pass unnoticed by most male narrators and historians. But the compulsion on the part of lower- and middle-class women, and some upper-class women too, to *get dressed,* indeed to get properly dressed for a boat-deck promenade *before moving to their lifeboat stations,* frequently killed them.

In his penetrating chronicle of the *Royal Charter* disaster of 1859, Alexander McKee pays especially close attention to the attire of women found washed ashore after the auxiliary steam clipper ran aground near Liverpool at the height of a hurricane. The clothing of the dead indicates that the women dressed frantically, many pulling on the clothes of their female cabinmates. More important, the clothing of the drowned reveals that some women had shed nightdresses and begun dressing as they normally would. Citing coroner descriptions, McKee writes that many of the dead drowned in corsets, steel-spring stays, and layers of outer clothing; many drowned wearing high-laced boots and gold-coin-filled money belts. McKee points out that some men drowned fully clothed too, but they had dressed and gone on deck when *Royal Charter* first got into trouble, while the women had stayed below in nightgowns. After the ship struck the rocks, the women had almost no time to get dressed, but dress they did.[45] Their clothing

offered them no chance at all in the surf pounding the shore a few yards beyond *Royal Charter,* and almost none shed clothing in an effort to rescue
themselves.

Victorian- and Edwardian-era women wore extraordinary amounts of
clothing, of course, and by today's standards those garments were quite complicated. Conrad mused about their attire in shipwreck situations. Mrs. Travers, the shipwrecked heroine of his 1919 novel *The Rescue,* not only winds
up wearing a Malay jacket and sarong but sandals over bare feet, clothing
that her husband condemns as "heathenish" costume. Without stockings,
luxuriating in filmy "native" clothing while her one European-style dress
dries, Mrs. Travers revels in finally dressing appropriately for the tropical climate, if not for her propriety-conscious husband. *The Rescue* slices as deeply
into the human nature of shipwreck as did *Lord Jim,* but it raises issues for
which Conrad is scarcely known. Perhaps half-clad women destroy discipline among male castaways, but perhaps the half-clad women enjoy a relief from mores they have long hated.[46] Shipwreck alerts Conrad's character to the restrictions women impose on themselves, including fears that
wearing male attire in the end makes them the equal of men.

Read with an eye for costume, shipwreck accounts reveal the unwillingness of so many women to abandon their traditional clothing. The chaos
of *Lusitania* after torpedoing is perhaps especially instructive, if only because
so many eyewitnesses and historians documented it. Women did not panic
even as they watched lifeboats swinging inboard, coming free from falls, and
rocketing down the forward-sloping deck into crowds of passengers. They
simply hung back, continuing to expect men—at first crewmen and officers, then fellow passengers—to assist them into the boats men must launch.
Gradually some realized that their voluminous skirts and coats would keep
them from boarding what lifeboats did swing free of the listing, bow-down
hull, and that they would have to jump into the sea in order to reach lifeboats
floating about haphazardly. Some men aboard *Lusitania* noted, almost ab-
sentmindedly, that ankle-length skirts kept women from walking about the
sloping decks: how were they then to jump into swung-out boats? Men simply abandoned most or all of their clothing, especially when they realized
that all lifeboats had been lowered or destroyed and that jumping remained
their last chance; but few women willingly followed their example. One un-
buttoned her skirt before jumping overboard, perhaps hoping that hitting
the water would pull it free, another mindlessly packed and unpacked a suit-
case, and most dithered as the stern of the ship swung higher and higher
until jumping overboard meant a sixty-foot drop.

One eyewitness beheld a first-class passenger systematically tearing off all his wife's clothing, then fastening a life belt around her, but most women who finally dove overboard, or floated free as the hull sank, found themselves in the water fully clothed, some even properly hatted. Only a group of younger women and teenage girls—perhaps because their survival swimming lessons routinely involved nudity and perhaps because their upper-class training had taught them to value life over modesty—tossed off most or all of their clothes beforehand. Men swimming toward lifeboats realized that the lightly dressed women did better than fully clothed women with life belts. One nonswimmer minus a life belt, wearing only a blouse and thin slip, survived by holding onto floating wreckage, something that made the drowning of fully clothed, life belt–wearing women all the more ironic.[47]

In the midst of the chaos, two facts became clear. First, life belts alone did not keep fully dressed women afloat for long: fully clothed, life belt–wearing women survived by hanging onto floating debris. Second, fully dressed women proved so drastically waterlogged that others had to struggle desperately to drag them into lifeboats. The huge, thick skirts kept many women from swimming to capsized lifeboats, from staying afloat until lifeboats could reach them, and from pulling themselves aboard lifeboats and wreckage.

Scrabbling from debris-strewn cabins along topsy-turvy decks, leaping from a rail to a lifeboat swinging below its davits, scrambling down rope ladders, even swimming a few feet or yards to the nearest lifeboat as its lurching occupants maneuvered it toward people nearby—all these things meant something vastly different for women than for men. Men slipping in leather-soled shoes thought nothing of discarding shoes and socks and moving barefoot toward safety. Most women hesitated to remove their shoes, let alone their stockings, and many slipped and fell as they tried to raise their hems above obstacles. Men who immediately shed suit coats, vests, neckties, and shirts in order to aid crewmen launching lifeboats never hesitated to remove their pants before diving overboard, but only a few upper-class women, so secure in their social position that they dismissed condemnation, followed this example. The Cunard Line underwrote emergency clothing expenses for the many survivors who left *Lusitania* literally naked, but only a handful were women.[48]

Some women, like those aboard the schooner *Pro Patria* when it struck a South Seas reef, survived almost in spite of themselves. James Norman Hall found them packing their clothing even as the schooner began to break up. No man had thought of packing anything, Hall recorded, and he dismissed the clothes packing as indicating the importance of such things to women.[49] But the women had nothing else to do in the immediate aftermath of that disaster; while male passengers dashed on deck half-naked to assess the situation and perhaps assist crewmen, they stayed below, dressed themselves normally, and then began packing. The women of *Pro Patria* behaved as so many other shipwreck-snared women did: they busied themselves while men arranged for boats.

Narratives of most other shipwrecks make one thing clear: in all but sedate shipwrecks like the *Titanic* or *Pro Patria* founderings, women passengers appeared to survive largely by abandoning all social proprieties before abandoning ship.[50] In cabins and on deck, then in the water alongside the sinking ship, and finally in a lifeboat, upper-class women chose autonomy over the bourgeois drawing-room and shipboard mores so prized by Conrad's Mr. Travers. For women, as Stacpoole demonstrates in *Blue Lagoon* and *The Beach of Dreams*, as Jack London explores in his 1911 novel *Adventure*, and as Conrad probes so relentlessly in *The Rescue*, surviving shipwreck demands a complete absence of prudery and a reliance on personal strength and nerve rather than on mores or men.[51] Women who expected special attention in a lifeboat because they were women soon learned the folly of their expectation. Women who intended to survive accepted the microcosm of the lifeboat without question.

"How far can you swim?" asked someone abruptly of Elizabeth Fowler, a woman jammed forward in an overcrowded, squall-beset twenty-six-foot-long lifeboat making land at night among the reefs of Barbados. Since the torpedoing of *West Kebar* ten nights earlier, Fowler had discovered one novelty after another—not least of which involved being the only woman aboard the lifeboat—and now she learned the dangerous combination of exhaustion and surf. As she relates in her 1944 narrative, while *West Kebar*'s master ordered everyone back into life jackets and the radio operator desperately signaled the blacked-out coast with flashlights, Fowler slowly realized that finding the islands had been simple compared with negotiating the reefs. Everyone aboard, all United States or Canadian seamen and passengers, cursed the inability of "the Limeys" to see the lifeboat rockets or the Morse code flashlight signals, and gradually the castaways fell silent. But Fowler could swim, and she answered the question casually, suddenly thinking about helping the only child in the boat, a boy whose father could not swim.

Standing Room Only: The Personal and Moving Record of a Woman's Experience during Ten Days in a Lifeboat with Thirty-Four Men after Their Ship Had Been Torpedoed by a German Submarine is the only book-length castaway narrative written by a woman, and one of the few wartime books that eschews the propaganda framework of so many newspaper and magazine articles. Fowler had the mind-set and time to work out many significances that brief news articles about and by women castaways ignore or slight, and while her book equals *Blue Hell, Lifeboat Number Seven,* and other first-rate World War II lifeboat narratives in its intensity of observation of equipment, weather, seamanship, and human emotion, it also unwittingly reveals Fowler's transformation from a tough-minded, freethinking young woman into a very capable castaway.

Nighttime attack had shattered the easygoing life aboard the small freighter steaming from Africa to the West Indies. From the beginning the passengers had divided into two groups, the homebound missionaries, all male, and three others, two men and Fowler. The young woman joined the off-duty navy gun crew to sunbathe by day—sometimes skipping rope or teaching herself to work her small portable typewriter—and to sit atop a hatch cover by night, singing songs to drown out the missionary prayer services. The wife of a medical doctor in the British colonial service, but herself an American citizen, Fowler had had enough of missionaries in West Africa and enough of their remarks about her shipboard behavior. She considered herself lucky to be aboard *West Kebar* at all, since two years earlier British authorities had radically curtailed the shipboard passage of civilian women.

The war had disrupted Fowler's family life substantially. Her four-year-old daughter had been evacuated from England to Canada, and Fowler had been able to visit her only by signing aboard a Norwegian tramp steamer as a crew member. She then returned to Africa to help her husband, then set out again a year later, alone, to Canada. *Standing Room Only* is a rich book, in part because its author remains slightly mysterious from beginning to end. Fowler not only liked men but moved among them easily, bantering and laughing, in part because her jungle life had made her confident and independent. While *West Kebar* waited in Freetown for a convoy to form, her male friends visited the ship with presents, a naval officer bringing the last bottle of Coca-Cola in West Africa, a Pan-American Airways pilot arriving with a box of chocolate.

Once aboard the freighter, however, she discovered that the men, though not the navy sailors, regarded her presence in the way Villiers had described. Even the ship's officers saw her as occupying a place a man might need more urgently, and as sort of a nuisance too. Try as she might, Fowler could do little to overcome the unfriendliness of the officers, something that bothered her increasingly as the weeks passed. She recognized that the officers saw her as unlikely to help herself in an emergency, and so mentally filed her away as a burden.[52] She made a concerted effort to give no trouble, to expect nothing special in her cabin accommodation or service, to be perfectly pleasant. Yet she chafed a bit under restrictions she understood but felt were less than fair, at least in wartime. For one thing, she had to wear dresses, even though by 1943 many women had begun wearing pants, and she absented herself from the male passengers whenever things got too ribald.[53]

The torpedo struck as Fowler was walking back to her cabin one night, away from the men singing lewd songs. She burst into her pitch-black cabin and grabbed her life belt, abandon-ship bag, and Burberry raincoat, not at all panicked but determined to prove her worth. She left the cabin and moved through the dark passage, counting her steps to each turn, and arrived quickly at her lifeboat station. All went well so long as she remained on her own, until she reached her station at Lifeboat Number 1. Then chaos, a chaos of men, enveloped her.

Most of the deck and all of the lifeboat had vanished. Only darkness and smoke remained, and a mass of twisted wire Fowler instantly recognized as both wireless antennae. Then she made a mistake. Spotting some male passengers, she asked what to do. One of the men grabbed her hand and dragged her, slamming her into twisted steel, and another began yelling for a light. When she opened her bag to get her flashlight, she spilled everything but the light, which some man snatched with a shouted "Good girl." Lifeboat Number 3 had been shattered by the blast, and men dragged Fowler, who had bent to gather up her abandon-ship equipment, away from her raincoat across to the port side, where she found the two lifeboats lowered and a cargo net slung down the hull. Another man grabbed her bag from her, flung it toward the lifeboat floating far below, then shoved her onto the net. She climbed down the net, then let go and tumbled into the tossing boat, half-expecting "a volley of applause."[54]

Instead she found "a different note. Grim. Intense." As ominous noises issued from the ship, the men disconnected the falls, shoved away from the hull, and began the work of getting out oars, shipping rudder, and picking up men diving from the ship. One of these turned out to be the captain. For twenty or thirty minutes, the *West Kebar* castaways tried to sort out equipment, before a second torpedo finished off the ship and a huge U-boat surfaced. "It was all too Hollywood," Fowler explains in her first use of cinema imagery to explain real-life events. "An artificial voice, straight from Hollywood, bellowed through a megaphone in carefully enunciated English, perfect and precise, 'What iss the name of your ship?' and 'Bring your boat alongside.'"[55] And then begins the most intriguing part of *Standing Room Only,* indeed the bulk of the book, as Fowler begins analyzing the behavior of men in distress.

The men make elbow room, run out oars, and row. "Out of the jam and confusion, there was an orderly discipline while the men bent to carry out the hated command. Our captain was silent. His silence crushed us." Discipline surged from the U-boat. The castaway men turned craven before the power of the U-boat commander, and Fowler suddenly perceived the authority implicit in power and what she considered the male willingness to obey it. She realized that no longer was she the lone woman aboard a steamship but the lone woman aboard the lifeboat, "an embarrassment to a boatload of men, a very symbol of bad and leery luck to sailors." She realized she knew nothing of what to expect. Almost from the moment the *West Kebar* master stepped from the U-boat back into the lifeboat and the submarine submerged, she began to learn and to change.[56]

Standing Room Only never mentions what happened to most of the missionaries, only two of whom seem to have arrived in the lifeboat, nor does it give its author's feelings about their loss. During the long waterborne moment—the lifeboat lay to a sea anchor for the rest of the first night—Fowler watched seasickness hit almost everyone aboard, and the men become selfish. Her account makes it clear how little the men did for Elizabeth Fowler during her lifeboat passage, and how little she needed their help anyway. Now and then she received special favors, almost all of them at the very beginning of the passage. The master of *West Kebar* ordered one of the six wartime-use blankets to be handed forward to her, but soon afterward a man dragged it from her and huddled in it himself. No one, not officer, crewman, or passenger, made him give it back. On another occasion, during a gale, some men tried to shove her into the shelter of a canvas lashed across the bow, but she doubted their motives: the air in the shelter stank, and they thrust their feet on top of her, hurting her so that she chose to sit up and endure the spray. But Fowler got very few other favors, and indeed the men spent her own resources on themselves, one covering his nakedness with her spare dress, the others using things spilled from her abandon-ship bag into the bilge and recovered during the passage. In a way, her efforts to cause no special trouble seem to have been abused by men unwilling to do much for anyone, especially for the woman the *West Kebar* officers left in the bow, as far as possible from themselves.

Urinating symbolized all her difficulties. "The first time the bucket appeared, I gritted my teeth and told myself not to be a fool," she recorded. "But the sound of that bucket being used by thirty-four men in turn was almost more than I could bear." Having to urinate into a pail produces some of the most thoughtful pages of *Standing Room Only* and describes functions male narrators habitually omit from their accounts. Fowler felt not only of a different sex but essentially alien, then "insecure and adrift," and the men make no effort to lessen her discomfort. "In between it was not so bad. We were all together, part of the boatload. But at such moments I was no longer one of them." For almost two days Fowler resisted using the pail and argued with herself about the restrictions society imposed on her as a girl. "But it was useless to rail against the Victorian doctrines inculcated into me that had changed a simple function of nature into something associated with shame." After she began using the pail, she reassured herself that in time it would become routine, but finally she accepted how much societal inhibitions were a part of her psyche. "Each time the struggle repeated itself, and I died a hundred deaths of shame and misery, despising myself for being unable to adapt myself to what, in theory, seemed so simple."[57]

Almost imperceptibly, her identity was eroding along with the femininity she ordinarily used to manage men. Yellowish green palm oil from the shattered cargo smothered everyone in the boat, and no amount of saltwater washing removed it or the soot of the explosion. Her dark-blue linen dress remained incredibly filthy, and her vomit-soaked hair defied all attempts to clean it. Days into the passage, one man said her slip showed, and without thinking Fowler grabbed it and threw it overboard, forgetting the desperate need for any sort of head-covering material. Her feet remained as black as her slip, and slowly she realized that the men around her had begun to treat her as a person, that having hands as black as the men's meant that she had been "accepted after an initiation." But still the men did not exactly see her as one of them, and indeed one night early in the voyage one man kissed her and she silently fought him off, terrified that men coming to her aid might hurt him.

Unlike Mary Cornish, Elizabeth Fowler had no one to tend. Fowler knew of the horror that had followed the 1940 sinking of *City of Benares,* a liner carrying evacuee children from Britain to America—the scores of children dead, the lifeboat passage of Cornish and her charges. But the sole child in the *West Kebar* lifeboat was a boy who spent the passage in the stern with his father and the officers, and while Fowler now and then went aft to tell him stories, she always had to return to her place in the bow. None of the men suffered injuries requiring first aid or any subsequent nursing, and in the end Fowler had nothing to do. Too weak to row, too ignorant of sailing to manage the lugsail next to her, she traveled as what she insisted she was not—a passenger.

But she listened and she watched, and *Standing Room Only* proves a marvelous portal on the creation of discipline aboard a lifeboat. From the beginning, Fowler scorns the religion of the two missionaries, a Catholic priest and a Protestant, the father of the little boy. Neither does anything for anyone, rowing included, even when asked. Fowler muses at some length on the place of religion in a lifeboat, but she devotes far more attention to the taciturn New England master navigating with a safety pin for a pair of dividers. He and his fellow officers strike her as the best of men (once away from the U-boat), saying very little by day and less by night, but confident and capable of quelling discontent and despair with a word. But *Standing Room Only* implies that the master of *West Kebar* never spoke to Fowler during the entire ten-day passage and that the first gentlemanly word she heard after the steamer sank came from the British officer commanding the torpedo boat that found the lifeboat entering the Barbados surf.[58]

Women swimming away from sinking ships or to freedom from Axis enemies figure in many World War II magazine articles, perhaps because earlier generations of women so feared the water that journalists still perceived swimming as an unusually perilous female activity.[59] But the lifeboat experience, even long before World War II, forced women to confront the prospect of swimming to lifeboats and thus the prospect of shedding social mores. Swimming—and urinating into a pail in front of men—epitomize the changes women must make in adapting to the essentially male discipline of the lifeboat. Any lifeboat that carried women sailed into a realm different from the all-male one, a realm in which officers sometimes handed women blankets not for warmth but from an understanding that they might use them for privacy. Even in *Reader's Digest* morale-boosting wartime stories like a 1943 one entitled "Which Was the Rescuer?" women castaways appear as fundamentally disoriented in male space.[60] On the other hand, the women often earned respect by announcing a special expertise like nursing and then immediately and efficiently tending to the wounded.[61]

When a torpedo sank *City of Cairo* in 1942, Betty Birchman, a registered nurse in Lifeboat Number 7, not only administered first aid but knew that the whale oil provided for frostbite might serve to alleviate the pain of sunburn and blisters. *Goodnight, Sorry for Sinking You,* Ralph Barker's 1984 study of that shipwreck and its subsequent lifeboat passages, devotes detailed attention both to the roles of women in the lifeboats making for St. Helena and to issues like their "fading of squeamishness" about urination and menstruation, subjects earlier historians slight. Barker relates that women contributed to lifeboat morale by such simple acts as passing around a bar of perfumed soap for the pleasure of smelling it, by cropping their hair, and by giving away their underwear for men to use as hats. In the one engine-equipped lifeboat, the compass had to be mounted in the bows, and there Mona Rooksby monitored it. The officer commanding the lifeboat silently enjoyed the cultured tone in which she called out course corrections, but he also soon trusted her readings and followed them without hesitation. But in the end, swimming divided women from men even in the *City of Cairo* boats. To escape the tropical heat and allay thirst, men dove overboard and swam in the calm sea, but women never left the boats.[62]

In the course of a lifeboat passage it often became clear that many men valued women less than they valued other men, and children even less than that. Women and children entered lifeboats almost as freight and found themselves devalued by mariners of all races. For white women, the cohesion of mariners despite race proved staggering in its implications.[63] Sybil Chapman fell asleep in a *City of Cairo* lifeboat and rested her head on the shoulder of the man wedged next to her. A moment later he pushed her away, muttering something about every man for himself.[64] His behavior put Chapman on her guard and made her acutely observant of the deterioration of civility thereafter. What she and some other castaway women discovered, especially in boats commanded by men whose worldview lacked the old age-of-sail code, proved disconcerting indeed. The whole framework of respect and camaraderie soon revealed itself as a veneer, and the much ballyhooed "law of the sea" based on "women and children first" devolved into a murky law of survival in which mariners of all races were pitted against passengers.

Only rarely did the public catch a glimpse of the reality beneath the veneer. In the Board of Trade inquiry into the foundering of *Lusitania*, Captain Turner, who called himself an "old-fashioned sailor man," condemned the way his own crew lowered—or tried to lower—lifeboats, and stated a blunt preference for the sailing-ship officers and crewmen of his youth.[65] But by the middle of the Great War "sailor man" meant more than a seaman proficient at seamanship. An old-time sailor man protected at least some of his passengers—the women and children the British and American public expected to be put first in lifeboats. But as incidents like the foundering of *Arctic* and *Titanic* demonstrated, the age-of-steam seaman did not automatically act as passengers trusted he might. All too frequently, the loyalty of seamen to each other thrust all passengers into a second-class position, and while self-preservation might unite seamen of different races and religions and nationalities, it often did so at the expense of passengers, perhaps especially women and children.

By World War II, the inexperience of far too many crewmen required men like the master of the *Express* lifeboat to quell disrespect, misbehavior, and what age-of-sail men would have called mutiny by the use of physical force. Whatever else women brought into the steamship-era lifeboat, then, they brought a perspective on problems of discipline and morale from an angle far different from that of men, crew and passengers alike. Women perceived the gulf between mariners and all other castaways, especially the ones mariners considered baggage.

BONES

WHEN *Lusitania* SANK, ONE FIRST-CLASS PASSENGER SAVED HER LIFE by shedding her clothes, diving from the listing liner, and swimming far from the funnels and radio antenna slowly toppling into the sea. After getting away from the capsizing liner, she spotted a swamped lifeboat drifting near her, swam through a mass of debris, and "using her gymnastic training" pulled herself into it.[1] "Castaway" would once have been the designation for a woman like this, though nowadays news media would call her a "victim" of disaster. "Survivor" designates her equally well, but with slightly unsettling connotations. The term implies living *over* something or someone, being superior to or above someone else, or perhaps rising and staying above disaster like shipwreck.

Just as the nakedness and near nakedness of many first-class *Lusitania* passengers usually passes unnoticed, so do many female observations about lifeboat discipline. Women like Elizabeth Fowler glimpsed something floating beneath the discipline ordering the *West Kebar* lifeboat that might cause some lifeboat occupants to survive over others. For all its measured grace, *Standing Room Only* clearly demonstrates that its author intended to survive, to be one of the castaways who lived. Women and children might serve as food for the strong. Seamen knew what passengers generally did not: cannibalism is part of the old open-boat code of conduct.

Such is the significance of Queequeg in Herman Melville's novel *Moby-Dick*. About to sail from New Bedford on a whaling voyage, Ishmael finds himself at an inn, sharing a bed with a tattooed Pacific Islands cannibal. But not every cannibal in early nineteenth-century New England hailed from the South Seas and advertised his social beliefs through extensive tattooing. Some had been born and raised on Nantucket, and other Yankees might pass them on the street without guessing that they had eaten human flesh, indeed that they had killed to obtain it. Away from the sea, anywhere inland from the alongshore zone where winter nights are still freighted with tales of U-boats and barratry, misdirected cargo and mutiny, radioactive powder and white slavery, cannibalism orders little contemporary thinking.[2] But now and then, in battened-down, gale-tossed boat shops along the Massachusetts coast, the word still causes eyes to narrow. Boat builders and fishermen toss some sticks in the wood-burning stove and speak of Nantucket.[3]

In 1820 a sperm whale deliberately rammed and sank the Nantucket whale ship *Essex* while most of its crew were off in whaleboats harpooning other whales in the pod. Melville remarks early in *Moby-Dick* that the mid-Pacific incident—about which he learned through tales, reading, and personal interviews—so affected him in his whaling days that he ordered his novel around it.[4] But about the aftermath of the *Essex* disaster Melville says nothing, although the character of Queequeg most definitely arose from that aftermath.

Nothing reveals the uselessness of whaleboats as long-distance lifeboats more clearly than the castaway passages of the *Essex* boats. Despite being heavily watered and provisioned from the hulk of the whaler; rerigged with two masts and equipped with extra sails, tools, and nails; and having gunwales raised six inches with cedar strakes carried to repair whaleboats damaged in hunting, the whaleboats performed poorly in their long Pacific passages. One whaleboat set off carrying six men; the other two each carried seven. All leaked at once, and early on a grampus attacked the master's boat, requiring the repair of at least one stove-in strake. Frequently strakes started from ribs, necessitating difficult repairs. After four weeks the three boats arrived at Elizabeth Island. A few days later, all but three men, who chose to stay behind, left for the South American mainland, having exhausted most of the food and water available on the islet. Their passage proved disastrous.

For two weeks the boats kept together. The second mate died, and his crew gave him a proper burial, sewing his body into a blanket weighted with a stone. Several days later, another man died in the same boat and also received a proper funeral. Then a storm separated the first mate's boat from the master's and the late second mate's, and caused all three boats to miss Easter Island. Eighteen days afterward, when another man died of convulsions, the castaways aboard the first mate's whaleboat dismembered the body, built a tiny fire on sand in the bottom of the boat, and roasted what flesh they did not place to dry in the rigging. The flesh gave them strength, and a few days later a British brig rescued them.

Aboard the other two boats things grew worse. Two men died in the second mate's boat, and the castaways in both vessels shared their flesh. Next came a storm in which the two craft separated, and the second mate's vanished forever. In the master's boat the men drew lots, and the fatal lot fell on Owen Coffin, a cabin boy and the master's nephew. The master gave Coffin a chance to renege, but the boy insisted on being killed, so after a second lottery to choose the executioner, a whaleman shot him. A few days later another man was shot to death, leaving only the master and the whaleman-turned-executioner alive in the boat. When the American whaler *Dauphin* found the two men, the whaleboat had been at sea sixty days since leaving Elizabeth Island and ninety-three days since *Essex* foundered. It had sailed four thousand miles from where *Essex* sank, but it figures in few tales of ship's boat passage making. Instead, it lives on in the annals of the macabre.[5]

In its basic version the story depends on the veracity of the pamphlet published by the first mate, Owen Chase, in 1821. Six of the *Essex* crewmen were African American, while the remaining fourteen were all friends and neighbors on Nantucket. The order in which men died perplexes any reader of what Melville chose to call the "authentic narrative," for with the exception of the second mate, who died in a fit early in the passage, the next men to die, of starvation, were all black. After them came young Owen Coffin (whose name Melville appears to remember in the name of the Spouter Inn landlord), then another African American. Since the second mate's boat went missing, no one knows about the order of death in it. Even assuming Chase's narrative is accurate, any reader must wonder at the early deaths of the African Americans: something must explain their surrender to starvation, and perhaps something untoward altered the chronology of deaths. Captain Pollard admitted not just to allowing the lottery that chose the cabin boy to die but to giving the boy the chance to escape the fate of the lottery.

What happened aboard the *Essex* boats still thrills contemporary readers even as it perplexes literary scholars intrigued by Melville's knowledge of the published account, his meeting with Chase's son, and his use of the story in *Moby-Dick* and, perhaps, in his rarely read *Mardi*.[6] Given the large number of mid-nineteenth-century shipwrecks that ended in cannibalism, Melville's interest surprises no reader of shipwreck narratives. But Conrad understood cannibalism as a secret twentieth-century phenomenon, now involving steamships. His gruesome short story "Falk" is about a steamship officer who survives by eating crewmen he has killed aboard his floating derelict; he then falls in love with a woman whose family thinks him a beast for his cannibalism ten years earlier. Conrad—but not Melville—sketched out the "mighty truth of an unerring and eternal principle" of survival that nowadays only jurists confront.[7]

Legally, the principle has existed for centuries in the aphorism "necessity knows no law." In broad terms, according to legal scholars, the principle means that when "a situation of real necessity exists, the parties are returned to a state of nature where might, not right, governs."[8] Not since Stacpoole, London, and the other naturalist writers of the early twentieth century have these concepts been presented in popular formats. Yet few Americans wholly forget them, for disasters—and apprehensions of disaster such as the vast computer-failure-spawned social breakdown once predicted for the year 2000—remind newspaper readers and television watchers that everyday order vanishes in fire and flood, plague and shipwreck.[9]

Only when order breaks down for extended periods of time do most victims of disaster begin to understand the old legal principle, and usually the first people to do so are women. Elizabeth Fowler glimpsed that understanding during the waterborne moment, when the U-boat surfaced next to the *West Kebar* lifeboat and the U-boat commander momentarily instilled a new kind of discipline. Sybil Chapman found it almost as soon in the *City of Cairo* lifeboat, when the man next to her shoved her away and declared his every-man-for-himself philosophy. Mary Cornish awoke to a new reality in the *City of Benares* lifeboat, as even the most optimistic and kind seaman behaved in ways that at first mystified her. The man treated illness with contempt, and despite his continual kindness to Cornish and the children, he alarmed her.[10] Women castaways, especially those who produced lengthy narratives, noticed the cracking social order at once, and often remarked on physical and emotional weakness as well. Perhaps because they saw themselves as dependent on the ordinary social order in which the strong help the weak, perhaps because they accepted weakness in ways men did not, the comments by women castaways illuminate the hierarchy of strength in the situations of real necessity that develop as days and weeks pass.

Throughout the nineteenth century, seamen viewed cannibalism as *ordinary* postshipwreck behavior. They were familiar with the tradition of drawing lots to see who would be killed and who would do the killing, and they understood that lotteries might happen only after mariners had eaten all the passengers first. European legal documents date the first recorded postshipwreck drawing of lots to sometime between 1629 and 1640, when an English crew in the Caribbean survived by killing and eating one of their number. They were properly accused of homicide by the proctor of the Dutch island of St. Martin, where they came ashore, and immediately and properly acquitted by the resident judge, who applied the doctrine of "inevitable necessity." Eating the bodies of people who had died naturally occurred so frequently as to be scarcely newsworthy throughout the eighteenth century, although the British *Gentleman's Magazine* often reported such behavior in its accounts of shipwreck. The drawing of lots attracted more attention, simply because it occurred less frequently and often under catastrophic conditions, and of course because it involved homicide. In the early nineteenth century, perhaps owing to the growth of the popular press in Britain, accounts of lot drawing became more frequent and detailed, as did accounts of cannibalism, like that aboard the derelict *Francis Mary* in 1826, when the fiancée of a man who died of exposure claimed the right to eat him, and thereafter became the self-appointed dismemberer of others who died naturally. Lurid popular-press accounts quoted survivors saying that human flesh tastes like beefsteak and human brains are delicious raw or cooked, and reported that aboard some drifting derelicts castaways preserved flesh for long-term use. In *Peggy,* wrecked in 1764, the crew pickled human flesh, and aboard *Tiger,* wrecked in 1766, the living smoked the flesh of the executed.

By the middle of the nineteenth century suspicions had begun to develop regarding the reported fairness of the lotteries. As jurist Alfred William Brian Simpson demonstrates in his *Cannibalism and the Common Law,* postshipwreck cannibalism-by-lottery almost always involved the strong eating the weak. In general, he reports, "the lot mysteriously falls on the obvious victim."[11] Throughout the eighteenth century, the lot fell almost invariably on any slave aboard, and after the abolition of slavery in the British Empire, on the weakest black man or boy. Moreover, as in the case of *Tiger,* the execution of an African did not always involve the casting of lots at all: sometimes the whites simply decided to kill and eat someone not white, and often white women instigated the procedure, perhaps knowing their own vulnerability.

If the derelict or boat carried passengers, however, the lot usually fell on them before anyone in the crew, white or black. "One cannot but be suspicious about the selection of a passenger," Simpson remarks. *Cannibalism and the Common Law* makes grim reading indeed. "Human nature being what it is," Simpson concludes, "the crew usually took priority in survival; women and children (and for that matter, male passengers) normally came last."[12]

Despite his careful research, Simpson admits that the ease with which castaways disposed of human remains makes analysis difficult. *Cannibalism and the Common Law* depends only on recorded cases stating that killing occurred, not on accounts alleging deliberate killing (which imply to any reader that the recorded accounts are merely the tip of an iceberg). Yet even recorded cases are untrustworthy, and he accepts little of the *Essex* story, for "the knowledge of what happened depends solely on these two accounts, both second-hand"—the sort of knowledge jurists suspect.[13] And after a celebrated British trial convened in 1884, mariners became very reluctant to speak of cannibalism at sea.

When the yacht *Mignonette* foundered in mid-Atlantic in 1884, the four crewmen headed for Australia took to its unprovisioned dinghy, and three survived—they claimed—only by killing the cabin boy, something the master did with his pocketknife. Found by the bark *Moctezuma* and delivered to Southampton, the three men were perfectly open about their actions, and at first no one, not even the family of the slain man, evidenced any hostility. Had Captain Thomas Dudley not asked for his pocketknife back "as a souvenir," the machinery of British law might not have started. But his request so unsettled a local constable, who had impounded the knife, that the officer begged instruction from his superiors. Eventually, the British crown tried all three men, having been goaded into motion by the penny newspapers, which harped on what they called "The Terrible Tale of the Sea" in a way that, paradoxically, brought public support for the three men. The machinery of British law broke down repeatedly, involving numbers of jurisdictions, barristers, magistrates, and judges, until finally the trial became one pitting common against enacted law, the sort that changes legal tradition and still entrances legal scholars.

Ten years before *Mignonette* sank, lots had been drawn aboard the derelict *Euxine,* resulting in the killing and eating of a boy. In the same year, aboard *Cospatrick,* survivors had certainly eaten the dead, and the dying had been hastened along their way by surgical bleeding. In 1878 an American coasting schooner bound from Charleston to Baltimore suffered severe damage in a storm and became the sort of floating derelict the Hydrographic Office eventually charted. Ten days after the storm, when the schooner had been at sea for forty-three days and provisions had run low, one African American shot another to death, after which the carcass was fried and eaten. Such cannibalism exacerbated feelings of moral indignation already aroused by examples of man-eating in the American West, including not just isolated individuals but the whole Donner party, trapped in a Rocky Mountain snowstorm.

In both Britain and the United States, popular opinion began to focus on outlawing cannibalism as a way of forestalling horrors like the *Euxine* catastrophe. When the crowded Australia-bound emigrant ship, freighted with linseed oil and hard liquor, caught fire and burned two thousand miles south of St. Helena, only 5 of the 479 people aboard survived. Lifeboats became abattoirs when passengers slaughtered and ate other passengers, as the inquiry convened in Singapore discovered. Public indignation against cannibalism also became a way of attacking the shipowner and mariner irresponsibility that resulted in disasters like the *Euxine.* Lifeboats for all—and lots of Board of Trade provisions—lessened the likelihood of cannibalism, or so went the argument.

Prosecuting the *Mignonette* survivors for something they and many seamen considered ordinary and proper behavior produced the predictable effect of causing castaways to fall silent about cannibalism. The last instance of confessed marine cannibalism occurred in 1889, when the steamship *Earnmoor* foundered in a hurricane and only eight of its crew of thirty-three managed to board a lifeboat. On the sixteenth day, discipline in the lifeboat collapsed, one man knifed another, and the castaways began talking of eating each other. A day later, when one man died in his sleep, the others prepared his body and consumed it, as they did again a day later when another man died. "One cannot but suspect that these deaths may not have been natural," Simpson remarks dryly.[14]

Simpson speculates that historians will find more evidence of cannibalism, perhaps during World War II. But he recognizes that fear of prosecution after about 1890 shifted accounts of lifeboat cannibalism from public realms, especially the legal one, to the annals of hearsay and folklore.[15] Certainly the cannibalism of the *Dumaru* castaways has entered into folklore, for all that one survivor told his tale to Lowell Thomas in 1930. According to Fred Harmon, everywhere on the Pacific Rim men fall silent at the *Dumaru* tale, and the story exists in hundreds of variations. Harmon had heard all sorts of versions of the lifeboat passage and its cannibalism, and nowadays *The Wreck of the Dumaru* endures as what Melville might call the "authentic" version. But Thomas's book begins with a long list of different versions of the story, and the list makes the published version as suspect as any Simpson examines.

Thirteen days after *Dumaru* blew up, the first man died aboard the Lundin lifeboat half-sailing, half-drifting across the Pacific. Graveyard Shaw, an African American who struck Harmon from the first as half-crazed, died from drinking saltwater, and the men pushed his body overboard without ceremony. Harmon remembers that moment as the point at which discipline collapsed into nightmare. The men worked with extraordinary cleverness to build a seawater evaporator from a bilge pump, and soon had a fire burning in a metal pail and a still producing small amounts of fresh water. But technological success did not forestall horror.

Harmon reported that the chief engineer died in a coma after raving for days, and he rebutted allegations that the other castaways killed him. The death divided the men along lines of nationality. Forward one Greek and the Filipino seamen surrounded the body and began dismembering it. Aft cowered the Americans, sickened, transfixed, and frightened by the knives and hatchets flashing a few feet away. Soon flesh bubbled in a pot above a fire in a pail. The mate and bosun moved forward, collected the hatchets, and forced the Filipinos to give up their knives, which they did without struggle; then the Americans began eating too. The flesh, they decided, tasted like veal.

Men began to die in a steady procession. The lifeboat atheist simply rolled over the low side of the boat and drowned himself, to the indignation of those who felt he should have died in the boat so his body might become food. The Greek seaman, crazed with thirst, got hold of a big clasp knife one night but was jumped on by several other men and lashed in a sort of kneeling position to the mast and a thwart. He died there two days later, followed the next day by two other men. At the beginning of the twenty-third day, seventeen men remained in the boat, but within hours more died, and the living began draining blood from the fresh corpses. Days and deaths later, as the boat drifted near island surf, the castaways kept cooking flesh despite one man's argument that they ought to throw the can and other evidence overboard.

No one in authority moved to punish the survivors when the story leaked out. A United States Navy inquiry into the reported suicide of a navy officer aboard the lifeboat simply accepted the testimony of those who claimed the lieutenant had jumped overboard. When two of the men volunteered that they had engaged in cannibalism, the navy officers recorded their remarks but did nothing more. No one in Manila seemed particularly intrigued by this dimension of the story. Relieved at not being prosecuted, the men boarded a Honolulu-bound ship which en route encountered a lifeboat with four men aboard. The irony of the situation was not lost on the former castaways.[16]

Discipline collapsed in the *Dumaru* boat in part because no one took charge of the lifeboat in the waterborne moment. Lieutenant Holmes of the United States Navy tried to take charge, but his effort ended in disaster the moment the trade winds shifted, and thereafter his authority declined with his health. His conflict with experienced seamen like the bosun only worsened a wretched situation, and his weak grasp of navigation caused morale to plummet, damaged as it was by the failure of the Lundin lifeboat to behave in traditional Board of Trade lifeboat ways. But Holmes's suicide marks a turning point aboard the *Dumaru* lifeboat, for in killing himself the self-proclaimed leader of the boat was turning his back on his crew. No wonder the navy inquiry ended so quickly. Holmes did not conduct himself in a manner befitting an officer, and the navy examiners avoided scandal. The survivors, and the press, might well argue that Holmes's incompetence condemned the men to a long lifeboat passage in the first place, and that his suicide created an absence of leadership that drove them first to despair, then to cannibalism.[17]

In 1941 Guy Pearce Jones published *Two Survived: The Story of Tapscott and Widdicombe, Who Were Torpedoed in Mid-Atlantic and Survived Seventy Days in an Open Boat.* No book better illustrates the difficulties confronting World War II writers trying to stimulate morale while dealing with open-boat reality. Tapscott and Widdicombe, along with five other men, abandoned the steamship *Anglo-Saxon* on August 21, 1940, after it was shelled by a German surface raider. Jones's account of the incident is simple and propagandistic. Finding the port Board of Trade lifeboat blown to pieces, Tapscott and Widdicombe and the others lowered the jolly boat (a small boat used for in-harbor hull maintenance) and lay off in the dark away from their sinking ship. In the morning the first mate, in charge of the boat, discovered three of the men badly wounded by gunfire, others slightly wounded by shrapnel, and the only water keg half-full. The jolly boat had been altered to serve as a lifeboat: it carried three flotation tanks, all the standard lifeboat equipment, and a lugsail rig. Nevertheless, from the beginning it performed less well than expected. Since no answer had come to the wireless operator's SOS, the mate set sail for the Leeward Islands, more than a thousand miles west, instead of bearing to windward for the Azores. The men had jumped aboard poorly equipped, except for the radio operator, who carried a case filled with tobacco and a book of Bible quotations, the pages of which the men read and then used to make cigarettes. Despite the use of first-aid kit's medicine and bandages, the wounded men soon smelled of gangrene, and the mate began to worry about their condition. In his rough log he wrote that making the Leeward Islands might be possible, but it would be terribly hard.

In the beginning, the mate successfully built cohesion, discipline, and morale, and even ran lotteries based on the day a rescue ship would find the eighteen-foot boat. But the dwindling water ration combined with blazing sun to weaken the wounded men, and on the eleventh day the wireless operator died. "Committed his body to the deep with silent prayer," the mate wrote in his log. Within two days he himself lingered near death, and all the old controversy between deck seamen and engine-room men exploded when the third engineer attempted to take command. Widdicombe, a deck seaman, insisted that he be put in charge of the boat, and in time the engineering officer accepted his demand. Almost at once another man, a gunner, committed suicide by pitching himself overboard.

Jones concluded that even if the castaways had been able to turn the boat to try to rescue him, the gunner had acted with free will. "Unwritten law gave a man in these circumstances, provided he was or appeared to be in his right mind, the inalienable right to choose his own way out of his suffering. The gunner had 'gone over the side.'" A deep pragmatism by then ordered the jolly boat castaways. Anyone who died meant one less castaway requiring a water ration.[18]

Two more suicides occurred on the thirteenth day, what Jones called "sailor's leave," the old term for seamen who walk away from docked ships, forgoing their pay. The mate struggled to his feet, removed his signet ring, asked Widdicombe to give it to his mother, then removed his coat. As he did so, the third engineer stood up and removed his coat too. Widdicombe and Tapscott knew what the two men intended, and Widdicombe asked the third engineer for his pants. The officer said no, that he might meet a mermaid below. The mate issued Widdicombe and Tapscott his final order: sail west, not southwest. Then the mate and the third engineer stood up on a thwart near the port gunwale, shook hands, and jumped. Stunned and saddened, Widdicombe and Tapscott recited the Lord's Prayer as the two officers, holding each other, floated astern.

A few days later, Widdicombe and Tapscott confronted a different sort of suicide. The third man left in the boat, Morgan, had been raving for two days, but his dementia cleared up and he relieved Widdicombe and Tapscott at the steering oar. The two men fell asleep, then awoke as the boat yawed wildly and Morgan went overboard. Tapscott followed him, grabbed him by the hair, and dragged him back aboard while Widdicombe succumbed to hysteria. From then on Morgan raved, and on the following day, with the boat nearly becalmed, he stood up, announced he was going down the street for a drink, and went overboard and down like a shot.

With all water gone, Tapscott and Widdicombe could scarcely see the suicides as increasing their own chances of survival, although the boat certainly floated slightly higher. The men began discussing suicide but determined to hold on as long as possible, and in order to do so began drinking urine. Then they opened the compass, poured out the alcohol, quaffed it, and plunged into oblivion. When they awoke they saw a squall ahead and soon were able to begin collecting rainwater from the sail. On the twenty-third day of their slow passage west, the cloudburst not only let them quench their thirst but provided water for six more days at the old ration. A few days later they encountered a rainstorm that lasted for four days, and undoubtedly that rain saved their lives. They began eating seaweed, ran out of water again, were nudged by a curious whale, and collected rainwater once more before drifting ashore, more dead than alive, on a beach in the Bahamas, some three thousand miles from where *Anglo-Saxon* went down.

For eight days the two men lay confined to a hospital room, suffering from hysteria and despondency, and their recovery proceeded slowly. Widdicombe became fit enough to travel home but died en route in February 1941, when a torpedo sank his ship, the liner *Siamese Prince,* off Scotland. Tapscott remained a convalescent in Nassau, and with Widdicombe now dead, he alone knew the entire tale of the jolly boat's passage.[19]

How "authentic," in the way Melville used the word, is *Two Survived?* Simpson might point out the possibilities for the two men to lie. Did they throw Morgan overboard to escape his ravings? Did they eat the dead? No one knows, and British authorities did not exactly struggle to find out. Suicide after suicide punctuated the passage of the *Anglo-Saxon* jolly boat. Had the two officers only endured another day, the rains might have given them enough water to last them to the Bahamas too. What caused them to give up?

Within a year, Richards and Banigan answered in *How to Abandon Ship,* in a chapter tersely entitled "Morale." The authors knew some version of the Tapscott and Widdicombe story, for they refer to it often, always as a bad example. The *Anglo-Saxon* jolly boat lacked full water breakers, something that spoke of dereliction of duty. It lacked an up-to-date medical kit containing "sulfanilamide, the miracle drug," which might have saved the life of the radio operator, whose mangled foot turned gangrenous. Some of the *Anglo-Saxon* castaways drank seawater, which turned them mad; others, including Tapscott, drank far too much rainwater when they finally obtained it. As Richards and Banigan clearly reveal, the men survived by luck.

Cigarettes, advises *How to Abandon Ship,* "steady" men, for they contain a magic no one quite understands. The chapter on morale begins with the recommendation that lifeboats carry a good supply of cigarettes pooled under the authority of the lifeboat commander and rationed fairly. In the dark of night, for the men on watch, cigarettes, courage, and confidence together make morale. Richards and Banigan introduce into the discussion of lifeboat survival a point that earlier writers ignore: that the loss of small, modern luxuries is a slayer of morale. Those who survive tend to be those who adapt best to such privation.

Richards and Banigan offer a few further suggestions for keeping despair at bay during a lifeboat passage. They advise readers to keep the oil lantern lit all night for the first couple of nights, since crewmen will not be used to genuine darkness. Men should stow some flavored water in every lifeboat to mask the staleness of keg water, and put books, magazines, playing cards, and even simple musical instruments in every boat. Liquor should be rationed out only after morale drops, they conclude, which is a virtual certainty unless the boat clips along at a good rate.

Crewmen used to a fifteen-knot steamship will despair in a becalmed boat, they caution. "Morale" ends with a lengthy discussion of suicide. The authors differentiate hysteria-induced suicidal feelings from another, more insidious type. A castaway suffering from hallucination has a physical illness that demands and rewards help, even if restraint proves the only assistance possible. The other sort of feeling, which makes men seemingly in their right minds go overboard, receives sustained attention from the authors, who contend that such men die because they turn their combative instincts against themselves. Unable to strike back at the enemy, their fierce desire to kill someone turns against themselves. Whatever the origin of such thinking— *How to Abandon Ship* lacks footnotes and bibliography—Richards and Banigan use it to argue that men can turn such impulses away from themselves. By busying themselves with seamanship duties, playing cards, and competing against wind and sea, determined and self-aware castaways can change their vengefulness into positive energy. In case they do not, however, Richards and Banigan follow this encouraging advice with the standard Church of England burial-at-sea prayer in italics.

In 1951, when Beito published his updated instructions on lifeboat navigation in *Navigation and Nautical Astronomy*, he took pains to focus on morale in a chapter otherwise devoted to navigational technique. He insisted that routine maintains morale. Ship's-boat and lifeboat narratives prove the worth of routine, he asserted, adding that "the story of Captain Bligh is perhaps one of the greatest illustrations of the value of patience and determination."[20] He also mandated a daily reading of the Bible and urged his readers to remember to pray. But Beito confused routine and drill with the sailing-ship-era spontaneity and adaptability autobiographers extolled. Moreover, he confused the outward forms of religiosity with the deep, lifelong faith that strengthened the *Hornet* castaways. Routine is not a substitute for patience, determination, or faith, and as his chapter deviates from the book's mathematical focus to emphasize a simplistic mix of routine and faith, any reader must wonder at how much wartime castaways kept to themselves the strength of traditional religion juxtaposed against pure selfishness.

More than a half century after the end of World War II, secrets like the "smart bomb" destruction of HMT *Rohna* or the sinking of the *Princess Sophia* continue to come to light. The story of Fritz Kuert reached readers only in 1994, for example, in Hans Herlin's *Survivor: The True Story of the Doggerbank*, and even now it remains very little remarked. As one of the 365 men aboard the German supply ship *Doggerbank* steaming in 1943 from Japan to Germany, Seaman Kuert felt great relief as his vessel neared the Azores well ahead of schedule, for the German admiralty had directed several U-boats to escort it from there to Bordeaux, in occupied France. Instead, U-43, one of seven U-boats operating southwest of the Azores and uninformed about the blockade runner, erroneously torpedoed *Doggerbank,* sending its men into rafts and a single small jolly boat almost exactly like the *Anglo-Saxon's.* The commander of U-43 was certain he had sunk a British freighter—*Doggerbank* had been launched in Scotland twenty years earlier and intentionally retained as British a silhouette as possible in order to slip past Allied ships and aircraft. Since *Doggerbank* sank without sending an SOS, no British warship or Allied merchant vessel knew anything about its foundering. *Doggerbank* simply vanished, and it was days before the German admiralty realized that its radio silence meant it had been sunk or captured.

Perhaps two hundred men died almost instantly as the three torpedoes struck the ship. Those who lived to abandon ship discovered little undamaged lifesaving equipment, and the few men in the jolly boat picked up equipment from the shattered lifeboats about them, finding oars and other gear floating amid the debris, along with the ship's mascot, a dog. Most of the men boarded a giant inflatable life raft, which they found so badly holed that they had to work continuously at its air pumps. Kuert found himself in the jolly boat, following the orders of an old fisherman named Boywitt, who knew how to lash oars together to make a mast and how to make a tarpaulin into a sail. Soon the jolly boat was negotiating the wreckage, bodies, and survivors; quickly its men picked up the *Doggerbank* captain and sailed away, searching for drifting lifeboats.

Kuert, a bosun on the *Doggerbank,* had already survived several sinkings, and he had a simple theory about lifeboat behavior that he related years later to Herlin: thinking killed. His philosophy came down to this: *"Keep your eyes open, remain alert by all means, but don't think, or, if you do, no more than a minute ahead."* Whatever his idea's worth, he again followed its dictates.[21]

At dawn of the second day, those on the jolly boat couldn't find the life raft. In near despair, they found only four swimming men, who claimed that when the jolly boat left to search farther away, they and the others on the raft thought the captain had abandoned them. For reasons best known to himself (or so the survivors maintained), the officer aboard the raft then removed his pistol from its waterproof pouch and shot himself. Within moments seamen grabbed the pistol, and seven more shots rang out as men dove from the raft, which quickly deflated. Only the four men who had kept on their life jackets survived. Even though some of the men aboard the raft might have been merchant seamen, not navy men, the collapse of discipline beginning with the suicide of a *Kriegsmarine* officer astonished everyone in the jolly boat.

Yet as the days passed, discipline collapsed in the jolly boat too. Forty hours after *Doggerbank* sank, carrying neither food nor water, it swung onto a course for South America, perhaps three weeks away. On the morning of the fifth day, one seaman almost capsized the boat when he tried to grab and eat the dog while others fought to protect it. In trying to defuse the situation, one seaman pointed out that the skinny little dog—who received its own ration of collected rainwater—would provide less food than one heavy-set seaman, and that any butchering ought to begin with something worth butchering. At that comment everyone fell silent. Once raised, Kuert implied to Herlin, the issue of cannibalism never vanished.

On the eighth day, one of the castaways simply stood up and stepped overboard. Apparently he had decided to kill himself before the rainwater ration had been distributed, and thus the decision meant one swallow more for the other six men. No one expected the suicide. The captain muttered that suicide is contagious and warned that a mass suicide might capsize the boat, but he made no effort to put about and rescue the man, whose life jacket kept him afloat.

The next day a gale capsized the little vessel, drowning two men and the dog and scattering much of the equipment. As Kuert crawled back into the swamped boat, he discovered that Captain Schneidewind had determined to kill himself, and nothing Kuert and the others said made any difference. The captain warned the men to remember his advice about navigating by stars, then pulled his pistol from its pouch, stood on the stern thwart and shot himself, leaving only Kuert and Boywitt behind to man the boat.

Slowly the two men sailed on. Kuert lashed together a crude rudder from fragments of thwarts, and he and Boywitt took one-hour turns at steering, dropping the sail at night to sleep. On the nineteenth day, Boywitt began drinking seawater, grew delirious, and died just as a tremendous rainstorm hit, filling the water-catching tarpaulin. Too weak to do more than hold the rudder, Kuert left Boywitt's body forward for four days, until its stench caused him to roll it overboard to the sharks following the boat. Immediately afterward a flying fish struck the sail and fell into the bilge. Kuert grabbed the fish, bit off its head, and sucked the blood from its gills, knowing he needed energy. Two days later, too weak to steer and once again horribly thirsty, he lay down on the dropped sail and dozed. A noise woke him, and he found the Spanish freighter *Campoamor* maneuvering alongside.

Like *Two Survived*, Herlin's *The Survivor* raises issues of leadership and morale against a backdrop of suicide, self-interest, and cannibalism. Both *Anglo-Saxon* and *Doggerbank* officers abandoned men under their command. Whatever gloss Jones applies to the *Anglo-Saxon* suicides proves too thin to mask the abandonment of duty. Hierarchy had simply collapsed, and different laws prevailed.

Unless Kuert lied, no one aboard the *Doggerbank* jolly boat followed through on the original suggestion of cannibalism. Even the dog was allowed to live. Some hidden precept kept the men from cannibalism and murder while allowing the option of suicide. But the first man to kill himself appears to have stepped overboard while out of his right mind, as witnessed by the fact that he leaped from the boat while still wearing his life jacket, and no one tried to stop him. His refusing the water ration meant, as Kuert explicitly notes, another swallow each for the rest. The suicide's companions may thus have implicitly conspired to let him kill himself, knowing that the jolly boat would float higher and that those remaining would have more to drink.

Selfishness sometimes masquerades as altruism. After the jolly boat capsized, Kuert first rescued his navigator-captain and then convinced old Boywitt to stiffen his resolve and get back aboard the swamped boat. Probably the strongest physically by that time, Kuert acted in a way that in retrospect makes one ever so slightly suspicious. He rescued his navigator and the only man experienced in small-boat handling, and allowed the others—including the dog—to drown.

Kuert believed in his own luck. He had survived other sinkings, and his narrative makes clear his eerie good fortune. After Boywitt succumbed to the temptation to drink seawater, it rained and rained; and after Boywitt died, a flying fish provided Kuert with energy-giving blood. Even his decision to lower the sail and rest for a while in its shade proved lucky, for it perhaps prolonged his energy just enough to keep him alive until *Campoamor* steamed over the horizon.

Kuert knew laws that prevented cannibalism but countenanced suicide, that permitted him to boost the useful men into the swamped boat in preference to less useful men, and that let him watch the latter swim away from the swamped boat. Kuert does not so much think as act in the most self-interested way imaginable, and in the end, he becomes the prototypical sole survivor.

Whatever Hans Herlin intended when he transcribed the story Fritz Kuert dictated into a tape recorder, he produced a book that increasingly unnerves the reader as it nears its close. Kuert himself, exchanged in a prisoner-of-war swap, fled after reporting to the German admiralty early in 1944 and finding himself about to be arrested. The twenty-six-year-old bosun spent the rest of the war hiding in a Hamburg garret, determined to make the world aware of German naval culpability in the deaths of his shipmates.

In the end, *The Survivor* focuses as much on the sinking of *Doggerbank* by U-43 as it does on the jolly boat passage. The book depicts Kuert as almost too efficient, too optimistic, too courageous, and perhaps it does so in order to make all the more powerful his singlehanded postwar effort to get answers from the admirals of the wartime German navy. *The Survivor* might as well have been entitled *The Witness,* and much of its power would erode if Kuert had been a lesser man than Herlin depicts him. The book reads like wartime propaganda and exposé combined.

Early in 1942 a news story appeared that took the United States by storm, in part because the nation desperately needed some good news. The federal government was obliged to manipulate the story a bit, as the British government did with the Dunkirk evacuation, in order to turn a defeat into a victory. On March 23, 1942, when *Time* magazine ran its version of the story of three navy airmen adrift in an inflatable raft, the American public met a brand-new standard of endurance and good behavior beyond anything connected with German supermen. Within a few months Robert Trumbull published *The Raft,* its dust jacket announcing it as "the story that generations of Americans will be telling their children to illustrate man's ability to master any fate." Reviewers adored the book; schoolchildren discovered it almost as quickly as adults. Immediately it became a best-seller. And almost at once it made all sorts of castaway behavior impossible to report.

The Raft is not strictly about lifeboats and the castaways making passages in them: it involves an inflatable raft and survival after a plane crash. Nonetheless, the book has a critical place in any analysis of lifeboat behavior, for since 1942 it has shaped subsequent castaway narrative. It was required reading in many junior and senior high schools by 1943, and during the 1950s it had an established identity on school reading lists as a book ordinarily read by boys obliged to write a book report. Well into the 1970s it held an honored place among adolescent reading materials, and even now it remains a fixture on any adventure or sea-story list, and among books that emphasize the indomitable human spirit.[22]

The Raft intrigues any reader of lifeboat narratives in part because Trumbull opens it with a brief summary of Bligh's open-boat passage and in part because Harold Dixon, the chief machinist's mate in charge of the two sailors aboard the raft, claimed that the castaways sailed the raft rather than drifted along. The men did the best they could, using a sea anchor and paddles made from shoes, to make the raft head west, toward an unnamed archipelago, but anyone accustomed to lifeboats would say they drifted. And anyone, accustomed or not to lifeboats, would marvel at their good fortune, for they not only collected rainwater frequently but caught fish, shot birds, and gathered floating coconuts, all in a timely manner. In the end the raft washed ashore on a tropical island.

The men did discuss cannibalism, Dixon admitted to Robert Trumbull, the *New York Times* Honolulu reporter. But Dixon claimed the idea came from one of the sailors, perhaps from the skinnier of the two who worried about being eaten and who on the twenty-eighth day suggested that all three men jump overboard. Although Dixon reported that the men had talked about which organs might provide the most sustenance, in retrospect, about two months later, he insisted that the topic really served more to pass time than to shape plans.

The Raft made superb wartime reading. The little raft became a symbol of the United States suddenly at war against enemies so powerful and far-flung that censors forbade the men from naming the island they reached, despite its being so remote that Dixon doubted anyone had ever heard of it. Stark naked, sunburned beyond description, emaciated and thirsty, the three United States Navy fliers stagger ashore, fall flat, then stand and walk inland, determined to let no Japanese soldier see an American crawl. With the publication of *The Raft*, the United States had a narrative whose message was straight out of Puritan New England: good men on a noble mission need not fear suicide or murder or cannibalism, for God will provide.[23]

Contemporary reviews of *The Raft* make clear the book's impact as a testimonial to American goodness. "One of the most gripping and in some ways the most inspiring story the war has produced thus far is contained in *The Raft*," raved a *Christian Science Monitor* writer.[24] The *New York Times* review called it "an astounding chapter, well told," and eight days later *Time* noted simply that "the special quality of this book is that it restores and documents by deed something that has long been lacking from men's books and minds—a sense of the therapeutic goodness of the unflagging will to live."[25] But the key concept surfaced most bluntly in an August 22 *New Yorker* review: "This is a far greater epic than Bligh's voyage, as the three Americans who performed this feat are better men than Bligh."[26] It is not suicide or murder or even cannibalism that concerned any contemporary reader of *The Raft*, but the extraordinary and implicit wartime belief that such evils befall only those castaways who somehow do not measure up to a vague but "higher" standard.[27]

Only rarely do seamen say much about open-boat behavior that contributes to making land, to surviving, perhaps because much of what they know proves unpalatable. Especially after the *Mignonette* prosecution and the *Dumaru* horrors, mariners became, as Simpson notes, closemouthed about much lifeboat behavior. Even books like *The Survivor*, published long after castaway horrors, bury important contradiction in appendixes. And books like Walter Gibson's 1953 *The Boat* pass almost unnoticed today.

Gibson first told his tale of open-boat chaos and cannibalism in a 1949 British newspaper article, and he finished his little book manuscript three years later to mark the anniversary of his escape from Singapore into hell. On March 2, 1942, a Japanese submarine torpedoed the overloaded Dutch freighter *Rooseboom* off Sumatra, killing about 300 people and leaving 135 in and about a single lifeboat the following morning. Given the number of people in the water holding onto lifelines, neither the master of *Rooseboom* nor the senior British army officers in command of the escaping troops thought of sailing or rowing anywhere.

At first everyone hoped to be rescued by the Royal Navy, and later even by the Imperial Japanese Navy. But instead the mass of humanity drifted for a month across a thousand miles of sea, the people in the water slowly dying, the Dutch merchant service officers killing themselves, crewmen murdering each other, and finally, three Javanese seamen killing and eating one of the three other survivors in the boat, an Englishman. Gibson said the Javanese had gone mad, and his grisly description of their offering organs to the other castaways seems obscene even today.[28] The following day the lifeboat drifted into surf breaking on the island of Sipora, a hundred miles from where *Rooseboom* had steamed out to sea a month before, and Gibson found himself a prisoner of war.

In *The Boat* the seeds of MacLean's *South by Java Head* are readily apparent. In very brief compass indeed, the narrative catches the whole fall of Singapore and subsequent evacuation chaos, the breakdown of order that ended in cannibalism. The book's themes of endurance, brutality, and cannibalism are the stuff of thrillers and Hollywood.

But unlike *The Raft*, *The Boat* appeared at exactly the wrong moment. The British were losing their empire, and British and American readers had lost their appetite for tales of the dark days of the Pacific War. Somehow *The Boat* strikes all the wrong notes. It is not about a lifeboat passage, or even about a life raft drift. It is about the slow festering of inhumanity in the glare of imperial collapse, and skirts along the far edge of Western notions of charity and endurance, love and hatred. It is about the rapid dissolution of a way of life and the passage of civilized people into a lawless realm. It is about the fragility of everyday order becoming cannibalism.

11

TRANSFORMATION

No ONE KNEW MUCH ABOUT THE REAL *Bounty* LONGBOAT PASSAGE after cinematographers got hold of the story, concluded Joseph Russell Stenhouse, author of *Cracker Hash: The Story of an Apprentice in Sail.* "The distortions of Hollywood history" wrenched Bligh and the longboat triumph into the afterglow of a mutiny with which cinematographers sympathized. *Cracker Hash* appeared in 1955, long after Stenhouse died in action in World War II, and decades after he served his apprenticeship aboard sailing ships and moved on to steam. Unlike most of his contemporaries, however, he left steam to become master of *Aurora,* the sailing ship that supported Ernest Shackleton's South Polar expedition. He returned from Antarctic adventure into the chaos of World War I, in which he commanded a heavily armed, three-masted schooner that preyed on U-boats. In subsequent years he commanded both steamships and sailing vessels, and wrote *Cracker Hash* in hours stolen from a second career as a "technical advisor to a film company." Unlike many seamen, Stenhouse had a firsthand understanding of cinema. He watched the celluloid remaking of classic tales into something filmmakers liked better.

Stenhouse prized the seagoing calm of long days and nights under sail. He scorned not only the short-term attention created by cinema and radio but also the way cinema and radio directed people's awareness away from natural phenomena, neighbors, and shipmates. Not many years earlier, he argued, people had devoted sustained attention to the world immediately about them, and perhaps knew fewer worries. They assembled their worldviews from firsthand knowledge, and they pondered that knowledge in long evenings at home and in long watches aboard sailing ships, often while doing routine physical tasks hospitable to reflection. *Cracker Hash* concerns the training of a man who valued enduring thoughtfulness over the short, choppy interludes showcased by early electric and electronic media.

At the end of his autobiographical fragment, Stenhouse concludes that cinemagoing satisfies a void created by modern forces. In the cinema house, viewers take in images of sailing ships and other phenomena they once saw outdoors and knew they might directly experience.[1] But cinema proves only a poor substitute, according to Stenhouse, whose work in making films like *Mutiny on the Bounty* convinced him that thriller writers had already recognized the incipient market among readers desperate for old-fashioned adventure, perhaps especially adventure at sea, as an indicator of what film might become.

"*Why not go to Memmert?* I said, in fun." So begins the crucial adventure in *The Riddle of the Sands.* Erskine Childers italicized the five words spoken by Carruthers, the narrator of his novel, for they are the foundation for all the subsequent action, surmise, and deadly public-policy message (the rise of German naval might) of the 1903 thriller. In the wake of Carruthers and his friend, Davies, hastily and furtively navigating fogbound German shallows in a lifeboat transformed into a small yacht, cruised other fictional secret agents, whether in substantial novels like Joseph Conrad's *The Secret Agent* of 1907, in mass-market thrillers like John Buchan's *The Thirty-Nine Steps* of 1915, or in hundreds of subsequent novels and screenplays. *The Riddle of the Sands* is the first British spy novel, and certainly the prototype of all subsequent spy novels, the model of secret service and adventure. Just as the little boat that sails half-charted waters is converted from a ship's lifeboat, so the novel itself is a sort of transformed lifeboat-passage narrative.

In the years after the Great War, lifeboat passage became not so much a mainstay of fiction as a sort of appendix to the whole sea-fiction genre. One example is H. M. Tomlinson's 1927 *Gallion's Reach,* a splendid novel opening in an obscure inlet of the Thames in London and progressing through a mid–Indian Ocean shipwreck into an orderly if difficult lifeboat passage. Tomlinson, a well-traveled writer of nonfiction books about the sea, understood not only the long waterborne moment following shipwreck but a great many details of lifeboat passage making, from the setting of lugsails to rudimentary navigation to the maintenance of morale and discipline.[2] Anyone familiar with Cecil Foster's *1700 Miles in Open Boats,* which appeared three years earlier, will quickly discern the origins of many details that enrich *Gallion's Reach.* Tomlinson's novel has its own secret agenda too, one somewhat similar to Childers's. *Gallion's Reach* focuses on a fact that Tomlinson thought too many Londoners ignored: steamships departing the Thames estuary might steam into shipwreck, and any passenger might find himself or herself thrust from a cozy cabin into a lifeboat. Sudden disjunction pulses through almost all lifeboat-passage fiction.

Fictional exploration of lifeboat passage making derives from the turn-of-the-century sail versus steam debate, and of course from open-boat fiction like Stephen Crane's 1897 short story "The Open Boat."[3] It includes novels written by experienced shipmasters like George H. Grant, whose 1938 *Take to the Boats* raises issues reexamined in World War II. Grant's novel of World War I involves a steamship sunk far west of Cape Frio in Africa by an unknown explosive device hidden in its forward hold in New York. Unlike the boats in *Gallion's Reach*, the three *Cumbercauld* lifeboats separate almost at once, enabling Grant to describe different fates following the waterborne moment. *Take to the Boats* provides far more detail than many readers may desire—its pages listing the equipment found in the starboard lifeboat could have been lifted directly from some Board of Trade Lifeboat Efficiency Examination textbook—and its threefold plotline is sometimes clumsy, but the novel summarizes much of what British merchant seamen learned in the Great War and after. It ranges over navigation and discipline, authority and provisioning, racism and the chasm separating seamen from below-deck crewmen. Yet like so much nonfiction, especially Cecil Foster's narrative of the *Trevessa* boats, *Take to the Boats* omits any mention of passengers, even the supercargo sort Tomlinson introduces in *Gallion's Reach*. By 1938 small freighters carried fewer and fewer passengers anyway, and passenger-liner shipwreck enticed fewer and fewer authors.

Passengers do figure in post–Great War shipwreck fiction, but only rarely do novels emphasize the foundering of passenger liners. Many of these works, like *Take to the Boats*, focus on the Great War experience, often in the propagandistic tone of Noyes's *Open Boats*, and most emphasize the foundering of small ships, usually freighters carrying a small complement of passengers. Stories set in the postwar period involve lifeboat lowering and at least the beginning of passage making, but by then radio communication struck most writers as likely to bring help to any disaster scene. *S.S. San Pedro*, for example, a gothic novel James Gould Cozzens published in 1930, prefigures actual ship founderings like that of *Morro Castle* in 1934 and *Lakonia* three decades later. But it devotes only a few pages to the willingness of white officers to get away first, infuses lifeboat lowering with racism—"they had more than thirty Negroes on board, and this, Miro recognized, was shameful," but not simply because the black men crowding the lifeboat were crew—and depicts one boat bereft of crew at all and filled entirely with women and children.[4] *San Pedro* is a cursed ship that steams almost free of the twentieth century, and the disasters that befall it originate in some malevolent force. In order for the novel to work, however, Cozzens must essentially dismiss radio communication. Indeed radio frustrated many other novelists by the late 1930s. It meant that ships no longer steamed in total isolation, that fire and other disaster brought ships rushing to help, and that launched lifeboats need only wait for other craft to arrive.

Novelists regarded radio as an intimation of the end of lifeboat passage making, and many, like Tomlinson, began to employ lifeboats as the means by which liners rescued passengers, if not crew, from foundering freighters and other small steamships. "Get the girl up here and make ready. That boat on the lee side is sound," barks the master of *Hestia*, the grubby little tramp steamer of Tomlinson's 1937 novel *Pipe All Hands*. "We can manage it better than the big thing out there for all I know."[5] The woman passenger must be freed from danger, and freed she is, as hundreds of passengers aboard the rolling liner watch the lifeboat creep through the towering seas. But in *Pipe All Hands* the lifeboat moves only a quarter mile between stopped ships, not across hundreds and thousands of miles of open sea. In the interwar years lifeboats became less and less important to passengers and novelists alike, and even in the gothic imagination of Cozzens, less and less real.

Nordhoff and Hall's 1933 *Men against the Sea* reawakened interest in presteamship open-boat passages, however, and books like Victor Slocum's 1938 *Castaway Boats* focused on them. In a series of brief chapters, Slocum chronicled great eighteenth- and nineteenth-century open-boat adventures, recounting the passage of Bligh's longboat, the *Saginaw*'s gig, and the *Essex* whaleboats, among others. At sea since childhood, witness to the rescue of open-boat castaways in the South Seas, and survivor of a small-boat passage himself after his father's bark foundered in Paranaguá Bay off Brazil in 1888, Slocum insisted on his authority to select illustrative tales. "The claim to the privilege of authorship in this field of maritime experience is based on a life spent very largely on the sea," he writes in his introduction.[6] But *Castaway Boats* opens with an indication of the profound influence of *The Riddle of the Sands,* and a suggestion that a new reader shapes what may be the object of all small-boat passage-making writing, fiction and nonfiction alike.

"The study of the small boat at sea may have other than romantic value; it may be the means of inculcating in the amateur sailor the spirit of self-reliance," Slocum maintains.[7] *Castaway Boats* is one of the first late-1930s books to rely on the book-buying power of yachtsmen, would-be yachtsmen, and readers interested in relics of the age of sail as portals on a better, less artificial past.

Slocum shapes his book cleverly, in search of the broadest possible readership. He first targets male readers, probably the male readers reviewers of *The Raft* subsequently identified, concerned that modern life (especially office work) somehow harmed everyone (especially men). Next he aims at the fast-growing group owning very small pleasure boats rather than large yachts, and emphasizes the singlehanded boating that so impresses Carruthers in the early chapters of *The Riddle of the Sands.* Whatever else it details, *Castaway Boats* focuses on the accomplishments of the small boats popular everywhere in depression-era United States coastal waters. Slocum insinuates that it is the small boater, not the owner of a large yacht, who has a kinship with the great passage makers of another era.

He also focuses on the antiurban, antisuburban, anti-fast-paced-living spirit that sent Nordhoff and Hall to the South Seas to write about transplanted Europeans and Americans who read no newspapers and find themselves discovering serenity.[8] Slocum argues that away from nerved-up cities and urban technology, men and women in small boats might rediscover the timeliness of nature and a sort of simple tranquillity wedded to excitement that contradicts the twentieth-century indoor passivity, effete entertainment, and technology Stenhouse scorned.[9]

Slocum knew or sensed the widening discontent sparked by high-technology inventions supplanting familiar, much-loved things while simultaneously eroding masculine independence, resourcefulness, and courage. As the editors of *The Rudder* discovered when a new magazine, *Motor Boating,* underscored the widening rift between sailboat aficionados and lovers of engine-driven boats, readers interested in "sea stories" turned toward the age of sail as the era of adventure, excitement, and romance vanquished by diesel-powered ships like *Morro Castle. Castaway Boats* is part of an emergent, backward-looking literature wider than age-of-sail autobiographies, like George Sorrell's 1928 *The Man before the Mast,* J. H. McCulloch's 1933 *A Million Miles in Sail,* or Pryce Mitchell's 1933 *Deep Water: The Autobiography of a Sea Captain,* so popular with depression-era readers. And it dovetails with nonfiction work like Ralph D. Paine's 1927 *The Ships and Sailors of Old Salem* and J. Ferrell Colton's 1937 *Last of the Square-Rigged Ships* even as it moves beyond the emergent interest in eighteenth- and nineteenth-century marine history that sparked a new scholarly journal, *The American Neptune,* in 1941.

Grant's *Take to the Boats* shares many similarities with *Castaway Boats,* some immediately evident. Both authors lived through the Great War, both commanded merchant ships, both learned of small-boat passage making in their years at sea, and both published books in 1938. This last is significant. By then the war clouds that blanket so much of Steinbeck's *The Log from the Sea of Cortez* had begun gathering, and authors like Grant and Slocum had begun to look back at the Great War and its lessons. "A good part of the fun of voyaging is to rediscover the places we have read of and have seen pictured in the atlas, and so verify romance and geography," Slocum writes of places drowsing since the end of the Great War.[10] But reexamining small-boat passages also forced a realization of the geographical import of headline news.

In 1938 small-boat passage making once again conjured up the specter of the lifeboat moving slowly away from the torpedoed steamship. Within six years such an image would grab the attention of United States cinemagoers and transform lifeboat imagery forever. In 1944 Alfred Hitchcock released his film *Lifeboat* and permanently altered popular thinking.

"You will come out of this movie knowing no more of what it is like in a lifeboat with a few people in mid-ocean than you did when you went in," sniffed *New Republic* film critic Manny Farber in January 1944.[11] But reviews in more popularly aimed magazines suggest that the public wholly embraced the Hitchcock imagery as realistic. "We didn't think it was possible for a 90-minute movie to have one locale and actually be moving cinema as well as one of the most exciting films of all filmdom," rhapsodized *Commonweal*.[12] *Newsweek* observed that "with his camera arbitrarily lashed to the mast, Hitchcock has not turned out the most satisfying job of his career as dean of cinematic melodrama; nevertheless, there is little doubt that twelve months from now, *Lifeboat* still will stack up as one of the most exciting films of the year."[13]

Only *Life*, in a January 31 photo review, worried that the mass of Americans might accept the screen imagery as gospel, and that the few voices raised in protest might be unheard. "*Lifeboat*, aside from a big storm and a surgical operation, is really Hitchcock's peculiar report on these survivors and the interplay of their personalities. Into the simple framework of a 26-ft. boat he has managed to squeeze prodigious suspense," notes the *Life* piece. But then, perhaps concerned by the growing ability of cinema to overwhelm photographic magazines, the reviewer wonders about the power of imagery most viewers will think authentic. Experienced castaways, he continues, will note the lack of bilge water, swollen lips, and competent seamanship, and will wonder how the characters look so energetic and clean for so long.[14] The *Life* review pinpoints exactly what Stenhouse predicted—the power of cinema to skew twentieth-century understanding.

Nazism, and the power of Nazism to triumph over disorganized opposition, focus the John Steinbeck short story that originally attracted Hitchcock's attention and that Jo Swerling adapted into a screenplay. The intent of the film is to demonstrate how Nazis wrest power from the weak and the divided, Manny Farber asserted. But the concept "leaves a number of points in doubt, proves itself by some rather flat-footed denouements, and is never essentially convincing." Viewers—again, as Stenhouse predicted—lack the will and opportunity to examine the film critically, however, for its pace sweeps them along.[15]

The setting, a big lifeboat on a sea empty until the final minutes of the film, demands and rewards notice, for it carries the old scenario baggage effortlessly, indeed slyly. As the film opens, Tallulah Bankhead sits alone in an ordinary ship's lifeboat, somehow having been lowered unaccompanied. No scene more clearly and cleverly demonstrates the scale of the lifeboat, and as other people climb aboard, including the captain of the U-boat that shelled the steamship, the lifeboat never seems crowded. Minute by minute, Hitchcock introduces the themes that have come to characterize lifeboat narratives. The castaways struggle to move the boat away from the suction they fear as their freighter sinks, and they worry about being at sea for weeks. They debate whether to throw the *Kriegsmarine* officer overboard, each of the British and American castaways adopting points of view any law school student or reader of lifeboat narratives would recognize instantly. "Whose law? We can make our own law," argues one, even as another claims that saving the captain is "the Christian way," and a third, putting his faith in marine law, insists that the German must be turned over to the authorities and "until such time we represent the authorities." The able seaman whose foot develops gangrene not only secretly drinks seawater but presents a dilemma when he wants to kill himself by jumping overboard. "It's murder," insists one castaway. "No it's not, it's execution," answers another. The quiet, African American steward puts his faith in God and his family, reciting the Twenty-third Psalm when the castaways bury a baby, and everyone except the German captain bemoans the lack of an officer or experienced lifeboatman to take charge of the boat. About all Hitchcock slights is cannibalism, and the film's placement of women aboard the boat and the development of subsequent romantic entanglements more than masks that omission. Its compass smashed and its food and water containers ruined, the lifeboat pushes the castaways into themselves and reveals how little they know of the sea, the stars, and—of course—human nature.[16]

Lifeboat profoundly disturbed some wartime viewers, who found the German captain the only capable person in the boat. "To survive, one must have a plan," he says near the end, mocking the aimlessness of the other castaways. Instead of analyzing the film as Farber did, most reviewers merely summarized it, perhaps because German capability seemed so brutally obvious, but some understood it as a condemnation of prewar American thinking.[17] By 1944 the Allies had cast off the dismay and disorganization in which Hitchcock revels, and which he almost burlesques; but suddenly cinema audiences had their faces rubbed in both—in an appalling scene of mob violence when the castaways commit murder.[18]

Americans tended to be far more critical of the war effort than popular memory recalls, and nowadays *Lifeboat* screens out of such context. Letters criticizing the British propaganda effort are everywhere in United States magazines, especially after British officialdom converted the retreat of the British Expeditionary Force into the glorious victory of the Dunkirk evacuation. The letters suggest that many readers of news magazines remained critical of all information from Britain long after United States entered the war, and they evince a healthy skepticism of United States official information too, and of Hollywood imagery.

In a September 1940 letter to the editors of *Time,* for example, a Pennsylvania reader, Merle B. McKaig, criticized a story reporting the torpedoing of *City of Benares,* the liner whose foundering put Mary Cornish into a lifeboat for days. The story lacked ordinary editorial objectivity, wrote McKaig. "Apparently you have accepted the British story whole cloth, without any examination of the details." McKaig argued that U-boats would have had difficulty attacking on very stormy nights, that a time bomb might have sunk the vessel, and that it might even have foundered as a result of a zigzagging collision in convoy. "An alert British propagandist could make excellent capital of the mishap—with a rigid sympathetic censorship holding up the news until the collective stories should hang together fair well." Point by point, *Time* rebutted McKaig's argument, but not especially effectively, perhaps because its editors recalled the part British officialdom had played in the *Lusitania* sinking.[19] As if to underscore the doubtful veracity of lifeboat-passage accounts, only a few months later, Huxley's *Atlantic Ordeal: The Story of Mary Cornish* reported that the gale had dropped before the torpedoing and that the seas had moderated enough for her to take a stroll on deck.[20]

More important, McKaig's criticism, however unwittingly, hit hard at the whole rationale of evacuating British children across an ocean essentially controlled by the *Kriegsmarine. City of Benares* left England carrying not only hundreds of children but a vast quantity of gold intended for the purchase of arms and munitions in the United States and Canada. Some military historians have voiced suspicions that the British government thought the gold might be safer in a liner known to be carrying children.[21]

More than a half century after the end of World War II, historians like Melanie Wiggins have begun tracing the skepticism exemplified by McKaig's letter to the editors of *Time.* Wiggins's *Torpedoes in the Gulf: Galveston and the U-Boats, 1942–1943* demonstrates the suspicion with which ordinary American citizens viewed everything from the Dunkirk evacuation to lifeboats coming ashore on the Texas coast to films like *Lifeboat*. "Say the Truth—And Don't Be Afraid of Death" began one 1942 *Fortune* article on the tanker sinkings just off the Atlantic coast. More than most present-day Americans know, United States citizens caught up in World War II viewed much war news with a skepticism born in World War I.

Hollywood began to color nonfiction lifeboat narratives as well. Elizabeth Fowler related that as the castaway men began quarreling, one became possessive of her and kissed her in the dark. After her rage subsided she began to see herself in Hollywood terms. "I had an irresistible desire to laugh. I was thinking of Joan Crawford as I had seen her in a movie. She was supposed to have been adrift in a lifeboat filled with men. Luckily for Miss Crawford, there had been only eight of them," Fowler wrote. "Her torn dress had been swathed about her body with all the allure of a sarong. Her hair had been a shining halo."[22] Fowler felt her own hair, sticky with palm oil and vomit and seawater, and looked at her oil-blackened hands and feet. Alone in a lifeboat jammed with thirty-four men, seated in the bow far from the taciturn New England shipmaster so intently focused on making his westing day after day, Fowler found the seamen around her becoming possessive, jealous, and sexually aroused, and she ruefully confronted the cinematic unreality. No longer was she the self-confident, saucy woman of the world drowning out the missionaries' evening service with ribald songs. Now her daily concerns included the reality of urinating into a pail and keeping men's hands off her.

In framing her predicament in Hollywood terms, Fowler alludes to a film that was closed by police in some United States cities and that never even opened in many others. The Catholic Legion of Decency condemned Frank Borzage's *Strange Cargo* before it premiered in 1940, lambasting not only what it considered the misrepresentation of passages from Scripture but the casting of Clark Gable in a cryptic, Christlike role. The eight men escaping from Devil's Island with Joan Crawford are more than convicts temporarily sharing a boat with a scantily clad woman. They represent both the world refugee crisis and the beginning of another war punctuated with steamship sinkings and lifeboat passages. The mysterious convict mystic who guides everyone in the boat and whose pronouncements so upset the Legion of Decency raises unsettling notions of leadership and the willingness of people to accept direction. One by one the men die, leaving Crawford alone with Clark Gable.[23] To Fowler, getting a rugged education in the realities of lifeboat existence, *Strange Cargo* offers up not only the potent absurdity of a lone woman controlling the passions of a boatload of men but the significant unlikelihood of that woman remaining perfectly coifed, attired, and groomed, day after day.

Any reader of Fowler's *Standing Room Only* must wonder at the Hitchcock film. She accurately identifies the automaton-like obedience of the *West Kebar* men as they follow the orders of the U-boat captain in the waterborne moment. She raises the same woman-of-the-world issues that fascinate Hitchcock, who in time reduces his two women characters to dependencies of men. She thinks in terms of *Strange Cargo,* and any viewer of *Lifeboat* must have recognized the similarities between the two films. And in the end she dismisses cinematic imagery. After her rescue, on a first-class train rolling north from Florida, Fowler realizes how much Hollywood skews the American vision of the war and especially skews the portrayal of American women caught up in war.

Nevertheless, after the war *Lifeboat* became a rear window on the past, on the whole war at sea, especially on the impact of ocean warfare upon civilians. In its dominance it occluded the power and message of books like Grant's *Take to the Boats* and Gibson's *The Boat* and profoundly influenced the way the United States population imagined, however inaccurately, all subsequent lifeboat narratives.[24] In later years, its intermittent appearance on television no doubt enhanced its impact.

Four years separate Hitchcock's *Lifeboat* from the publication of Gavin Douglas's manual *Seamanship for Passengers.* In those four years the world of ocean travel changed so dramatically that lifeboats slipped further and further from passenger consciousness. "Wars accomplish few goods, but among them are improved survival appliances," Douglas muses near the close of his book, in a chapter entitled "Lifeboats and the Passenger." "From the last two wars seamen have profited by the grim tests of lifeboats from torpedoed and bombed ships," he argues. "The day of a ship's lifeboat being a dump for unwanted oddments of the ship, of carelessness in tending them and a reluctance to move them from their chocks, is gone."[25] But the day of liner travel had begun to end too, and with it much of the significance of *Lifeboat.*

Seamanship for Passengers emphasizes something different from so many how-to-travel-by-liner books. It focuses on obedience to instruction, especially during lowering and the waterborne moment. Passengers must do as crewmen order, Douglas advises. "In the boat, sit down and be quiet. Every unnecessary word spoken confuses orders."[26] Being a hero, even politely ushering someone else into a boat after being told to board, invariably causes delay and possible disaster. Following orders makes abandoning ship vastly more effective; disobeying orders is a quick way to get killed or to kill others. World War II produced better marine safety appliances, Douglas implies, but it did nothing to correct certain social weaknesses most apparent during and after shipwreck.

Obedience sans officers and certificated lifeboatmen preoccupied Douglas, who advanced a unique argument born in his reading of World War II experience. In a lifeboat lowered and released but missing seamen, the castaways must stream the sea anchor and trim the boat, moving themselves sternward until the boat rides with its bow higher than its stern. They must then choose a leader, who will sit aft at the tiller and steer while others row. Then and only then should they unpack the portable radio and begin cranking, sending the automatic distress signal and homing beacon before settling themselves to await the aid that is almost certain to arrive. Douglas presumed a longish waterborne moment, one ending with the arrival of steamships, perhaps preceded by aircraft, but he knew another scenario too.

If help does not come, the castaways must make for land, using the pilot charts and remembering the approximate position in which their ship sank. Douglas's book explains rudimentary navigation using charts and Boy and Girl Scout techniques, and explains how to find the North Star and the sword of Orion in the event the compass proves inoperative. The book details the simple downwind seamanship the traditional lifeboat rewards, and even explains foul-weather seamanship. Beyond that, it sketches the human component of lifeboat existence, outlining the rationing of food and water, first aid for the injured, and even ways morale rises and falls. It ends with instructions for bringing the big boat into the surf and onto the beach, instructions borrowed from the Royal National Lifeboat Institution. "Do not be concerned about the boat being smashed, but go from it as fast as you can," concludes the instruction on lifeboat passage making. "It has done its job—as it always will do if it is cared for and handled sensibly."[27]

Among the other things *Lifeboat* taught cinemagoers, its first minutes showcased the roominess of the traditional lifeboat. Bankhead sat in solitary splendor, surrounded by space, and even with ten people aboard, the Hollywood lifeboat provided plenty of elbowroom. Whatever else the thousands of war surplus lifeboats meant to passersby who saw them ranged in urban ship-breakers' yards and junkyards, they meant the possibility of obtaining large, well-built boats for almost nothing. Between 1946 and about 1955, would-be adventurers in the United States and Great Britain began buying lifeboats in large numbers and converting them into yachts. The traditional lifeboat's strength shamed yachts and other postwar boats built of first-generation plywood, and its possibilities for conversion seemed endless. The postwar lifeboat conversion fad actually began at least a decade earlier, when depression-mired boaters found disused lifeboats to be an inexpensive way of getting onto the water, if not in style, at least in safety and comfort, and throughout World War II authors reminded boatless readers that peace would cast thousands of surplus boats into a buyer's market. It did.

Surplus boats enticed Americans as early as the 1870s. In the aftermath of the wholesale destruction of the Union merchant marine by Confederate raiders, and the shift to wood- and iron-hulled steamships like *Arctic*, many observers found harbors filled with aged, disused sailing vessels equipped with still serviceable jolly boats, launches, and whaleboats. For ordinary Americans eager to get on the water, the port of New Bedford, slumbering in the twilight of the whaling era, offered special opportunities. In its harbor filled with inactive sailing ships and punctuated everywhere with whaleboats and other castoff boats, the anonymous author of "Cheap Yachting," an 1873 *Harper's New Monthly Magazine* article, spent two weeks rowing and finally located an inexpensive boat.[28] For decades afterward, low-budget amateur mariners sought out ship's boats, and eventually lifeboats.

As a series of economic downturns struck the United States merchant marine after World War I, ship's boats became a favorite means of cheap yachting. The slump in shipping after the war deposited many lifeboats on the beach, and the depression beached many more. British writers—perhaps simply because they had the example of alongshore fishing boats like those R. Thurston Hopkins described in his 1931 *Small Sailing Craft*, perhaps because by 1935 Britain already knew the Converted Cruiser Club as a champion of converted lifeboats—were the first to detail how an ordinary Board of Trade lifeboat might become another *Riddle of the Sands* converted beauty.

C. E. Tyrrell Lewis published *Lifeboats and Their Conversion* in 1935, and his advice went through several subsequent printings, including one in 1943. The book inspired American Carl D. Lane to deal with conversions at length in his 1941 *Boatowner's Sheet Anchor: A Practical Guide to Fitting Out, Upkeep, and Alteration of the Small Yacht* and served as a model for other, more detailed books, especially two published in 1951 in Britain: Michael Verney's *Practical Conversions and Yacht Repairs* and John Lewis's *Small Boat Conversion*.

C. E. Tyrrell Lewis sketches the history of the *Trevessa* boats as an example of what lugsail-driven boats can do. In his view, except for the patent boats, and, perhaps, the largest of the steel lifeboats equipping new liners, Board of Trade lifeboats offer perfect opportunities to those unable to afford expensive small sailboats and engine-driven cruisers. "If you have ideas of your own with regard to the rig or accommodation, that you wish to try out: or you do not mind appearances, but want seaworthiness: or if you enjoy boat tinkering for its own sake—then under any of these latter circumstances it is worthwhile buying a lifeboat hull, and spending the leisure hours of six to twelve months in making a cruiser of her," he advises. His text explains everything from locating an inexpensive ship's lifeboat to arranging railroad and road transport to installing an engine. It emphasizes the relative ease with which the amateur builder can approach the entire project, since work proceeds with the hull right side up. No special tools are required, only space a bit removed from other houses in order to forestall complaints about hammering and other noise. Conversion work tends to be fun, he insists; all of it proceeds in a sort of waking dream, since so many changes happen so quickly and relatively easily. But here, he warns, the builder must stop and ponder the uses to which the converted lifeboat will be put. Should the cabin have full standing headroom or should it be low-ceilinged, like the cozy, seaworthy lifeboat in *The Riddle of the Sands*? How should the engine be installed? What sort of rig should the vessel carry?[29]

Such questions fascinated not only British builders but American ones too, and in 1941 Carl Lane confronted them squarely. He warned that no matter how fine a conversion might be, the converted lifeboat would always be a lifeboat. While lifeboats could be bought for ten to several hundred dollars, conversion was not really an inexpensive way of acquiring a boat, for a lot of time and some money for materials must go into the job. No matter how powerful the engine, no converted lifeboat would go fast under power, and all rolled in a seaway, although false keels and steel keel shoes snubbed the rolling. Lane argued against conversion, but admitted cheerfully that he had done a dozen conversions himself and that hundreds were done every year in the United States alone. Having argued against the idea, he went on to explain how to make it a reality.

Ten years taught amateur boaters a lot about lifeboat capabilities and at least something substantial about conversion. Writing in 1951, Michael Verney had at his fingertips all sorts of precise information about lifeboats from sixteen to thirty feet in length. His *Practical Conversions and Yacht Repairs* explains the difficulties of converting steel lifeboats, even of discovering if they are sound, and extols the virtues of diagonally planked, traditional, Board of Trade wood boats. His book focuses on cabin planning, perhaps because by 1951 women had begun boating in large numbers and enjoyed the comfort and privacy of cabins. Lovers separated by war intended to enjoy themselves together in peace, including at sea; if they had children, they intended to bring them along too. The key to pleasant accommodation, Verney emphasizes, is limiting headroom and planning bunk and galley arrangements carefully.

While Lane and Verney certainly expanded in some ways on C. E. Tyrrell Lewis's thinking, John Lewis authored the premiere book on conversion in 1951. *Small Boat Conversion* opens with a step-by-step recounting of Lewis's own conversion of a twenty-six-foot lifeboat into an auxiliary ketch, an account that details even problems like the miring of the delivery truck in beach sand and the effort involved in removing the keel bolts with "hammer, punch, and monkey wrench." Unlike his predecessors, however, Lewis insisted that the word *conversion* connoted too many conflicting possibilities. Lifeboats might be converted into all sorts of pleasure craft if only visionaries would ask the right question early on. Once a would-be converter decided whether he wanted a roomy river cruiser or an offshore one, and figured out if he had the spare time to do the work himself, conversion rewarded anyone making a small initial outlay of money and then making a lot of low-skilled effort.[30]

In the pages of *Yachting, The Rudder, Motor Boating,* and other magazines, the cabin of *The Riddle of the Sands*'s converted lifeboat became the cabin of choice in the years following World War II, and the converted lifeboat a model of comfort and safety. Until about 1960, the converted lifeboat pleased thousands of amateur shipwrights and seamen. What Hitchcock emphasized in film, the converted lifeboat proved in actuality. It provided plenty of elbowroom for those going fishing overnight, for honeymooners wanting privacy, for parents and children off for a week, for teenagers out for a long day's sail, even for the family dog. It offered space for a tiny enclosed toilet if its owners wanted. It offered space for a tiny galley fitted with any sort of stove, coal-burning and otherwise, and lockers galore for holding the extra sweaters and beach chairs and fishing tackle and swimsuits and ice chests that featured in impromptu cruises. Forward it provided storage for spare anchors and warp, and extra sails and line. It featured a large cockpit aft, a small forward deck especially useful while anchoring and mooring, and a cabin top excellent for sunbathing.

It also proved remarkably true to its origins, especially in heavy weather. While passengers might not like running for home in a rising sea, even in a thunderstorm or squall, everyone aboard knew that the boat would behave well. However it rolled, it rolled only so far before rolling the other way, its stability a hallmark of its earlier existence. Following seas split against its pointed stern, and head seas splayed from its bows, sending spray far aft but never green water. Perhaps few families took these vessels into heavy weather, but even those who converted lifeboats into river cruisers lacking mast and sails knew that in emergency conditions their boats would endure.

Seaworthiness meant more in the decade immediately following World

War II than it did later. Before the great suburbanizing of the United States coastal regions, most converted lifeboats sailed from urban locations, and their owners confronted the wakes of steamships and warships moving in confined waters. Moreover, oncoming bad weather meant running for home, for away from urban harbors lay little accommodation for recreational small craft. In the decades before marinas opened in many small harbors and provided nearby emergency docking, a change in weather meant lighting oil-burning lamps, fishing a compass from a locker, donning oilskins, and heading toward home. In such moments conning a converted lifeboat became, if not a pleasure, at least as comforting as relaxing in the stove-heated cabin, for the converted lifeboat behaved well under duress.

In squalls and in the wake of ever larger oil tankers and aircraft carriers, the converted lifeboat offered its owners another, more subtle pleasure. It had a history along with a time-tested hull. Sometimes the history appeared plainly in some identifying mark, even the name of the steamship to which the lifeboat once belonged. Sometimes the history loomed vaguely, announcing itself only in cryptic letters and numbers chiseled into the stem or keel. And now and then history became romance, when the amateur shipwright discovered a bullet hole or even the spent bullet itself, an eerie reminder of a ship once strafed from the sky. Whatever the converter of a lifeboat thought of history, even the history of his or her own vessel, the history of all lifeboats gave rise to a remark heard all alongshore in the 1950s and '60s. Anyone who joked at a converted lifeboat, anyone who asked how such a slow-moving boat got anywhere, heard the stock answer flung back. "Hey, this boat's been to Europe and back." And so it had, secure in its davits and chocks of course, but nonetheless ready to be launched into mid-ocean seas.

Then suddenly the converted lifeboats disappeared. By 1965 there were far fewer than there had been a decade earlier, and by 1970 they had almost vanished. Plywood boats designed for outboard engines supplanted them within a decade. Small boats fitted for water-skiing and other high-speed activities proved far easier to run up on a sandy beach and easier to trailer behind automobiles. Larger plywood boats, V-hulled and carrying inboard engines designed for square-sterned craft, offered spacious cabins and flying bridges, if not much stability in storms. By the middle 1960s, production boats built by Lyman and Chris-Craft had well-nigh vanquished the converted lifeboats, and within a few years, fiberglass boats by Boston Whaler and other firms vanquished the plywood-and-mahogany craft. Only in the small boatyards deep in the salt marshes could anyone find the last of the converted lifeboats, some serving as summertime houseboats, most decaying on cradles tangled in bayberry and poison ivy.[31]

Car ferry *Ashtabula* with steel lifeboat. Egg-shaped vessel with capacity of forty persons. (Courtesy of the Mariners' Museum, from the R. G. Skerrett Collections)

More than changing fashion vanquished the converted lifeboat, though. The supply of traditional lifeboats shrank, then almost vanished itself.

By 1960 the traditional lifeboat remained traditional mostly in hull shape and rig, but not necessarily in material. Steel supplanted wood, and steel, Carl Lane insisted, made for difficult conversions. The steel lifeboat is the curse of the amateur boating world, he warned as early as 1941, for only an expert can tell if it has been twisted in a way impossible to repair. Even men trained in metalworking should avoid boats formed on great hydraulic presses, and even expert welders often give up on skewed hulls.[32] All too often, a steel lifeboat seemed fine only until its new owner tried to set it level and plumb on blocks. Then it displayed a pronounced warp unnoticed as it lay on the ground in some junkyard.[33] If a steel lifeboat remained fair-lined, it still posed problems to the amateur builder intent on installing wood decks and a wood cabin, for attaching deck beams meant first fabricating complicated brackets, then welding them to the steel hull. Fitting an engine meant first fitting brackets to carry the wood engine bed, then cutting a hole in the steel skin of the hull, neither a simple job and both likely to be botched. When the amateur builder first started the engine and heard its roaring echo magnified by a metal hull, he might collapse in despair. Now and then a military-trained welder installed a steel deck and a steel cabin, and quickly learned that he had not only a top-heavy boat but one with condensation running down the interior sides of the cabin and soaking everything.

Something else happened around 1960 that shifted attention from all lifeboats, steel-hulled and otherwise. Travel by sea suddenly seemed outdated and trivial. Air travel transformed sea-survival devices, and more and more ships began carrying motorized, cocoonlike lifeboats, enclosed inflatable rafts, and other appliances that evolved from wartime experience and the long-distance-bomber years of the Cold War. The few great liners that struggled against airline competition carried engine-driven steel lifeboats, some of which carried a hundred people each, and small vessels began carrying boxed inflatable rafts. With fewer wood lifeboats for sale and with passenger attention fixed on inflatable rafts that in time evolved into pleasure craft like the outboard-engine-driven inflatable boats built by Zodiac, the old Board of Trade ship's lifeboat became an anachronism, something in reserve aboard many vessels carrying high-tech survival equipment based on the simple notion that rescuers always arrive soon after shipwreck.

For two generations of Americans born after World War II, *lifeboat* connotes something from cinema, especially from the Hitchcock film. Even readers of *The Riddle of the Sands* now miss the significance of *Dulcibella*, the staunch boat owned by Davies, a man with very little money but with a fierce desire to voyage in a boat he has converted from lifeboat to cruiser. While *The Raft* endures as a stock high school reading-list book, narratives like Foster's *1700 Miles in Open Boats* and West's *Lifeboat Number Seven* have vanished into some library limbo as deserted and unfrequented as so many parts of the oceans after steam supplanted sail. Noyes's *Open Boats* gathers dust in the general history section of my university library, alongside Lauriat's denunciation of the *Lusitania* disaster. Hollywood history skews any reading of Nordhoff and Hall's *Men against the Sea,* and the brilliant if atypical navigational material of Gatty's *The Raft Book* is ignored in books aimed at yachtsmen and professional mariners alike. Almost no one, including naval architects, knows of John Lewis's *Small Boat Conversion,* and only a very few yachtsmen know much about double-ended vessels at all. The traditional lifeboat floats in a sort of twilight now, lit intermittently by flashes of more Hollywood history.

Despite their reputation for not performing well in fire and for getting twisted while stowed, steel-hulled lifeboats outfitted with traditional rigs made port everywhere during the latter years of World War II. (Courtesy of National Archives)

CONCLUSION

ON THE HORIZON, THE ORANGE HYPHEN IN THE BLUE BEGINS TO foreshorten. I watch it, my sunglassed eyes narrowed against the blinding sun. The orange hyphen vanishes suddenly, blocked by the big rudder head jutting up over the stern of the lifeboat.

It is a hot day in Massachusetts Bay just east of Boston Harbor, and a rare one. Windless. The lifeboat is becalmed, its lugsail lowered and stowed on its yard, the tiller lashed amidships, the twelve-foot-long ash oars pulling slowly but regularly. The old wisdom, at least along the south coast of Massachusetts, holds that a man produces a quarter horsepower when rowing. I sit, pull on the oar with both hands, and think. Does the old wisdom mean a man rowing an ordinary pulling boat, one oar in each hand? Does it mean that together, my wife, Debra, and I produce a half horsepower? The lifeboat moves steadily but excruciatingly slowly across the glassy sea. It is exactly low tide. In ten, perhaps fifteen minutes, the incoming tide will produce a current that will help us along. Debra, hidden behind sunglasses herself, glances over her shoulder now and again at Boston Harbor light, judging our progress. I close my eyes and pull.

Since Debra saw the first lifeboat delivered on a trailer truck and unloaded between our henhouse and vegetable garden, she has wondered about the deeper significances of lifeboat passage making. That lifeboat, all twenty-six feet of it, now sits in the barn cellar, minus its cabin and other accoutrements installed in a mid-1950s conversion. Stripped of its plywood elements, it proved in need of years of part-time repair that I might or might not find in retirement. The next lifeboat, slightly smaller, arrived on a slightly smaller truck. It needed work—a whole strake removed years earlier in an abortive repair had never been replaced, one gunwale needed replacing, two thwarts lay broken, many strakes needed refastening, and the keel bolts had to be replaced in a job as filthy and exhausting as John Lewis described in

Small Boat Conversion—but the work progressed, month after month, even on winter weekends. Debra started lifting weights when we found the first pair of antique oars. Every other evening, the barbell and dumbbells went up and down. At another antique shop we found another pair of oars, and our teenage twin sons began muttering mutinously. A gloomy chandlery produced another pair. Adam and Nathaniel grumbled about hiring the high school weight-lifting team. A Green Harbor spar maker and rigger arrived in the barn, measured and fussed in growing delight, and returned with a rigging model. The new-made lugsail arrived from a Virginia sail maker but did nothing to stop weight-lifting and murmuring, and eventually new spars arrived too. Oh yes, insisted the elderly men stopping by to see the project in the barn, it will sail. But always take the oars. Always.

Debra glances at me, keeping perfect pace, smiles slightly. "I just want you to know I can't do this forever," she says. I nod. "You know, every other man on this coast has an engine." I nod again, suggest drifting a while and having a Coke, open my eyes.

I see the hyphen. Moving fast, the bone of white water in its teeth, the Boston Harbor Pilot Boat is bearing down on us. I look over my shoulders, stare around 360 degrees, and see nothing, absolutely nothing that explains the tearing burst of speed. The pilot boat is an all-weather, well-nigh cylindrical craft capable of coming alongside any oceangoing vessel in almost any seas. Now and then, it rescues distressed mariners. I know it has powerful engines, but I have never thought of it moving fast. I begin to think about collision and wonder about the competency of the people behind the smoked-glass windows of the wheelhouse. Then I realize.

The pilot boat crew has seen a lifeboat pulling slowly into the outer reaches of Boston Harbor. Not a converted lifeboat. Not a boat that resembles a lifeboat. But a real lifeboat, double-ended, one big hook jutting up forward, another jutting up aft, its rudder chained to the hull.

Before I can think what to do, what to say, the orange boat throttles back, slows, and veers slightly. A door opens, and a man carrying a bullhorn steps out.

Debra slides her oar inboard and waves, a big, long-armed, happy wave.

Several more men burst from the pilot house, stare, then slowly wave back. The pilot boat slows, almost dead in the water. I see what they see.

A petite, buzz-cut-haired waving woman, her yellow-and-green bikini two wisps against her skin, her smile as dazzling as the sun.

The men wave once more and step inside to air conditioning. Instantly the pilot boat accelerates and swings away, a long foaming wake behind it.

Debra slides her oar outboard, takes off her sunglasses, and says, "Those guys have an engine. I bet they'd give me a lift."

I promise her an engine. Ten minutes later we see the first riffle that announces a breeze, and twenty minutes later we are curled up on the stern seat, the lugsail set and drawing beautifully, our nameless lifeboat sailing east. Debra holds the sheet and tiller, squirms out onto the gunwale, grins, and says, "Well, let's see what speed we *can* make, okay?" I nod, and wonder again at my experiment.

I wonder a lot. I wonder at the looks people give the boat, at the comments people make, at what a lifeboat means now.

In Boston Harbor the lifeboat causes comment. Everyone knows it is a lifeboat. Little children point and yell; older people stare and turn to talk about it. One pleasure boat after another comes alongside for a look, some of the biggest ones scaring me as they approach and some of the fastest ones scaring me even worse. Speedboats crewed by gaggles of guys come alongside, their occupants pointing, shaking their heads. The shouts are the same, pretty much. "Great boat," is the usual one. "How old is it?"

Despite its glossy paint, Stockholm-tarred rigging, new lugsail, and oiled spars, the boat strikes everyone as a relic. Yet it is not a museum piece, for all it turns so many heads. We wear no period costume. In swimsuit or faded blue shirts and khakis, in high-tech white sunscreen fabric or tiny Tango Rose bikini, we might be any man and woman in a Boston Whaler or fiberglass sloop. We have no castaway look about us. Forward are the most high-tech of storage boxes, a big York Box filled with everything from a compass to flares, an Otter Box for my camera and Debra's cell phone and pager, other boxes for Debra's scuba-diving gear, a self-consuming portable toilet. The boat is no Flying Dutchman, sailed from some 1930s past into the new millennium.

But in some ways it affronts people too, especially when it rides gracefully through the confused waves and currents of the estuary mouth. That a boat small compared with the big fiberglass cabin cruisers around it rides so smoothly seems to irritate the occupants of boats not meant to cross oceans. Even the Coast Guard, patrolling aboard inshore-waters-only boats, scrutinizes the old lifeboat; inspecting it close up, as one Coast Guard officer told me, might lead to embarrassing comparisons of safety equipment.

Other times the lifeboat strikes me as a sort of goddess for whom doors are opened and drawbridges raised. One summer day a friend and I sailed into an East Boston marina, having consumed our supply of Coca-Cola. Made fast against the piers floated yacht after yacht, some on passage from

Florida to Maine. The marina dockhands, all young men, spotted the lifeboat, gathered in a group, waved, and shouted, "Come on, we'll make a place for you." Muscling aside two towering yachts, they made a place indeed, exclaiming over the boat, selling us the Coca-Cola we craved, then providing free ice for the coolers. Above the ice machine a sign announced ice at five dollars a bag. The young men laughed and told me ice is free "to a boat like that." A few hours later, moving south, we encountered *Pride of Baltimore*, the topsail schooner owned by the city of Baltimore as a sort of world-traveling ambassador. As we passed each other on parallel courses, its helmsman sung out, "What a beautiful boat!" and its crew waved wildly. But then again, *Pride of Baltimore* ships a crew fond of old ways.

Maybe the lifeboat is not so much beautiful as it is glamorous in the old meaning of the term, something touched by *glamourie*, by enchantment.

People over forty-five or so pay the lifeboat a sustained attention that frustrates me. Only rarely can I stop, come about, sail near some wharf or urban promenade and ask what they think, and even then I am self-conscious at seeming to beg for compliments. In Boston Harbor, among the docked steamships, I see crewmen look up from their jobs and stop and stare hard at the lifeboat, then turn and glance at the lifeboats slung in davits against the day the high-tech lifesaving devices fail, then glance back and wave. Sometimes older men watch the lifeboat nose into a beach or dock, then shout, "Here come the last *Titanic* survivors." Onlookers laugh. Do they think of shipwreck, of the wartime years when lifeboats focused the attention of so many children, of Hitchcock's film?

Younger people see the boat as something else. Over and over they ask if it is one of the lifeboats from the 1998 film *Titanic*. Until I learned that the film's lifeboats had been sold, supposedly for $10,000 each, I could not imagine how anyone acquired a prop from a film. I saw the film twice, sulking about its lack of verisimilitude and critical because in the final scene the lifeboat probing among bodies seemed a bit low in the water.

Now and then elderly men approach and stare and sometimes turn away, shaking their heads. Only a rare one speaks. What they say troubles me. I am no expert in collecting oral history, and Yankee alongshore men incline toward a taciturnity hard to crack. They know times to talk, especially about World War II, and times to be silent, and old age encourages silence. But when they talk to me about the lifeboat I listen, knowing that asking questions is a privilege belonging to men older than I.

"That's a good boat in an oil fire." I look up at the old man staring down at me from the finger pier. It is March, cold and raw and windy, and I have lashed the lifeboat to its custom-built trailer and backed it inch by inch down the launching ramp, hoping to find a devious leak in its stern. I look up from the bilge at the old man bundled against the cold and wait, nodding my interest. He considers, then speaks again. "They gave us steel lifeboats, in '43. Said wood boats didn't belong in tankers. First thing out we get hit, abandon, and the steel boat floats around in the burning oil like a frying pan, red hot. A wood boat burns and doesn't heat up and it gets you away from the fire. Next time out, I got a gasoline tanker with wood boats."

I ask softly what happened that trip. Wrong question. He shrugs, says "Made Liverpool," and rubs his running nose. The left side of his face is pale, scarred from a fire long ago. "Take care of that boat, 'cause it will take care of you," he says and walks off, shaking his head. I see the water pooling around my feet, and grab the pencil and paper I brought to sketch the whereabouts of the leak. As fast as I can, I write down what I think he said.

I ask around. Steelmakers tell me they use wood, usually two-by-fours, to skim the dross floating atop molten steel: anything metal melts. The two-by-four burns, but lasts long enough for the job. Perhaps fire twisted the metal lifeboats that vexed postwar conversion experts, but nowadays not many amateur boaters think to take a wood lifeboat through an oil or kerosene or gasoline fire. Once one starts thinking about it, fire proves surprisingly absent from contemporary survival-after-shipwreck guidebooks mandating lifesaving appliances made of fiberglass and other plastics.

Any sort of fire, perhaps especially an electric one, can ignite the diesel fuel that powers almost every oceangoing vessel. If the fire does not ignite the fuel, a flare most certainly can. In the dark, aboard an inflated raft, castaways may well toss overboard a flare straight into the middle of fuel, igniting it and destroying their raft and any other rafts caught in the spreading blaze.

So ships still cross oceans carrying a traditional lifeboat, not as the primary survival vessel perhaps, but one with a special purpose. If the crew of the ship must rescue castaways, especially people caught in fire, all the high-tech rafts prove almost useless. Into the inferno must go a boat capable of withstanding high seas and fire together, the boat no raft yet replaces. Descended from the traditional lifeboat, that boat endures as a memorial to the artifact in my barn.

For 150 years, oared, lugsail-rigged Board of Trade lifeboats saved lives in all sorts of circumstances. Now one of the last boats sails around on summer days, making me smile and think and learn.

It performs splendidly. It can be trusted to behave itself. Designed to care for its shocked and disoriented passengers during the waterborne moment when they can do little for themselves, to be sailed by landsmen if necessary, to make a passage in the wake of the *Bounty* launch, to manage gales and surf, it sails now as an antique, but it sails and rows exquisitely.

It whispers of times when they—whoever they are—do not come to help.

The lifeboat, long ago secured beneath davits and canvas cover but now sailing on a July afternoon, reminds everyone that things go wrong. Ships sink. Skyscrapers collapse.

It speaks honestly and bluntly of a way out of trouble that depends on no one but its castaway occupants.

NOTES

INTRODUCTION

1. Sim, *Malayan Landscape*, 248.
2. Gilmour, *Singapore to Freedom*, 32, 42–44.
3. Tomlinson, *Malay Waters*, 41–42; see also 51.
4. Jones, *Two Survived*, xiii.
5. Jones, *Two Survived*, x.
6. Coates and Morrison, *The Sinking of the Princess Sophia*, xiii.
7. Clark, *When the U-Boats Came to America*, 6–7, 15, 19, 53–54, 68, 70–74, 78–79.
8. Wiggins, *Torpedoes in the Gulf*, esp. 134–36, 156, 124–25, 149–51.
9. Dobson, Miller, and Payne, *Cruellest Night*, 14–15, 154–56, 190–96. See also Sellwood, *The Damned Don't Drown*.
10. Schoen, *Untergang*; and Lass, *Die Flucht*.
11. Grattidge, *Captain*, 174–75.
12. Villiers, *And Not to Yield*, 19–20.
13. Trew, *Death of a Supertanker*, 94.
14. Center, *Practical Guide*, ix.
15. Turner, *War in the Southern Oceans*, 54–63, 186.
16. Richards, *Tug*, 136–40.
17. Faith, *Mayday*, 7, 19, 51, 72, 156. On the sinking of *Sun Vista*, see "Sunken Ocean Liner was Boston Mainstay," *Quincy (Mass.) Patriot Ledger*, May 28, 1999. Apparently only because the ship once regularly called at Boston did the newspaper see fit to note its destruction.
18. Quoted in a Calvin Woodward Associated Press wire-service story dated July 20, 1999.
19. Monsarrat, *Three Corvettes*, 63–64.

CHAPTER I. Missing

1. Villiers, *Posted Missing*, 198; see also 161–69; and Hadfield, *Sea-Toll of Our Time*.
2. Villiers, *Posted Missing*, 28–36.
3. Conrad, *The Mirror of the Sea*, 56–59.
4. Villiers, *Posted Missing*, 168.
5. Conrad, *Mirror of the Sea*, 63–66; quotation on 64.
6. Hadfield, *Sea-Toll of Our Time*, 66–67.
7. It still does, especially in countries bordering the Indian Ocean. See, for example, Jenkins, *Scend of the Sea*.
8. McCulloch, *A Million Miles in Sail*, 97–98.
9. McCulloch, *Million Miles in Sail*, 99–100, 118, 149–52.
10. McCulloch, *Million Miles in Sail*, 156.
11. McCulloch, *Million Miles in Sail*, 150–54.
12. McCulloch, *Million Miles in Sail*, 170.
13. Munro, *The Roaring Forties and After*, 97.
14. Munro, *Roaring Forties and After*, 42, 96–104; quotation on 103.
15. Clements, *A Gipsy of the Horn*, 50, 45, 117–18, 161; quotation on 118.
16. Clements, *Gipsy of the Horn*, 118.
17. Clements, *Gipsy of the Horn*, 118.
18. Clements, *Gipsy of the Horn*, 108, 152–54, 172, 175; quotation on 154.

19. Leavitt, *Wake of the Coasters,* 94–95.
20. Ashley, *The Ashley Book of Knots,* 487–510.
21. Smith, in *The Arts of the Sailor,* suggests that the decline in rigging and ropework skills began early in the nineteenth century and was "well underway" by the middle of the century: "With the passing of the merchant sailing ship the rope-and-canvas sailor was headed for oblivion, cast adrift on a mechanized sea and master of an almost obsolete craft" (iii).
22. Dana, *Two Years before the Mast,* 187, 235.
23. Dana, *Two Years before the Mast,* 197.
24. Conrad, *Lord Jim,* esp. 140–42.
25. Making, *In Sail and Steam,* 135.
26. Making, *In Sail and Steam,* 135.
27. Making, *In Sail and Steam,* 145.
28. Making, *In Sail and Steam,* 191–92, 205, 219.
29. Making, *In Sail and Steam,* 242–43, 252.
30. Making, *In Sail and Steam,* 207–29, 240–65.
31. McCullough, *Million Miles in Sail,* 250–51.
32. Cooper, "Blue Lagoon."
33. Munro, *Roaring Forties,* 16.
34. Stacpoole, *The Beach of Dreams,* 84–85.
35. Stacpoole, *Beach of Dreams,* 96–97.
36. Stacpoole, *Beach of Dreams,* 274.
37. Stacpoole, *Beach of Dreams,* 276–77. Other authors explored the significance of castaway couples marooned together; see, for example, Morgan Robertson's short story "Primordial"; and W. Clark Russell, *Marooned.*
38. Conrad, *Last Essays,* 77–78.
39. Conrad, *North Sea,* 17.
40. McCulloch, *Million Miles in Sail,* 179.
41. On this period, see Kern, *The Culture of Time and Space.*
42. Ellis, *Round Cape Horn,* 293. United States Coast Guard officer-cadets still train aboard the sailing ship *Eagle.* The distinction between their training and that of cadets at the United States Naval Academy rewards analysis.
43. On this superiority, see Ellis, *Round Cape Horn,* 72–75, 90–93.

CHAPTER 2. Crumbs
1. "Yachtsmen—Keep Watch for These Jap Ships."
2. "This Is How the United States May Be Invaded," esp. 17.
3. "Nazi Fifth Column," esp. 51.
4. Wiggins, *Torpedoes in the Gulf,* 107, 84, 124–25. For a general survey of the Caribbean campaigns, see Kelshall, *The U-Boat War in the Caribbean.*
5. Rogers, *Hazards of War,* 157.
6. Rogers, *Hazards of War,* 82.
7. Lea's *The Valor of Ignorance* and McFee's *North of Suez* deserve careful scrutiny by intellectual historians, especially those of colonialism and the origins of the state of Israel.
8. Jones, *Short Voyages,* 55–62.
9. "The U.S. Reckons Up Its Losses," 26–27. This is a hard headed, chilling article that deserves wider currency today.
10. *The Raft* remains intriguing reading, especially in its suggestion that the navy fliers expected to find the Japanese south of the New Hebrides.
11. Kelshall, *U-Boat War,* 46–47.
12. Steinbeck, *The Log from the Sea of Cortez,* 46.
13. Steinbeck, *Log from the Sea of Cortez,* 47, 122–23.
14. Tomlinson, *Tide Marks,* 30–35. The Admiralty Hydrographic Office's *Red Sea and Gulf of Aden Pilot* provides fourteen pages on the island, which it spells Sokotra (402–15).
15. Tomlinson, *Gifts of Fortune,* 40–41.
16. Tomlinson, *Gallion's Reach,* 91.
17. Tomlinson, *Gifts of Fortune,* 41.
18. Tomlinson, *Tide Marks,* 80.
19. Steinbeck, *Log,* 288.
20. Melville, *Moby-Dick,* 319, 174.
21. Melville, *Moby-Dick,* 461.
22. Melville, *Moby-Dick,* 236, 318–19.

23. Conrad, "Youth," 31. On cargo coal fires, see Rowan, *Coal.*

24. Conrad, "Youth," 35, 37.

25. Conrad, "Youth," 38–39, 40.

26. Making, *In Sail and Steam,* 34.

27. Lords Commissioners, *The China Sea Directory,* 1:4–8.

28. Lords Commissioners, *China Sea Directory,* 1:5.

29. United States Hydrographic Office, *East Indies Pilot,* 1:138.

30. United States Hydrographic Office, *Sailing Directions for Sunda Strait and Northwest Coast of Borneo and Off-Lying Dangers,* 54–55.

31. United States Hydrographic Office, *Sailing Directions for Sunda Strait,* 54.

32. Tomlinson, *Malay Waters,* 48–50.

33. McCulloch, *Million Miles in Sail,* 97, 107–8.

34. Blain, *Home Is the Sailor,* 42; see also 48–49.

35. United States Hydrographic Office, *Sailing Directions for Malacca Strait and Sumatra,* 98–100.

36. Blain, *Home Is the Sailor,* 97.

37. United States Hydrographic Office, *Sailing Directions for Malacca Strait,* map following 60.

38. United States Hydrographic Office, *The Coast of Brazil,* 1:335–37.

39. Turner, Gordon-Cumming, and Betzler, *War in the Southern Oceans,* 227–28.

CHAPTER 3. **Bligh**

1. Hall, *The Tale of a Shipwreck,* 123. The book also appeared as "From Med to Mum," *Atlantic Monthly* 153 (March–July 1934), 257–68, 404–14, 568–78, 708–18, 856–79.

2. Hall, *Tale of a Shipwreck,* 115–16.

3. Hall, *Tale of a Shipwreck,* 117–19.

4. Hall, *Tale of a Shipwreck,* 122–23.

5. Hall, *Tale of a Shipwreck,* 126–27.

6. Hall quoted in Briand, *In Search of Paradise,* 90; see also 91, 124, 285.

7. Hall, *Tale of a Shipwreck,* 136–39; quotation on 137.

8. Fussell, *The Great War and Modern Memory,* offers many insights into the changes wrought by World War I on its participants.

9. Hall and Nordhoff, *Faery Lands of the South Seas,* esp. 6–7, 222–24.

10. Twain, *Letters from Hawaii,* 135–60.

11. Twain, "My Debut as a Literary Person," 88. See also Irvine, "The Lone Cruise," 571–77.

12. All quotations from the *Hornet* diaries are from Brown, ed., *Longboat to Hawaii.*

13. Twain, "Forty-three Days in an Open Boat," 111–13.

14. Twain, "My Debut as a Literary Person," 87–88.

15. Brown, *Longboat to Hawaii,* 147.

16. Conrad, *Notes on Life and Letters,* 197.

17. Munro, *Roaring Forties,* 75–76, 102.

18. Quoted in Daiches, *Robert Louis Stevenson,* 92.

19. Conrad, *Notes on Life and Letters,* 250–51.

20. Hall and Nordoff, *Faery Lands,* 265–68; quotation on 265–66.

21. Nordhoff and Hall, *Men against the Sea,* 6, 60; quotation on 172.

22. Nordhoff and Hall, *Men against the Sea,* 224.

23. Bligh, *The Bligh Notebook,* 23–24. See also Danielsson, *What Happened on the Bounty.*

24. Delano, *Narrative of Voyages,* 118–25. See also Edwards and Hamilton, *Voyage of HMS Pandora,* 156–58.

25. Stevenson, *In the South Seas,* 151–53, 157–59.

26. Stevenson, *In the South Seas,* 151.

27. Gauguin, *Noa Noa,* 136–37.

28. Cubbins, *The Wreck of the Serica,* 8, 14, 19, 23, 32, 36–37.

29. Knight, *Modern Seamanship,* 323–38; and Nares, *Seamanship,* 112–15.

30. Read, *The Last Cruise of the Saginaw,* 61–62, 101–12.

31. Wood, "Escape of the Confederate Secretary of War," 111–16.

32. Worsley, *Shackleton's Boat Journey*, 171, 175.

33. Conrad, *Notes on Life and Letters*, 245.

CHAPTER 4. **Boats**

1. Benson, *The Log of the El Dorado*, 18–19.

2. Benson, *Log of the El Dorado*, 20–22.

3. Benson, *Log of the El Dorado*, 18–19, 20–22, 24–25, 27, 34–35, 40, 43, 31, 29.

4. Collier, *The Sands of Dunkirk*, 204–5.

5. "British Rear Guard," esp. 35. See also "Battle of Flanders."

6. Chatterton, *The Epic of Dunkirk*, 127–28.

7. Collier, *Sands of Dunkirk*, 154, 246–50.

8. R. Lewis, *History of the Life-Boat*, 79.

9. Lewis, *History of the Life-Boat*, 77–79.

10. Dawson, *Britain's Life-Boats*, 91–94. See also Cameron, *The Life-Boat and Its Work*.

11. King, *The Coast Guard Expands*.

12. On the Humane Society efforts, see, for example, R. B. Forbes, *Notes* and *The Life Boat*.

13. See, for example, Kimball, *Organization and Methods;* and King, *Coast Guard Expands*.

14. Lincoln, *Rugged Water*, 26–27. Lincoln's novel seems as realistic as Dalton's nonfiction *The Life Savers of Cape Cod*, which appeared a year earlier. On contemporary rescue boats, see Middleton, *Lifeboats of the World*.

15. Willoughby, *Rum War at Sea*, is a useful introduction to the issues of Prohibition.

16. Brann, *The Little Ships*, 5 (quotation), 106–22.

17. Divine, *Nine Days of Dunkirk*, 101, 150, 153. See also Holman, *The Little Ships*.

18. Mercator, "Atlantic Steam Navigation," 433–35.

19. United States Navy, *Boat Book*, 12–13.

20. United States Navy, *1940 Bluejacket's Manual*, 312–13. For general background, see Scheina, *United States Coast Guard Cutters*. See also United States Navy, *The Power Boat Book*.

21. Morrill and Martin, *South from Corregidor*, 56.

22. Morrill and Martin, *South from Corregidor*, 70–71.

23. Morrill and Martin, *South from Corregidor*, 77, 100–101, 146, 154, 183, 210–11, 224–27.

24. Morrill and Martin, *South from Corregidor*, 72, 4.

25. Divine, *Dunkirk*, 135, 164–65; Buster, *Return*, 251, 243.

26. Divine, *Dunkirk*, 135.

27. Ballantyne, *The Lifeboat*, 145.

28. United States Navy, *1940 Bluejacket's Manual*, 568–72.

29. Masefield, *Nine Days Wonder*, 50.

30. D. Williams, *Retreat from Dunkirk*, 84–86.

31. Buster, *Return via Dunkirk*, 252–55.

32. Divine, *Nine Days of Dunkirk*, 101.

33. Drew, *Amateur Sailor*, 168.

34. Drew, *Amateur Sailor*, 173–74.

35. Drew, *Amateur Sailor*, 180.

36. Drew, *Amateur Sailor*, 181–82.

CHAPTER 5. **Board of Trade**

1. West, *Lifeboat Number Seven*, 181, 176.

2. West, *Lifeboat Number Seven*, 7–8, 154.

3. See West, *Lifeboat Number Seven*, 19, on the genre of lifeboat-passage narrative.

4. West, *Lifeboat Number Seven*, 35.

5. Hopkins's *Small Sailing Craft* is a very important book, one sometimes ignored by contemporary students of traditional fore-and-aft rigs.

6. Mansfield, "The Shetland Sixareen."

7. Chapelle, *American Small Sailing Craft*, 217–21. On seagoing canoes, see Beck, *The American Indian as a Sea-Fighter*. See also N. L. Forbes, *Suggestive Curves*. Few Americans now know the canoe as an artifact of the seafaring abilities of coastal tribes in Maine and the Pacific Northwest; in the latter region, tribes traded with Hawaii and Polynesia.

8. Simmons, "That's a Matinicus Double-Ender."

9. P. H. Spectre, "As Alike as Two Peas in a Pod?" 47. See also his "Slick as a Pod"; and Gardner, *Building Classic Small Craft*, 129–45.

10. Chapelle, *American Small Sailing Craft*, 277–91.

11. Chapelle, *American Small Sailing Craft*, 288.
12. Chapelle, *American Small Sailing Craft*, 190–91.
13. H. J. James, *German Subs in Yankee Waters*, 101–9.
14. Clark, *When the U-Boats Came*, 233–42.
15. Mercator, "Atlantic Steam Navigation," argues that "our men-of-war are not furnished, either with a whale boat or coble; our ships of war have not a boat *fit to take a beach in any surf whatsoever*" (431–32).
16. "Boats, Life Boats, and Buoys," 93.
17. Blake, *The Lifeboat*, 3–5.
18. Layton, *Ship's Lifeboats*, 15–16.
19. Sutton, *A Story of Sidmouth;* see also Cornish, *Scenery of Sidmouth;* Homeland Association, *Sidmouth;* R. Lane, *Old Sidmouth;* Gosling, *Around Seaton and Sidmouth;* and Barry and Gosling, *Around Sidmouth*.
20. Reynolds, *A Poor Man's House*, 117, 175–78.
21. Ellis, *Round Cape Horn*, 128.
22. Leather, *Spritsails*, 190–203; and Reynolds, *Poor Man's House*, 148.
23. Foster, *1700 Miles*, 53.
24. Rough, "Ships Boats."
25. Foster, *1700 Miles*, 37–38.
26. Blake, *Lifeboat*, 14–15, 13.
27. Layton, *Ship's Lifeboats*, 14–20.
28. Blake, *Lifeboat*, 13–14. See also Layton, *Ship's Lifeboats*, 22.
29. Foster, *1700 Miles*, 113; Blake, *Lifeboat*, 14.
30. Layton, *Ship's Lifeboats*, 22.
31. Blake, *Lifeboat*, 41.
32. Blake, *Lifeboat*, 49–52; Layton, *Ship's Lifeboat*, 68–73.
33. Knight, *Modern Seamanship*, 168, 171, 247.
34. Turpin and MacEwan, *Merchant Marine Officers' Handbook*, 496–98. See also United States Coast Guard, *Manual for Lifeboatmen*.
35. Hauge and Hartmann, *Flight from Dakar*, 117.

CHAPTER 6. **Davit**

1. Clark, *When the U-Boats Came*, 43–46. Clark gives the name of U-151 captain as Von Henchendorf; James, *German Subs*, 17, gives the name as Von Nostitz und Janckendorf but offers no source. Here I follow Clark. In his account James, 34–36, also wholly ignores the panic.
2. C. Simpson, *Lusitania*, 40–41, 71.
3. Low, *Submarine at War*, esp. 45–46.
4. Low, *Submarine at War*, 69–70.
5. Simpson, *Lusitania*, 238–39. The incidents are not mentioned in Chatterton's *Q-Ships* but are documented in Spindler, *La guerre sousmarine*, 2:322–29.
6. James, *German Subs*, 36–42; and Clark, *When the U-Boats Came*, 52–57.
7. Clark, *When the U-Boats Came*, 79–80.
8. Clark, *When the U-Boats Came*, 55–80; and James, *German Subs*, 56–69.
9. James, *German Subs*, 137.
10. James, *German Subs*, 126–41.
11. Mulville, *Schooner Integrity*, 53.
12. James, *German Subs*, 140–43.
13. Reynolds, *Alongshore*, 102, 89.
14. Russell, *The Mate*, 253, 258–59.
15. Conrad, *Last Essays*, 76.
16. Conrad, *Last Essays*, 76–77.
17. Murdoch, *A True Account*, 2–4, 6, 8–9.
18. Conrad, *North Sea*, 36–37.
19. Brown, *Women and Children Last*, esp. 67–94; quotation on 81. For a treatment of a subsequent similar collision, see Weiss, *Personal Recollections*.
20. Strong, *Diary*, 1:191.
21. Strong, *Diary*, 1:192; his full account of the sinking of *Arctic* is on 186–94.
22. "Monthly Record," 830.
23. Abbott, "Ocean Life," 61–66.
24. "Ship-Masters and Mariners," esp. 112–13.
25. "Loss of the *Arctic*," 107–8.
26. "Disasters at Sea," esp. 114–16.
27. Brown, *Women and Children Last*, 147.
28. "Nautical Club," esp. 443–46.
29. *Life-Saving Inventions*, esp. 26–27.
30. S. F. Walker, "Lifesaving at Sea," esp. 403. Patented davits appear in many

turn-of-the-century issues of *International Marine Engineering.*

31. Welin, *Lifeboats on Ocean-Going Ships,* 28–35, 36. See also "Improvements in the Welin Quadrant Davit."

32. Welin, "Paper Read," 45.

33. Francey, "RMS *Titanic,*" 46.

34. *Titanic Disaster Hearings,* 182.

35. *Titanic Disaster Hearings,* 188.

36. *Titanic Disaster Hearings,* 195.

37. McFee, *More Harbours of Memory,* 144.

38. McFee, *More Harbours of Memory,* 150.

39. Blocksidge, *Ships' Boats,* 421–22.

40. Hendry, *The Ocean Tramp,* 74.

41. Hendry, *Ocean Tramp,* 74–75.

42. *Tait's Seamanship,* 54. See also F. B. Williams, *On Many Seas,* 101–5.

43. Grant, *Consigned to Davy Jones,* 273–74, 281.

44. Grant, *Consigned to Davy Jones,* 281.

45. Grant, *Consigned to Davy Jones,* 172, 198, 282–86; quotation on 286.

46. Simpson, *Lusitania,* 160–65.

47. Hoehling and Hoehling, *Last Voyage,* 125.

48. Simpson, *Lusitania,* 164–68.

49. Bernard, *Cock Sparrow,* 158.

50. Droste, comp., *Lusitania Case,* 151–52, 175.

51. Lauriat, *The Lusitania's Last Voyage,* 17–18, 20. For another example of round-bar davits impaling lifeboats, see McCoy, *Nor Death Dismay,* 133.

52. Bernard, *Cock Sparrow,* 160–67.

53. Lauriat, *Lusitania's Last Voyage,* 79–80.

54. Simpson, *Lusitania,* 232.

55. Simpson, *Lusitania,* 88.

56. Kenworthy quoted in Simpson, *Lusitania,* 129.

CHAPTER 7. Waterborne

1. West, *Lifeboat,* 25.

2. West, *Lifeboat,* 27–28.

3. West, *Lifeboat,* 32–33.

4. Gilchrist, *Blue Hell,* 19.

5. *Titanic Disaster Hearings,* 195–96.

6. *Titanic Disaster Hearings,* 196, 201–2.

7. Brown, *Women and Children Last,* 121–22.

8. Strong, *Diary,* 1:356.

9. Lauriat, *Lusitania's Last Voyage,* 88–90.

10. Venables, "The Torpedoing of the *Assyrian,*" 24–25, 28–29. See also Foss, *Shoot a Line,* 24–32.

11. Fox, "A Ship Sinks," 69.

12. Page, "All in the Same Boat," 176–77.

13. Some of the directions appear to have been printed broadsides posted at wharves and distributed to seamen. See also T. Lane, *Merchant Seamen's War,* 226–27; and United States Coast Guard, *Suggestions concerning Tank Vessel Operation during Wartime,* esp. 1–5. The latter warned tankermen to expect uncovered lifeboats to contain 6–12 inches of oil after torpedo-caused explosion.

14. Bombard, *The Voyage of the Heretique,* x.

15. Gibbons, *How the Laconia Sank,* 19–20.

16. Gibbons, *How the Laconia Sank,* 20–22.

17. Venables, "Torpedoing of the *Assyrian,*" 26. See also Hayden, *Wanderer,* 190.

18. Gibbons, *How the Laconia Sank,* 32–33.

19. Gibbons, *How the Laconia Sank,* 24–29.

20. Layton, *Ship's Lifeboats,* 26–27, 32–33.

21. Noyes, *Open Boats,* 67–69.

22. Ellsberg, *Under the Red Sea Sun,* 20.

23. Ellsberg, *Under the Red Sea Sun,* 21–23. For another account of unpreparedness, see Klitgaard, *Oil and Deep Water,* 9–10, 18, 92–95, 121–22.

24. Mence, *Practical Seamanship,* esp. 192–96. Many merchant seamen spent years in sailing vessels without using boats; see, for example, Nordhoff, *The Merchant Vessel.* In his 1940 *Watch Below,* McFee notes that "boats were rarely provisioned and usually were all dried up and leaking at the moment of need" (172–73).

25. Ellis, *Round Cape Horn,* 124.

26. Richards and Banigan, *How to Abandon Ship,* 55, 7, 3. See also Bunker, *Heroes in Dungarees,* esp. 50–51, 114–15, 178–81.

27. Richards and Banigan, *How to Abandon Ship,* 20.

28. Richards and Banigan, *How to Abandon Ship,* 20, 45.

29. Richards and Banigan, *How to Abandon Ship,* 46.

30. Shaw, *White Sails,* 11–12.

31. Shaw, *White Sails,* 23. Shaw's recounting of the *Dovenby* rescue (11–23) is the most detailed account of ship-to-ship rescue I have discovered. The fact that Shaw's boat made two round trips perhaps accounts for his precision. See also Shaw, *Merchant Navy at War,* for the account of a board examining a young cadet by asking him "how he would approach a boat to a waterlogged wreck to ensure the safety of the boat and the crew and the people he hoped to rescue" (58).

32. Mulville, *Schooner Integrity,* 55–56.

33. Mulville, *Schooner Integrity,* 58–77. See also Nicholl, *Survival at Sea,* 44–47.

34. Hayler, Keever, and Seiler, *Cornell Manual,* 39–40.

35. Hayler, Keever, and Seiler, *Cornell Manual,* 40, 45–46, 47.

CHAPTER 8. **Passage**

1. Foster, *1700 Miles,* 138.

2. Foster, *1700 Miles,* 23.

3. Foster, *1700 Miles,* 33–40, 45, 74–75, 53; quotation on 33.

4. Foster, *1700 Miles,* 61, 66. See also Sobel, *Longitude.*

5. Stafford, "No Man Alone," 29–31.

6. "These Sights Save Lives."

7. T. Lane, *Merchant Seamen's War,* 225.

8. Roskill, *A Merchant Fleet in War,* 210–14.

9. Roskill, *Merchant Fleet in War,* 214–16.

10. Roskill, *Merchant Fleet in War,* 210–20.

11. McCulloch, *Million Miles,* 90–91.

12. McCulloch, *Million Miles,* 90–100, 107–8, 146–48, 105.

13. Barnes, *When Ships Were Ships,* 420–21.

14. Barnes, *When Ships Were Ships,* 423–24.

15. Barnes, *When Ships Were Ships,* 424–33, 262–65.

16. Lane, *Merchant Seamen's War,* 239.

17. Layton, *Ship's Lifeboats,* 59.

18. C. D. Lane, *The Boatman's Manual,* 321.

19. Ogg, *Compasses and Compassing,* 10–16; quotation on 15.

20. Turpin and MacEwen, *Merchant Marine,* 29.

21. Lane, *Boatman's Manual,* 307–16.

22. Turpin and MacEwen, *Merchant Marine,* 717.

23. Richards and Banigan, *How to Abandon Ship,* 51.

24. Richards and Banigan, *How to Abandon Ship,* 57, 62–67. See also United States Coast Guard, *Manual for Lifeboatmen and Able Seamen,* 28.

25. Chapman, "Life in a Lifeboat"; see also the story related in Lee, *Landlubber's Log,* 19–20.

26. Gilchrist, *Blue Hell,* 28–29.

27. Gilchrist, *Blue Hell,* 42–45, 50, 63, 76–80.

28. Manning, *U.S. Coast Survey,* 32–36. This important book offers wonderful insights into the controversies regarding navigation.

29. E. Hayden, "The Pilot Chart of the North Atlantic Ocean," 267–68, 269–70. *Science* approvingly excerpted parts of Hayden's article. See also United States Hydrographic Office, *Supplement to Pilot Chart of the North Atlantic for December, 1883,* esp. 10–12.

30. Hayden, "Pilot Chart," 274–75.

31. Hayden, "Pilot Chart," esp. 267–70, 450–55.

32. Walker, *An Unsinkable Titanic,* 27–29.

33. United States Coast Guard, *Manual for Lifeboatmen and Able Seamen,* 29.

34. "Well-Furnished Lifeboat." As late as 1945, writers told merchant seamen to buy a map of the world; see, for example, Lee, *Landlubber's Log,* 66.

35. De Hartog, *A Sailor's Life,* 94–96.

36. Lane, *Merchant Seamen's War,* 224–35. See also Saunders, *Valiant Voyaging.*

37. For an example of anti-German rhetoric, see H. Russell, *Sea Shepherds.* See also Roskill, *War at Sea,* 3:245; and Cameron, *The Peleus Trial.* On British machine-gunning of German lifeboats being lowered, see Kerr, *Business in Great Waters,* 58–61.

38. B. Edwards, *Fighting Tramps,* 117–23.

39. Lane, *Merchant Seamen's War,* 222.

40. Lane, *Merchant Seamen's War,* 225.

41. Liverpool Libraries, *Battle of the Atlantic,* 29.

42. Roskill, *War at Sea,* 3:245; and Cameron, *Peleus Trial.*

43. Lane, *Merchant Seamen's War,* 220–27.

44. Kerr, *Business in Great Waters,* 59–60. See also Saunders, *Valiant Voyaging,* 154.

45. Lane, *Merchant Seamen's War,* 239–40. I have been unable to find the *Times* letters Lane cites; microfilm copies of December 19, 1942, and January 5, 1943, are perhaps incomplete.

46. Gilchrist, *Blue Hell,* 45.

47. L. Thomas, *The Wreck of the Dumaru,* 80–105.

48. Merrien, *Lonely Voyagers,* 10–11.

49. Borden, *Sea Quest,* esp. 25, 164, writes that boats less than twenty-five feet in length are perfect for the Pacific, but he is speaking about yachts, not Lundin boats.

50. Sterling, *Small Boat Navigation,* 104–8.

51. Gatty, *Raft Book,* 4–6.

52. Gatty, *Raft Book,* 94.

53. Gatty, *Raft Book,* 117, 9, 17–18, 15–16, 36–38, 45, 55–56, 53, 57–69. Now and then yachtsmen use Gatty's techniques, perhaps because they learned of them while in World War II military service. See, for example, E. Bradford, *The Wind Off the Island,* esp. 99: "The breeze was coming. It brought the land with it—fruit and hot earth and all the Sicilian summer." The difference in sophistication between *The Raft Book* and other wartime manuals, say the War Shipping Administration's *Safety for Seamen,* is appalling; see, for example, the approach of the latter to nighttime navigation, 47–49. Much of Gatty's thinking has been buttressed by subsequent researchers; see, for example, Riley, ed., *Man across the Sea.*

54. American Associaton of Scientific Workers, *What to Do Aboard a Transport,* 152–54. World War II training of United States forces sea and air navigators may have been less than perfect; see, for example, McClendon, *The Lady Be Good,* which focuses on an April 1943 navigat-

ing error aboard a B-29 bomber only recently found in the middle of the Sahara.

55. Bagshaw, *Coasting Sailorman,* 156–57.

56. Beito, *Navigation and Nautical Astronomy,* 593–99; quotation on 594–95.

57. D. Robertson, *Sea Survival,* 59, 85–87. On inflatable rafts after World War II, see Nicholl, *Survival at Sea.*

58. During the war, United States Army and Navy fliers were equipped with cloth charts for use in inflatable rafts. See charts OPNAV-16-V #S109 of the Division of Naval Intelligence. Information on the charts advised downed airmen to rig a drogue, sit tight, and await rescue. The 1943 *How to Survive on Land and Sea* provided additional information to pilots of one-man aircraft.

CHAPTER 9. **Baggage**

1. Roskill, *Merchant Fleet in War,* 180–81.

2. Cornell and Hoffman, *American Merchant Seaman's Manual,* 179.

3. Richards and Banigan, *How to Abandon Ship,* 26.

4. Stafford, "No Man Alone," 31.

5. Bone, *Bowsprit Ashore,* 13.

6. See, especially, Gibson, *The Boat.*

7. *Titanic Disaster Hearings,* 429; see also 211, 226–27, 351, 371–72, 439–40, 484–85.

8. Biel, *Down with the Old Canoe,* 49–58.

9. Howells, *The Myth of the Titanic,* 100–101, 95–97.

10. Conrad, "Some Aspects," 594–95.

11. Howells, *Myth of the Titanic,* 101–2.

12. Howells, *Myth of the Titanic,* 95–96.

13. Lord, *A Night to Remember,* 98.

14. Richards and Banigan, *How to Abandon Ship,* 46.

15. Arzt, *Marine Laws,* 702.

16. Cornell and Hoffman, *American Merchant Seaman's Manual,* 585. See also Turpin and MacEwen, *Merchant Marine,* 21–23.

17. McCallum, *Journey with a Pistol,* 16–19.

18. American Association of Scientific Workers, *What to Do Aboard a Transport,* 256–60.

19. Cornell and Hoffman, *American Merchant Seaman's Manual* (5th ed., 1964), 257.

20. Gilchrist, *Blue Hell*, 18, 52, 102.

21. West, *Lifeboat*, 90–91.

22. Richards and Banigan, *How to Abandon Ship*, 84–85.

23. Gilchrist, *Blue Hell*, 29.

24. Gilchrist, *Blue Hell*, 34, 28, 36, 12.

25. Lee, *Landlubber's Log*, 54.

26. Kerr, *Business in Great Waters*, 118–20.

27. Conrad, *Lord Jim*, 61–71, 101.

28. Conrad, *Lord Jim*, 89.

29. Williams, *On Many Seas*, 254–55.

30. Kerr, *Business in Great Waters*, 152–53, 148.

31. Foster, *1700 Miles*, 143, 129–30.

32. Merrien, *Lonely Voyagers*, 3–4.

33. Foster, *1700 Miles*, 75–76.

34. Woodard, *Narrative*, xxi.

35. Woodard, *Narrative*, xxxiii–xl; quotation on xxxix.

36. Woodard, *Narrative*, 233–37, xxxi–xl.

37. Turpin and MacEwen, *Merchant Marine*, 2.

38. Wooldridge, *The Wreck of the Maid of Athens*, esp. 126–30. In some ways, mid-nineteenth-century castaways seem more concerned with remaining "proper" than depicting themselves as brave.

39. Villiers, *The Way of a Ship*, 107.

40. Villiers, *Way of a Ship*, 107–8.

41. Villiers, "Hoodoos Still Haunt," 8.

42. James, *Vital Reserves*, 56–57.

43. See, for example, Wilson, *American Jitters;* and Chase, *Men and Machines*.

44. Woods, *The Broadway Limited*, 8.

45. McKee, *The Golden Wreck*, esp. 100–104.

46. Conrad, *The Rescue*, 268–76. This brilliant novel is unjustly ignored by contemporary literary critics and by readers of what anthologists have decided are "Conrad's sea stories."

47. Hoehling and Hoehling, *Last Voyage*, 104–44.

48. Hoehling and Hoehling, *Last Voyage*, 170.

49. Hall, *Tale of a Shipwreck*, 102–3.

50. On women shedding clothes and swimming, see Chopin, *The Awakening*, 46–49.

51. London, *Adventure*, 90, 134, 178–79. Chopin, *The Awakening*, 19. On differences in swimming ability between upper- and middle-class women and girls after 1890 and the difference between upper-class swimming costumes and middle-class bathing suits, see Stilgoe, *Alongshore*, 334–66.

52. Fowler, *Standing Room Only*, 130, 6–7, 31, 9.

53. See, for example, Bowman, *Slacks and Callouses;* and Taylor, *Six Iron Spiders*. The latter, published in 1942, describes the abandonment of skirts and dresses by Cape Cod women involved in civilian defense.

54. Fowler, *Standing Room Only*, 13–15.

55. Fowler, *Standing Room Only*, 27–28.

56. Fowler, *Standing Room Only*, 37.

57. Fowler, *Standing Room Only*, 84–85, 59.

58. Fowler, *Standing Room Only*, 59–61, 76, 92, 127, and passim.

59. See, for example, Gellhorn, "Hatchet Day for the Dutch."

60. Runbeck, "Which Was the Rescuer?" esp. 25.

61. How much this is a stock literary device dating from the early nineteenth century I do not know. See, for example, Jane Austen's *Persuasion* for stunning insights into the capabilities of women.

62. Barker, *Goodnight*, 118–20.

63. On this cohesion, see, for example, McGuiness, *Sailor of Fortune*, 44–47.

64. Barker, *Goodnight*, 121.

65. C. Simpson, *Lusitania*, 227.

CHAPTER 10. Bones

1. Hoehling and Hoehling, *Last Voyage*, 71.

2. See, for example, Internet sites using the word. For decades I have asked students in my introductory course if they have ever met a murderer or a cannibal. Almost without exception, they say no. Only rarely does a student raise his or her hand and say, very hesitantly, "Well, not that I know of."

3. In my boyhood in the 1950s I learned that the *Essex* disaster should not be spoken of publicly because it opened on *other* Nantucket behavior.

4. Melville, *Moby-Dick*, 179.

5. The facts in the *Essex* case are not easily sorted. See Slocum, *Castaway Boats*, 79–102; and O. Chase, *Narrative*.

6. See Vincent, *The Trying Out of Moby-Dick*, esp. 46–48. Nowadays few college students realize that professors used to assign *Moby-Dick* as an introduction to Melville's thinking on a host of appallingly difficult subjects ranging from racism (in *Mardi* and *The Confidence Man*) to incest (in *Pierre*). In the same way, too few readers of Conrad's *Heart of Darkness* ever reach the shoal waters of *The Rescue*.

7. Conrad, *Typhoon*, 235. It is perhaps significant that Conrad's narrator encounters Falk as the master of a modern steam tug. In juxtaposing Falk against a sailing-ship master who loathes him, Conrad intensifies his scrutiny of the fundamental morality of the steamship seaman.

8. Alan Dershowitz, introduction to A. W. B. Simpson's *Cannibalism*, i.

9. Piers Paul Read's *Alive* is one recent explication of cannibalism after a 1972 airplane crash in the Andes; the book seems to be popular among high school students. On January 18, 2000, the *Quincy (Mass.) Patriot Ledger* reported that an alleged serial murderer on Cape Cod had reported murdering, then eating his women victims, a report that was greeted with horror and fascination at the general store and gasoline station I frequent.

10. Huxley, *Atlantic Ordeal*, 59.

11. Simpson, *Cannibalism*, 49–51. The cases of cannibalism form the core of Simpson's book.

12. Simpson, *Cannibalism*, 97.

13. Simpson, *Cannibalism*, 82–83, 111–12; see also 113.

14. Simpson, *Cannibalism*, 139–40.

15. Simpson, *Cannibalism*, 126–28, 164, 163–86, 115–17, 260–61.

16. Thomas, *Wreck*, 3, 7–8, 104–15, 144, 157–62, 176–83, 258–59.

17. Thomas, *Wreck*, 1–5.

18. Jones, *Two Survived*, 62.

19. Jones, *Two Survived*, 55, 88–89, 95–98, 106–12, 116, 120, 124–25, 133–34, 141–43, 148–50, 192–93.

20. Beito, *Navigation and Nautical Astronomy*, 596.

21. Herlin, *Survivor*, 81.

22. Many high schools post their summer reading lists on the Web: *The Raft* appears on the list published by the Hanover, New Hampshire, high school, for example.

23. Trumbull, *The Raft*, 56–57, 79, 28, 29–30, 104, 147–49, 172, 185, 190.

24. R. R. M., "The Raft."

25. Stevens, "The Raft"; and "The Raft," *Time*.

26. "The Raft," *New Yorker*.

27. See, for example, Kenneth Roberts's 1955 novel *Boon Island* on this point.

28. Gibson, *The Boat*, esp. 60.

CHAPTER 11. **Transformation**

1. Stenhouse, *Cracker Hash*, 34–39, 46, 65.

2. Tomlinson, *Gallion's Reach*, esp. 106–30.

3. Crane, *The Open Boat*, esp. 57–58. Crane is particularly effective in describing the malaise that hits even strong-minded people when "willy-nilly, the firm fails, the army loses, the ship goes down."

4. Cozzens, *S.S. San Pedro*, 80–83.

5. Tomlinson, *Pipe All Hands*, 310–12.

6. Slocum, *Castaway Boats*, 17.

7. Slocum, *Castaway Boats*, iii.

8. Nordhoff and Hall, *Faery Lands*, 68–70.

9. Slocum, *Castaway Boats*, 12. See also Kern, *The Culture of Time and Space*.

10. Slocum, *Castaway Boats*, 262–63.

11. Farber, "Among the Missing."

12. "Water Water Everywhere," 374.

13. "Hitchcock's Hand," 66.

14. *"Lifeboat,"* 77.

15. Farber, "Among the Missing."
16. Farber, "Theatrical Movies."
17. "Hitchcock's Hand."
18. "Water Water Everywhere," 374. "Lifeboat."
19. McKaig, "Suspicion."
20. Huxley, *Atlantic Ordeal*, 24–25.
21. Entering *City of Benares* into any Internet search engine proves remarkably revealing of current conspiracy theory and continuing suspicion of British war activities once classified with the "D" for censored.
22. Fowler, *Standing Room Only*, 76–77.
23. Walsh, *Sin and Censorship*, 166–67. *Strange Cargo* is a difficult film to find, and when one does, one must be careful to ascertain whether the version is the original or the later one, altered by MGM to earn the approval of the Legion of Decency. Anyone remotely intrigued by the place of small boats in Hollywood films quickly realizes the issues in-the-boat situations raise for directors wary of condemnation; see, for example, *The African Queen* (1951), starring Humphrey Bogart and Katharine Hepburn.
24. I have asked many older Americans about *Lifeboat*. Almost everyone I ask, including men who served in the military during World War II, saw it, and only those intimate with the sea and small boats remember finding anything odd about its visual imagery. Lifeboats figure in other World War II films, especially *San Demetrio* (1943) and *Western Approaches* (1944), both British.
25. Douglas, *Seamanship for Passengers*, 130–31.
26. Douglas, *Seamanship for Passengers*, 132–33; see also 141.
27. Douglas, *Seamanship for Passengers*, 137–38.
28. "Cheap Yachting," 1–3.
29. C. E. T. Lewis, *Lifeboats and Their Conversion*, 17–25, 71, 46–63.
30. J. Lewis, *Small Boat Conversion*, 12–15, 35–40, 90–91.
31. These paragraphs derive from my boyhood alongshore, including the dinnertable remarks of my boatbuilder father, John F. Stilgoe, who tackled lifeboat conversions as cheerfully as he built skiffs and replaced garboard strakes in lobster boats grounded out at low tide.
32. Lane, *Boatowner's Sheet Anchor*, 88.
33. In the dim reaches of my memory squats a steel lifeboat in a junkyard owned by Hyman Rome, a man who permitted boys to wander among his treasures and visualize inventions. He kept the boat as a playpen for very little children accompanying their fathers. Once when I asked why the boat never sold, he glanced at it and said, "Twisted." Later, in the car on the way home, my father explained what that meant.

BIBLIOGRAPHY

Abbott, John S. C. "Ocean Life." *Harper's New Monthly Magazine* 5 (June 1852), 61–66.

Admiralty Hydrographic Office. *Red Sea and Gulf of Aden Pilot.* London: Potter, 1900.

American Association of Scientific Workers. *What to Do Aboard a Transport.* Washington, D.C.: Science Service, 1943.

Ansel, Willits D. *The Whaleboat: A Study of Design, Construction and Use from 1850 to 1970.* Mystic, Conn.: Mystic Seaport, 1983.

Arzt, Frederick K. *Marine Laws: Navigation and Safety.* Stony Brook, N.Y.: Equity House, 1953.

Ashley, Clifford W. *The Ashley Book of Knots.* New York: Doubleday, 1944.

Bagshaw, H. A. E. *Coasting Sailorman.* Ware, Hertfordshire: Chaffcutter, 1998.

Ballantyne, R. M. *The Lifeboat: A Tale of Our Coast Heroes.* London: McIntyre, 1886.

Barker, Ralph. *Goodnight, Sorry for Sinking You: The Story of the S.S. City of Cairo.* London: Collins, 1984.

Barnes, William Morris. *When Ships Were Ships.* New York: Boni, 1930.

Barrow, John. *The Mutiny and Piratical Seizure of HMS Bounty.* [1834] London: William Tagg, 1872.

Barry, L., and Gerald Gosling. *Around Sidmouth.* St. Mary's Mills, Gloucestershire: Chalford, 1994.

"Battle of Flanders: Most of the BEF Escapes the German Trap." *Life* 8 (June 10, 1940), 31.

Beaver, Paul. *U-Boats in the Atlantic: A Selection of German Wartime Photographs.* Cambridge: Patrick Stephens, 1979.

Beck, Horace P. *The American Indian as a Sea-Fighter in Colonial Times.* Mystic, Conn.: Marine Historical Association, 1959.

Beito, Edwin A. *Navigation and Nautical Astronomy.* [1926] 10th ed. Annapolis: United States Naval Institute, 1951.

Bell, Robert W. *In Peril on the Sea: A Personal Remembrance.* Garden City, N.Y.: Doubleday, 1984.

Benson, N. P. *The Log of the El Dorado.* San Francisco: Barry, 1915.

Bernard, Oliver P. *Cock Sparrow: A True Chronicle.* London: Jonathan Cape, 1936.

Bevan, David. *Drums of the Birkenhead.* Cape Town: Purnell, 1972.

Biddle, Tyrrel E. *Amateur Sailing in Open & Half-Decked Boats.* London: Norie & Wilson, 1886.

Biel, Steven. *Down with the Old Canoe: A Cultural History of the Titanic Disaster.* New York: Norton, 1996.

Bladow, Janet. "Swept Away: Castaways May Need More than Food and Shelter." *Omni* 12 (September 1990), 20–22.

Blain, William. *Home Is the Sailor: The Sea Life of William Brown, Master Mariner and Penang Pilot.* New York: Sheridan House, 1940.

Blake, Harold S. *The Lifeboat: Its Construction, Equipment and Management: A Guide to the Board of Trade Lifeboat Efficiency Examination.* Glasgow: Brown, Son & Ferguson, 1933.

Bligh, William. *The Bligh Notebook.* Ed. John Bach. London: Allen & Unwin, 1987.

Blocksidge, Ernest W. *Ships' Boats: Their Qualities, Construction, Equipment, and Launching Appliances.* New York: Longmans, 1920.

"Boats, Life Boats, and Buoys." *Nautical Magazine,* February 1847, 93.

Bombard, Alain. *The Voyage of the Heretique: A True Chronicle.* New York: Simon & Schuster, 1953.

Bone, Alexander. *Bowsprit Ashore.* Garden City, N.Y.: Doubleday, 1933.

Borden, Charles A. *Sea Quest.* Philadelphia: Smith, 1967.

Bowman, Constance, and Clara Marie Allen. *Slacks and Callouses.* New York: Longmans, 1944.

Bradford, Ernle. *The Wind Off the Island.* London: Grafton, 1960.

Bradford, Gershom. *Mariner's Dictionary.* New York: Weathervane, 1952.

Brann, Christian. *The Little Ships of Dunkirk.* Cirencester: Collectors' Books, 1990.

Briand, Paul. *In Search of Paradise: The Nordhoff-Hall Story.* New York: Duell, Sloan & Pearce, 1966.

"British Rear Guard Fights Out of Flanders and Wades into Sea Toward Home and England." *Life* 8 (June 24, 1940), 34–35.

Brooke, Geoffrey. *Alarm Starboard! A Remarkable True Story of the War at Sea.* Cambridge: Patrick Stephens, 1982.

Brown, Alexander Crosby. *Women and Children Last: The Loss of the Steamship Arctic.* New York: Putnam's, 1961.

———, ed. *Longboat to Hawaii: An Account of the Voyage of the Clipper Ship Hornet of New York Bound for San Francisco in 1866.* Cambridge, Md.: Cornell Maritime Press, 1974.

Buchheim, Lothar-Gunther. *Die U-Boot-Fahrer: Die Boote, die Besatungen und ihr Admiral.* Munich: Bertelsmann, 1985.

Bullen, Frank T. *The Cruise of the Chahalot.* New York: Appleton, 1899.

Bunker, John. *Heroes in Dungarees: The Story of the American Merchant Marine in World War II.* Annapolis: Naval Institute Press, 1995.

"Burning of the Clipper *Hornet.*" *Harper's Weekly Magazine* 10 (September 29, 1866), 614–16.

Buster, Gun. *Return via Dunkirk.* London: Hodder & Stoughton, 1940.

Butler, John A. *Strike Able Peter: The Stranding and Salvage of the USS Missouri.* Annapolis: Naval Institute Press, 1985.

Cameron, John. *The Life-Boat and Its Work.* London: Royal National Life-Boat Institution, 1911.

———. *The Peleus Trial.* London: Hodge, 1948.

Center for the Study and Practice of Survival. *A Practical Guide to Lifeboat Survival.* Trans. David S. Jeffs and David Keating. Annapolis: Naval Institute Press, 1997.

Chadwick, F. E. *Ocean Steamships: A Popular Account of Their Construction, Development, Management, and Appliances.* New York: Scribner's, 1891.

Chambliss, W. C. "Recipe for Survival: What to Do When You Have to Abandon Ship." *Collier's* 113 (March 25, 1944), 22–24.

Chapelle, Howard I. *American Small Sailing Craft: Their Design, Development, and Construction.* New York: Norton, 1951.

Chapman, Chanler A. "Life in a Lifeboat." *Life* 13 (September 28, 1942), 17–18.

Chase, Owen. *Narrative of the Most Extraordinary and Distressing Shipwreck.* [1821] New York: Corinth, 1963.

Chase, Stuart. *Men and Machines.* New York: Macmillan, 1929.

Chatterton, E. Keble. *The Epic of Dunkirk.* London: Hurst, 1940.

———. *Q-Ships and Their Story.* Boston: Lauriat, 1923.

"Cheap Yachting." *Harper's New Monthly Magazine* 48 (June 1873), 1–16.

Chopin, Kate. *The Awakening.* [1899] New York: Avon, 1972.

Clark, William Bell. *When the U-Boats Came to America.* Boston: Little, Brown, 1929.

Clements, Rex. *A Gipsy of the Horn: The Narrative of a Voyage round the Horn in a Windjammer Twenty Years Ago.* Boston: Houghton, Mifflin, 1925.

Coates, Ken, and Bill Morrison. *The Sinking of the Princess Sophia: Taking the North Down with Her.* Toronto: Oxford University Press, 1990.

Cobbett, William. *Rural Rides.* [1830] Hammondsworth: Penguin, 1967.

Collier, Richard. *The Sands of Dunkirk.* New York: Dell, 1961.

Colton, J. Ferrell. *Last of the Square-Rigged Ships.* New York: Putnam's, 1937.

Conrad, Joseph. *Last Essays.* Ed. Richard Curle. Freeport, N.Y.: Books for Libraries Press, 1970.

———. *Lord Jim.* [1900] Garden City, N.Y.: Doubleday, 1920.

———. *The Mirror of the Sea.* [1905] Garden City, N.Y.: Doubleday, 1924.

———. *The North Sea on the Edge of War.* London: author, 1919.

———. *Notes on Life and Letters.* Garden City, N.Y.: Doubleday, 1924.

———. *The Rescue.* [1919] Garden City, N.Y.: Doubleday, 1924.

———. "Some Aspects of the Admirable Inquiry." *English Review* 11 (July 1912), 581–95.

———. *Typhoon and Other Stories.* Garden City: Doubleday, 1924.

———. "Youth." Pp. 26–45 in *Great Sea Stories,* ed. Alan Villiers. New York: Dell, 1959.

Cooke, Kenneth. *Man on a Raft* [orig. *What Cares the Sea?* 1960]. New York: Berkley, 1965.

Cooper, F. T. "Blue Lagoon." *Bookman* 27 (August 1908), 579.

Copplestone, Bennet. *Dead Men's Tales.* Edinburgh: Blackwood, 1926.

Cornell, Felix M., and Alan C. Hoffman. *American Merchant Seaman's Manual: For Seamen by Seamen.* [1938] 2nd ed., New York: Cornell Maritime Press, 1942. 5th ed., Cambridge, Md.: Cornell Maritime Press, 1964.

Cornish, Vaughan. *Scenery of Sidmouth.* Cambridge: Cambridge University Press, 1940.

Cozzens, James Gould. *S.S. San Pedro.* New York: Harcourt, 1930.

Crane, Stephen. *The Open Boat and other Stories.* London: Heinemann, 1898.

Cubbins, Thomas. *The Wreck of the Serica: A Narrative of 1868.* London: Dropmore, 1950.

Cugle, Charles H. *Examination Guide for Lifeboat Men and Seamen.* New York: Dutton, 1937.

Daiches, David. *Robert Louis Stevenson and His World.* London: Thames and Hudson, 1973.

Dalton, J. W. *The Life Savers of Cape Cod.* Boston: Barta, 1923.

Dana, Richard Henry. *Two Years before the Mast.* [1840] New York: Doubleday, 1973.

Dangerous Voyage Performed by Captain Bligh, with a Part of the Crew of His Majesty's Ship Bounty in an Open Boat, Over Twelve Hundred Leagues of the Ocean in the Year 1789, The. Dublin: Napper, 1824.

Danielsson, Bengt. *What Happened on the Bounty.* New York: Rand McNally, 1964.

Davis, Richard Harding. *Stories for Boys.* New York: Scribner's, 1891.

Dawson, A. J. *Britain's Life-Boats: The Story of a Century of Heroic Service.* London: Hodder & Stoughton, 1923.

De Hartog, Jan. *The Call of the Sea.* New York: Athenaeum, 1966.

————. *A Sailor's Life.* [1955] London: White Lion, 1976.

De Kerchove, Rene. *International Marine Dictionary.* New York: Van Nostrand, 1947.

Delano, Amasa. *Narrative of Voyages and Travels . . . in the Pacific and Oriental Islands.* Boston: House, 1817.

"Disasters at Sea, Reported for the Past Month." *Monthly Nautical Magazine* 1 (November 1854), 114–19.

Divine, A. D. *Dunkirk.* London: Faber & Faber, 1945.

Divine, David [A. D. Divine]. *The Nine Days of Dunkirk.* New York: Norton, 1959.

Dixon, C. C. "The Sargasso Sea." *Geographical Journal* 66 (November 1925), 434–42, 480.

Dobson, Christopher, John Miller, and Ronald Payne. *The Cruellest Night: Germany's Dunkirk and the Sinking of the Wilhelm Gustloff.* London: Hodder & Stoughton, 1979.

Douglas, Gavin. *Seamanship for Passengers.* London: Lehmann, 1949.

Drew, Nicholas. *Amateur Sailor.* London: Constable, 1944.

Droste, C. L., comp. *Lusitania Case.* Richmond, Va.: Dietz, 1915.

"Editor's Easy Chair." *Harper's New Monthly Magazine* 10 (December 1854), 119–20.

Edwards, Bernard. *Blood and Bushido: Japanese Atrocities at Sea, 1941–1945.* Upton-Upon-Severn: Self Publishing Association, 1991.

————. *The Fighting Tramps: The Merchant Navy Goes to War.* London: Robert Hale, 1989.

Edwards, Edward, and George Hamilton. *Voyage of HMS Pandora Dispatched to Arrest the Mutineers of the Bounty in the South Seas.* Ed. Basil Thomson. London: F. Edwards, 1915.

Ellis, Fred W. *Round Cape Horn in Sail.* South Croydon, Surrey: Blue Book, 1949.

Ellis, Frederick D. *The Tragedy of the Lusitania.* Philadelphia: National, 1915.

Ellsberg, Edward. *Under the Red Sea Sun.* New York: Dodd, Mead, 1946.

Faith, Nicholas. *Mayday.* London: Macmillan, 1998.

Farber, Manny. "Among the Missing." *New Republic* 110 (January 24, 1944), 160.

————. "Theatrical Movies." *New Republic* 110 (February 14, 1944), 210–11.

Ferber, Edna. "Lifeboat." *Scholastic* 42 (February 1, 1943), 25–26.

Forbes, N. Lyles. *Suggestive Curves.* Salem, Mass.: Peabody-Essex Museum, 2000.

Forbes, Robert Benett. *The Life Boat and Other Life-Saving Inventions.* Boston: Williams, 1880.

————. *Notes on Some Few of the Wrecks and Rescues during the Present Century.* Boston: Little, Brown, 1889.

Forester, C. S. *The African Queen.* New York: Little, Brown, 1935.

Forrester, Larry. *Fly for Your Life.* London: Muller, 1956.

Foss, Denis. *Shoot a Line: A Merchant Mariner's War.* Yeoril, Somerset: Linden Hall, 1992.

Foster, Cecil. *1700 Miles in Open Boats: The Story of the Loss of the S.S. Trevessa in the Indian Ocean and the Voyage of Her Boats to Safety.* [1924] London: Rupert Hart-Davis, 1952.

Fowler, Elizabeth. *Standing Room Only: The Personal and Moving Record of a Woman's Experience during Ten Days in a Lifeboat with Thirty-four Men after Their Ship Had Been Torpedoed by a German Submarine.* New York: Dodd, Mead, 1944.

Fox, C. "A Ship Sinks." In *Touching the Adventures,* ed. J. Lennox Kerr. London: Harrap, 1953.

Francey, John B. "RMS *Titanic:* Survival Equipment." *Titanic Communicator* 20 (Spring 1996), 46–47.

Fussell, Paul. *The Great War and Modern Memory.* New York: Oxford University Press, 1975.

Gann, Ernest. *Twilight for the Gods.* New York: Sloane, 1956.

Gardner, John. *Building Classic Small Craft.* Camden, Maine: International Marine, 1977.

Gatty, Harold. *The Raft Book: Lore of the Sea and Sky.* New York: Grady, 1943.

Gauguin, Paul. *Noa Noa: The Tahiti Journal of Paul Gauguin.* Trans. O. F. Theis. San Francisco: Chronicle, 1994.

Gaul, Gilbert. "Jamaica." *Century* 45 (March 1893), 682–88.

Gellhorn, Martha. "Hatchet Day for the Dutch." *Collier's* 113 (March 25, 1944), 27, 59–60.

Gibbons, Floyd P. *How the Laconia Sank.* Chicago: Daughaday, 1917.

Gibson, Walter. *The Boat.* Boston: Houghton, Mifflin, 1953.

Gilchrist, Derek C. *Blue Hell.* London: Heath Cranton, 1943.

Gilmour, Oswald W. *Singapore to Freedom.* London: Burrow, 1943.

Gosling, Ted. *Around Seaton and Sidmouth.* Phoenix Mill, Gloucestershire: Sutton, 1991.

Grant, George H. *Consigned to Davy Jones: My Third Voyage in the Half Deck of a British Tramp Steamer.* Boston: Little, Brown, 1934.

———. *Take to the Boats.* Boston: Little, Brown, 1938.

Grattidge, Henry. *Captain of the Queens.* New York: Dutton, 1956.

Gray, Edwin A. *The U-boat War.* London: Pen and Sword, 1994.

Hadfield, R. L. *Sea-Toll of Our Time: A Chronicle of Maritime Disaster during the Last Thirty Years Drawn from Authentic Sources.* London: Witherby, 1930.

Hadley, Peter. *Third Class to Dunkirk.* London: Hollis and Carter, 1944.

Hall, James Norman. *The Tale of a Shipwreck.* Boston: Houghton, Mifflin, 1934.

Hall, James Norman, and Charles Bernard Nordhoff. *Faery Lands of the South Seas.* New York: Harper, 1921.

Halsted, Ivor. *Heroes of the Atlantic: The British Merchant Navy Carries On.* New York: Dutton, 1942.

Hamilton, John. *War at Sea, 1939–1945.* Poole: Blandford Press, 1986.

Hauge, Eiliv Odde, and Vera Hartmann. *Flight from Dakar.* Trans. F. H. Lyon. London: Allen and Unwin, 1954.

Hayden, Everett. "The Pilot Chart of the North Atlantic Ocean." *Journal of the Franklin Institute* 125 (April and June 1888), 265–78, 447–62. Also excerpted as "Pilot Chart of the North Atlantic Ocean." *Science* 12 (August 10, 1888), 65–66.

Hayden, Sterling. *Wanderer.* New York: Norton, 1963.

Hayler, William B., John M. Keever, and Paul M. Seiler. *The Cornell Manual for Lifeboat-*

men, Able Seamen, and Qualified Members of the Engine Department. Centreville, Md.: Cornell Maritime Press, 1984.

Hendry, Frank C. *The Ocean Tramp.* Illus. Frank H. Mason. London: Collins, 1938.

Henty, G. A. *Under Drake's Flag: A Tale of the Spanish Main.* New York: Lupton, 1902.

Herlin, Hans. *Survivor: The True Story of the Doggerbank.* Trans. John Brownjohn. London: Cooper, 1994.

"Hitchcock's Hand Steers *Lifeboat* Safely through Film's Troubled Sea." *Newsweek* 23 (January 17, 1944), 65–66.

Hoehling, A. A., and Mary Hoehling. *The Last Voyage of the Lusitania.* New York: Holt, 1956.

Holman, Gordon. *The Little Ships.* London: Hodder and Stoughton, 1943.

Homeland Association. *Sidmouth and the Neighbourhood.* London: Homeland Association, 1926.

Hopkins, R. Thurston. *Small Sailing Craft.* London: Allan, 1931.

Hotchkiss, Charles F. *On the Ebb: A Few Long-Lines from an Old Salt.* New Haven, Conn.: Tuttle, 1878.

Hough, Richard. *Captain Bligh and Mr. Christian: The Men and the Mutiny.* New York: Dutton, 1972.

———. *The Great War at Sea, 1914–1918.* Oxford: Oxford University Press, 1983.

How to Survive on Land and Sea: Individual Survival. Annapolis: Naval Institute, 1943.

Howells, Richard. *The Myth of the Titanic.* London: Macmillan, 1999.

Huxley, Elspeth. *Atlantic Ordeal: The Story of Mary Cornish.* New York: Harper, 1941.

"Improvements in the Welin Quadrant Davit." *International Marine Engineering* 15 (August 1910), 333–35.

Irvine, Legh H. "The Lone Cruise of the *Hornet* Men." *Wide World* 5 (September 1900), 576–77.

Jackson, C. *Forgotten Tragedy: The Sinking of HMT Rohna.* Annapolis: Naval Institute Press, 1997.

James, Henry J. *German Subs in Yankee Waters.* New York: Gotham, 1940.

James, William. *On Vital Reserves.* New York: Holt, 1922.

Jenkins, Geoffrey. *Scend of the Sea.* London: Collins, 1971.

Jones, Guy Pearce. *Two Survived: The Story of Tapscott and Widdicombe, Who Were Torpedoed in Mid-Atlantic and Survived Seventy Days in an Open Boat.* Introduction by William McFee. New York: Random House, 1941.

Jones, Stephen. *Noank: The Ethereal Years.* Noank, Conn.: Noank Historical Society, 1988.

———. *Short Voyages: Forays along the Littoral.* New York: Norton, 1985.

Jones, William H. S. *The Cape Horn Breed: My Experience as an Apprentice in Sail in the Full Rigged Ship British Isles.* New York: Criterion, 1956.

Kayle, Allan. *Salvage of the Birkenhead.* Johannesburg: Southern Book, 1990.

Keegan, John. *The Mask of Command.* New York: Viking, 1987.

Kelshall, Gaylord T. M. *The U-Boat War in the Caribbean.* [1988] Annapolis: Naval Institute Press, 1994.

Kent, Rockwell. *Voyaging Southward from the Strait of Magellan.* [1924] New York: Grosset and Dunlap, 1968.

Kenworthy, Joseph. *The Freedom of the Seas.* London: Liveright, n.d.

Kern, Stephen. *The Culture of Time and Space, 1880–1918.* Cambridge, Mass.: Harvard University Press, 1983.

Kerr, George Fleming. *Business in Great Waters: The War History of the Peninsula & Orient.* London: Faber and Faber, 1951.

Kiley, Deborah Scaling. *Albatross: The True Story of a Woman's Survival at Sea.* Boston: Houghton, Mifflin, 1994.

Kimball, Sumner Increase. *Organization and Methods of the United States Life-Saving Service.* Washington, D.C.: GPO, 1894.

King, Irving H. *The Coast Guard Expands, 1865–1915: New Roles, New Frontiers.* Annapolis: Naval Institute Press, 1996.

Klitgaard, Kaj. *Oil and Deep Water.* Chapel Hill: University of North Carolina Press, 1945.

Knight, Austin M. *Modern Seamanship.* [1901] Rev. ed., New York: Van Nostrand, 1941.

Kurzman, Dan. *Left to Die: The Tragedy of the USS Juneau.* New York: Pocket Books, 1994.

Lane, Carl D. *The Boatman's Manual: A Complete Manual of Boat Handling, Operation, Maintenance, and Seamanship.* New York: Norton, 1942.

———. *Boatowner's Sheet Anchor: A Practical Guide to Fitting Out, Upkeep, and Alteration of the Small Yacht.* New York: Norton, 1941.

Lane, Reginald. *Old Sidmouth.* Exeter, Devon: Devon, 1990.

Lane, Tony. *The Merchant Seamen's War.* New York: Manchester University Press, 1990.

Lass, Edgar Gunther. *Die Flucht, Ostpreuseen 1944–45.* Bad Nauheim: Podzun, 1964.

Lauriat, Charles E. *The Lusitania's Last Voyage.* Boston: Houghton Mifflin, 1915.

Layton, C. W. T. *Ship's Lifeboats: A Handbook for the Board of Trade Examination for Certificates in Lifeboat Efficiency.* Glasgow: Brown, 1938.

Lea, Homer. *The Valor of Ignorance.* [1909] New York: Harper, 1942.

Leather, John. "The Humble Lug." *Boatman* 10 (November/December 1993), 66–74.

———. *Spritsails and Lugsails.* London: Granada, 1979.

Leavitt, John F. *Wake of the Coasters.* Middletown, Conn.: Wesleyan University Press, 1970.

Lee, Norman. *Landlubber's Log: 25,000 Miles with the Merchant Navy.* London: Quality, 1945.

Lewis, C. E. Tyrrell. *Lifeboats and Their Conversion.* [1935] London: Witherby, 1943.

Lewis, John. *Small Boat Conversion.* London: Rupert Hart-Davis, 1951.

Lewis, Richard. *History of the Life-Boat and Its Work.* London: Macmillan, 1874.

Liddle, Peter H. *The Sailor's War, 1914–1918.* Poole: Blandford, 1985.

"*Lifeboat:* Hitchcock Throws Eight People and the Nazi Who Torpedoed Them Together in an Open Boat." *Life* 16 (January 31, 1944), 77–81.

Life-Saving Inventions. Washington, D.C.: GPO, 1868.

Lincoln, Joseph C. *Rugged Water.* New York: Burt, 1924.

Liverpool Libraries. *Battle of the Atlantic.* Liverpool: Picton, 1993.

London, Jack. *Adventure.* New York: Regent, 1911.

Lord, Walter. *A Night to Remember.* New York: Bantam, 1955.

Lords Commissioners of the Admiralty. *The China Sea Directory, Containing Directions for the Approaches to the China Sea and to Singapore by the Straits of Sunda, Banka, Gaspar, Carimata, Rhio, Varella, Durian, and Singapore.* London: Hydrographic Office, Admiralty, 1867.

"Loss of the *Arctic.*" *Griffiths and Bates Nautical Magazine* 1 (November 1854), 105–11.

"Loss of the *Wairarapa*." *Borderland* 2 (January 1895), 69.

Low, Archibald Montgomery. *The Submarine at War.* London: Hutchinson, 1941.

MacLean, Alistair. *South by Java Head.* New York: Doubleday, 1958.

Making, Victor Leslie. *In Sail and Steam: Behind the Scenes of the Merchant Service.* London: Sidgwick & Jackson, 1937.

Manning, Thomas G. *U.S. Coast Survey vs. Naval Hydrographic Office: A 19th-Century Rivalry in Science and Politics.* Tuscaloosa: University of Alabama Press, 1988.

Mansfield, Kathy. "The Shetland Sixareen." *The Boatman* 34 (May 1996), 19–24.

"Maps: Global War Teaches Global Cartography." *Life* 13 (August 3, 1942), 57–65.

Masefield, John. *The Nine Days Wonder: The Operation Dynamo.* New York: Macmillan, 1941.

Mather, Cotton. *Thoughts for the Day of Rain.* Boston: Green, 1712.

McCallum, Neil. *Journey with a Pistol: A Diary of War.* London: Gollancz, 1959.

McClendon, Dennis: *The Lady Be Good: Mystery Bomber of World War II.* Fallbrook: Aero, 1982.

McCoy, Samuel Duff. *Nor Death Dismay: A Record of Merchant Ships and Merchant Marines in Time of War.* New York: Macmillan, 1944.

McCulloch, John Herries. *A Million Miles in Sail.* New York: Dodd, Mead, 1933.

McFee, William. *More Harbours of Memory.* Garden City, N.Y.: Doubleday, 1934.

———. *North of Suez.* Garden City, N.Y.: Doubleday, 1930.

———. *Watch Below: A Reconstruction in Narrative Form of the Golden Age of Steam, When Coal Took the Place of Wind and the Tramp Steamer's Smoke Covered the Seven Seas.* New York: Random House, 1940.

McGuiness. *Sailor of Fortune.* Philadelphia: Macrae, 1935.

McKaig, Merle B. "Suspicion." *Time* 36 (September 21, 1940), 8.

McKee, Alexander. *The Golden Wreck.* London: Souvenir, 1986.

Melville, Herman. *Moby-Dick, or, The Whale.* [1851] Ed. Harrison Hayford and Hershel Parker. New York: Norton, 1967.

Mence, C. S. *Practical Seamanship for Use in the Merchant Service.* London: George Philip, 1919.

Mercator. "Atlantic Steam Navigation." *Nautical Magazine* 6 (June [?] 1840), 429–40.

Merrien, Jean. *Lonely Voyagers.* Trans. J. H. Watkins. New York: Putnam's, 1954.

Methley, Noel T. *The Life-Boat and Its Story.* London: Sidgwick & Jackson, 1912.

Middleton, E. W. *Lifeboats of the World: A Pocket Encyclopedia of Sea Rescue.* New York: Arco, 1978.

Minnaert, M. G. J. *Light and Color in the Outdoors.* [1937] Trans. Len Seymour. New York: Springer, 1974.

Mitchell, Pryce. *Deep Water: The Autobiography of a Sea Captain.* Boston: Little, Brown, 1933.

Monsarrat, Nicholas. *Three Corvettes.* London: Mayflower, 1972.

"Monthly Record." *Harper's New Monthly Magazine* 7 (November 1854), 830.

Morgan, Lael. *The Woman's Guide to Boating and Cooking.* Freeport, Maine: Bond Wheelwright, 1968.

Morrill, John, and Pete Martin. *South from Corregidor.* New York: Simon and Schuster, 1943.

Mulville, Frank. *Schooner Integrity.* New York: Sheridan House, 1979.

———. *Terschelling Sands.* London: Conway Maritime, 1968.

Munro, D. J. *The Roaring Forties and After.* London: Sampson Low, 1929.

Murdoch, James. *A True Account of that Terrible Catastrophe, the Burning of the Magnificent Packet Ship, Ocean Monarch.* Lowell, Mass.: Huntoon, 1848.

Nares, George S. *Seamanship.* Exeter: Wheaton, 1862.

"Nautical Club." *Nautical Magazine* 30 (August 1861), 435–45.

"Nazi Fifth Column and Communist Allies Are Active in Mexico." *Life* 8 (July 10, 1940), 51–52.

Nicholl, G. W. R. *Survival at Sea: The Development, Operation, and Design of Inflatable Marine Lifesaving Equipment.* London: Coles, 1960.

Nordhoff, Charles. *The Merchant Vessel.* New York: Dodd, Mead, 1895.

Nordhoff, Charles, and James Norman Hall. *Men against the Sea.* [1934] Boston: Little, Brown, 1962.

Noyes, Alfred. *Open Boats.* Edinburgh: Stokes, 1917.

Odyssey of a Torpedoed Transport, The. Trans. Grace Fallow Norton. Boston: Houghton, Mifflin, 1917.

Ogg, Robert D. *Compasses and Compassing.* Portland, Maine: Danforth, 1977.

Otis, James. *The Castaways; or, On the Florida Reefs.* New York: Burt, 1888.

Page, Michael F. "All in the Same Boat." In *Touching the Adventures,* ed. J. Lennox Kerr. London: Harrap, 1953.

Paine, Ralph. *The Ships and Sailors of Old Salem.* Boston: Lauriat, 1927.

Parkin, Ray. *Out of the Smoke.* London: Hogarth, 1960.

Patenall, Henry. "Madagascar via Lifeboat." *Travel* 83 (July 1944), 22–27, 33.

Peary, Matthew. "Notes in Regard to Safety on Steamers at Sea." *Hunt's Merchants' Magazine* 32 (March 1855), 379–80.

Pease, Howard. *The Tattooed Man.* [1926] Reprint, New York: Comet, 1947.

Phillips, G. Purssey. *Dark Seas Remember.* London: Stanley Paul, 1943.

Piddington, Henry. *Sailor's Horn-Book for the Law of Storms.* London: Norgate, 1889.

Qualtrough, E. F. *The Sailor's Handy Book and Yachtsman's Manual.* New York: Scribner's, 1892.

R. R. M. "The Raft." *Christian Science Monitor* 34 (August 31, 1942), 18.

"Raft, The." *New Yorker* 18 (August 22, 1942), 59.

"Raft, The." *Time* 40 (August 31, 1942), 103.

Raynal, F. E. *Wrecked on a Reef; or, Twenty Months among the Auckland Isles.* London: Nelson, 1874.

Read, George H. *The Last Cruise of the Saginaw.* Boston: Houghton, Mifflin, 1912.

Read, Piers Paul. *Alive.* New York: Avon, 1974.

Redifer, Albert E. *Lifeboat Manual.* New York: Cornell Maritime Press, 1944.

Reynolds, Stephen. *Alongshore: Where Men and Sea Face One Another.* New York: Macmillan, 1910.

———. *A Poor Man's House.* London: Lane, 1909.

Richards, Phil, and John J. Banigan. *How to Abandon Ship.* New York: Cornell Maritime Press, 1942.

Riley, Carroll L., ed., et al. In *Man across the Sea: Problems of Pre-Columbian Contacts.* Austin: University of Texas Press, 1971.

Roberts, Kenneth. *Boon Island.* New York: Doubleday, 1955.

Robertson, Dougal. *Sea Survival: A Manual.* New York: Praeger, 1975.

Robertson, Morgan. "Primordial." In *Where Angels Fear to Tread.* New York: McClures, 1898.

Rogers, Stanley. *Hazards of War.* London: Harrap, 1944.

Roskill, S. W. *A Merchant Fleet in War.* London: Collins, 1962.

———. *War at Sea, 1939–1945.* 3 vols. London: HMSO, 1959–61.

Rough, David. "Ships Boats: With Copper or Zinc Cylinders." *Nautical Magazine* 8 [?] (June 1848), 322.

Rowan, Thomas. *Coal: Spontaneous Combustion and Explosions Occurring in Coal Cargoes: Their Treatment and Prevention.* London: Spon, 1882.

Royal National Life-Boat Institution. *Stories of the Life-Boat.* London: RNLI, 1940.

———. *Stories of the Life-Boat: A Further Selection.* London: RNLI, 1956.

Royal Navy. *Admiralty Manual of Seamanship.* 3 vols. London: HMSO, 1972.

Runbeck, M. L. "Which Was the Rescuer? A Woman and a Little Boy Adrift in a Lifeboat." *Reader's Digest* 43 (November 1943), 23–25.

Russell, Herbert. *Sea Shepherds.* London: Murray, 1941.

Russell, W. Clark. *Alone on a Wide, Wide Sea: An Ocean Mystery.* London: Chatto & Windus, 1892.

———. *Marooned.* Chicago: Rand McNally, 1889.

———. *The Mate of the Good Ship York; or, The Ship's Adventure.* Boston: Page, 1902.

Rutter, Owen, ed. *Voyage of the Bounty's Launch as Related in William Bligh's Despatch to the Admiralty and the Journal of John Fryer.* London: Golden Cockerel Press, 1934.

Saunders, Hilary St. G. *Valiant Voyaging: A Short History of the British India Steam Navigation Company in the Second World War.* London: Faber & Faber, 1949.

"Say the Truth—And Don't Be Afraid of Death." *Fortune* 25 (April 1942), 108.

Scheina, Robert L. *United States Coast Guard Cutters and Craft of World War II.* Annapolis: Naval Institute Press, 1982.

Schoen, Heinz. *Untergang die Wilhelm Gustloff.* Baden: Pabel, 1960.

Sellwood, A. V. *The Damned Don't Drown: The Sinking of the Wilhelm Gustloff.* London: Wingate, 1973.

Shaw, Frank H. *Merchant Navy at War.* New York: Stanley Paul, 1943.

———. *White Sails and Spindrift.* London: Stanley Paul, 1944.

"Ship-Masters and Mariners." *Griffiths and Bates Nautical Magazine* 1 (November 1854), 111–14.

Sigsbee, Charles D. *The Maine: An Account of Her Destruction in Havana Harbor.* New York: Century, 1899.

Sim, Katherine. *Malayan Landscape* [1946]. Singapore: Donald Moore, 1969.

Simmons, Walter J. "That's a Matinicus Double-Ender, Mabel." *Small Boat Journal* 1 (May 1980), 19–23.

Simpson, Alfred William Brian. *Cannibalism and the Common Law: The Mignonette Case.* Introduction by Alan Dershowitz. Birmingham, Ala.: Notable Trials Society, 1990.

Simpson, Colin. "Lusitania." *Life* 73 (October 13, 1972), 58–80.

———. *Lusitania.* London: Longmans, 1972.

Slocum, Victor. *Castaway Boats.* New York: Furman, 1938.

Smith, Hervey Garrett. *The Arts of the Sailor.* [1953] Mineola, N.Y.: Dover, 1990.

Smyth, William H. *Sailor's Word Book: An Alphabetical Digest of Nautical Terms.* London: Conway, 1996.

Sobel, Dava. *Longitude.* New York: Walker, 1995.

Sorrell, George. *The Man before the Mast.* Ed. C. Fox Smith. London: Methuen, 1928.

Spector, Ronald H. *The Eagle against the Sun: The American War with Japan.* New York: Free Press, 1985.

Spectre, Peter H. "As Alike as Two Peas in a Pod?" *Boatman* 31 (November/December 1993), 46–55.

———. "Slick as a Pod." *Maine Boats & Harbors* 35 (July 1995), 34–39, 67.

Spindler, Arno. *La guerre sous-marine.* 2 vols. Trans. Rene Jouan. Paris: Payot, 1934.

Stacpoole, H. De Vere. *The Beach of Dreams.* New York: John Lane, 1919.

———. *The Blue Lagoon.* Philadelphia: Lippincott, 1908.

Stafford, Edward P. "No Man Alone: 41 Sailors in a Very Small Boat." *Oceans* 17 (May–June 1984), 28–32.

Steinbeck, John. *The Log from the Sea of Cortez.* [1941] New York: Penguin, 1966.

Steinmuth, Hans. *Lusitania.* Stuttgart: Deutsches Verlag, 1915.

Stenhouse, Joseph Russell. *Cracker Hash: The Story of an Apprentice in Sail.* London: Percival Marshall, 1955.

Sterling, F. W. *Small Boat Navigation.* [1917] New York: Macmillan, 1942.

Stevens, Charles. *A Sailor Boy's Experience.* Nepanee, Ontario: author, 1892.

Stevens, H. A. "The Raft." *New York Times* 94 (August 23, 1942), 3.

Stevenson, Robert Louis. *In the South Seas.* New York: Scribner's, 1896.

Stilgoe, John R. *Alongshore.* New Haven: Yale University Press, 1994.

———. *Borderland: Origins of the American Suburb, 1820 to 1939.* New Haven: Yale University Press, 1988.

Strong, George Templeton. *Diary.* Ed. Allan Nevins and Milton Halsey Thomas. New York: Macmillan, 1952.

Sutton, Anna. *A Story of Sidmouth.* Exeter, Devon: Townsend, 1959.

Tait's Seamanship for Board of Trade Examinations. Glasgow: Brown, 1912.

Taylor, Phoebe Atwood. *The Six Iron Spiders.* [1942] Woodstock, Vt.: Countryman, 1992.

"These Sights Save Lives." *Messing About in Boats* 13 (January 1, 1996), 6.

"This Is How the United States May Be Invaded." *Life* 8 (June 24, 1940), 16–18.

Thomas, Gordon, and Max Morgan Witts. *Shipwreck: The Strange Fate of the Morro Castle.* New York: Dell, 1973.

Thomas, Lowell. *The Wreck of the Dumaru: A Story of Cannibalism in an Open Boat.* New York: Collier, 1930.

Thomas, R. *Interesting and Authentic Narratives of the Most Remarkable Shipwrecks.* New York: Strong, 1837.

Titanic Disaster Hearings. Ed. Tom Kuntz. New York: Pocket Books, 1998.

Tomlinson, H. M. *Gallion's Reach: A Romance.* [1927] London: Rupert Hart-Davis, 1955.

———. *Gifts of Fortune and Hints for Those About to Travel.* New York: Harper, 1926.

———. *Malay Waters: The Story of Little Ships Coasting Out of Singapore and Penang in Peace and War.* London: Hodder and Stoughton, 1950.

———. *Pipe All Hands.* New York: Harper, 1937.

———. *Tide Marks: Being Some Records of a Journey to the Beaches of the Moluccas and the Forest of Malaya in 1923.* New York: Blue Ribbon, 1924.

Townend, William. *Long Voyage.* London: Chapman and Hall, 1943.

Treanor, Thomas Stanley. *Heroes of the Goodwin Sands.* London: Religious Tract Society, 1892.

Trew, Antony. *Death of a Supertanker.* New York: St. Martin's, 1978.

Trumbull, Robert. *The Raft.* New York: Holt, 1942.

Turner, L. C. F., H. R. Gordon-Cumming, and J. E. Betzler. *War in the Southern Oceans, 1939–1945.* Cape Town: Oxford University Press, 1961.

Turpin, Edward A., and William A. MacEwen. *Merchant Marine Officers' Handbook.* New York: Cornell Maritime Press, 1944.

Twain, Mark. "Forty-three Days in an Open Boat." *Harper's Magazine* 34 (December 1866), 104–13.

———. *Letters from Hawaii.* Ed. A. Grove Day. New York: Appleton, 1966.

———. "My Debut as a Literary Person." *Century Magazine* 59 (November 1899), 104–13.

"United States." *Harper's New Monthly Magazine* 9 (November 1854), 830.

United States Coast Guard. *Manual for Lifeboatmen.* Washington, D.C.: GPO, 1973.

———. *Manual for Lifeboatmen and Able Seamen.* Washington, D.C.: GPO, August 1944.

———. *Safety Hints.* Washington, D.C.: GPO, 1944.

———. *Suggestions concerning Tank Vessel Operation during Wartime.* Washington, D.C.: USCG, 1945.

United States Hydrographic Office. *The Coast of Brazil.* 2 vols. Washington, D.C.: GPO, 1873.

———. *East Indies Pilot.* 2 vols. Washington, D.C.: GPO, 1924.

———. *Pilot Chart of the North Atlantic, August, 1891.* Washington, D.C.: GPO, 1891.

———. *Sailing Directions for Malacca Strait and Sumatra.* 2 vols. Washington, D.C.: GPO, 1933.

———. *Sailing Directions for Sunda Strait and Northwest Coast of Borneo and Off-Lying Dangers.* Washington, D.C.: GPO, 1935.

———. *Supplement to Pilot Chart of the North Atlantic for December, 1883.* Washington, D.C.: GPO, 1883.

United States Navy. *Boat Book.* Washington, D.C.: GPO, 1920.

———. *1940 Bluejacket's Manual.* Annapolis: United States Naval Institute, 1940.

———. *The Power Boat Book.* Washington, D.C.: Bureau of Naval Personnel, 1951.

"U.S. Reckons Up Its Losses on the Scoreboard of War." *Life* 13 (August 5, 1942), 26–27.

Venables, William H. "The Torpedoing of the *Assyrian.*" In *Touching the Adventures,* ed. J. Lennox Kerr. London: Harrap, 1953.

Verney, Michael. *Practical Conversions and Yacht Repairs.* [1951] London: Murray, 1957.

Villiers, Alan. *And Not to Yield: A Story of the Outward Bound.* London: Hodder and Stoughton, 1953.

———. "Hoodoos Still Haunt the Sea." *New York Sunday Herald Tribune,* March 27, 1932, 8–12.

———. *Posted Missing: The Story of Ships Lost without Trace in Recent Years.* New York: Scribner's, 1956.

———. *The Way of a Ship.* New York: Scribner's, 1953.

Vincent, Howard P. *The Trying Out of Moby-Dick.* [1949] Carbondale: Southern Illinois University Press, 1967.

Walker, J. Bernard. *An Unsinkable Titanic: Every Ship Its Own Lifeboat.* New York: Dodd, Mead, 1912.

Walker, Sidney F. "The Boat Problem." *International Marine Engineering* 15 (September 1910), 358–60.

———. "Lifesaving at Sea." *International Marine Engineering* 15 (October 1910), 403–7.

Walsh, Frank. *Sin and Censorship: The Catholic Church and the Motion Picture Industry.* New Haven: Yale University Press, 1996.

War Shipping Administration. *Safety for Seamen.* [Washington, D.C.]: War Shipping Administration, [1943].

"Water Water Everywhere." *Commonweal* 39 (January 28, 1944), 374–75.

Weiss, N. *Personal Recollections of the Wreck of the Ville-Du-Harve and the Loch-Earn.* New York: Randolph, 1875.

Welin, Axel. *Appliances for Manipulating Lifeboats.* London: W. Johnson, 1912.

———. *Lifeboats on Ocean-Going Ships and Their Manipulation.* London: W. Johnson, 1912.

———. "Paper Read at a Representative Gathering of Shipowners . . . May 25, 1910." Glasgow, 1910.

"Well-Furnished Lifeboat." *New Yorker* 19 (May 15, 1943), 13–14.

West, Frank Lawrence. *Lifeboat Number Seven.* London: Kimber, 1960.

Wiggins, Melanie. *Torpedoes in the Gulf: Galveston and the U-Boats, 1942–1943.* College Station: Texas A&M University Press, 1995.

Williams, Douglas. *Retreat from Dunkirk.* New York: Brentanos, 1940.

Williams, Frederick Benton. *On Many Seas: The Life and Exploits of a Yankee Sailor.* Ed. William Stone Booth. New York: Macmillan, 1897.

Willoughby, Malcolm Francis. *Rum War at Sea.* Washington, D.C.: GPO, 1964.

Wilson, Edmund. *American Jitters; a Year of the Slump.* New York: Scribner's, 1932.

Wood, John Taylor. "Escape of the Confederate Secretary of War." *Century Magazine* 47 (November 1893), 110–23.

Woodard, David. *The Narrative of Captain David Woodard and Four Seamen, Who Lost Their Ship While in a Boat at Sea. . . .* Ed. William Vaughan. London: J. Johnson, 1804.

Wooldridge, Emily. *The Wreck of the Maid of Athens; Being the Journal of Emily Wooldridge, 1869–1870.* Ed. and illus. Laurence Irving. New York: Macmillan, 1953.

Woods, Katherine. *The Broadway Limited: 1902–1927.* Philadelphia: Pennsylvania Railroad Company, 1927.

Worsley, F. A. *Shackleton's Boat Journey.* [1940] New York: Norton, 1977.

Worth, Claud. *Yacht Navigation and Voyaging.* London: Potter, 1927.

"Wreck of the Life-Boats, The." *Harper's Weekly* 31 (January 22, 1887).

Wrightman, Frank A. *The Wind Is Free.* New York: Duell, 1955.

"Yachtsmen—Keep Watch for These Jap Ships." *The Rudder* 58 (January 1942), 26–27.

INDEX